Modeling and Analysis of Reservoir System Operations

Ralph A. Wurbs

For book and bookstore information

http://www.prenhall.com

Prentice Hall PTR, Upper Saddle River, NJ 07458

Library of Congress Cataloging-in-Publication Data

Wurbs, Ralph Allen.
    Modeling and analysis of reservoir system operations / Ralph A.
Wurbs.
      p. cm.
    Includes bibliographical references and index.
    ISBN 0–13–605924–4 (hardcover : alk. paper)
    1. Reservoirs—Computer simulation. 2. Reservoirs—United States.
I. Title.
TD395.W87   1996
628.1 32—dc20   95–52377
                                            CIP

Editorial/Production Supervision: Betty Letezia
Acquisitions Editor: Bernard Goodwin
Manufacturing Manager: Alexis R. Heydt
Cover Design and Illustration: Design Source

The publisher offers discounts on this book when ordered in bulk quantities.
For more information, contact:

Corporate Sales Department
PTR Prentice Hall
One Lake Street
Upper Saddle River, NJ 07458
Phone: 800-382-3419
FAX: 201-236-7141
e-mail: corpsales@prenhall.com

Printed in the United States of America
10 9 8 7 6 5 4 3 2 1

ISBN 0-13-605924-4

Prentice-Hall International (UK) Limited, London
Prentice-Hall of Australia Pty. Limited, Sydney
Prentice-Hall of Canada, Inc., Toronto
Prentice-Hall Hispanoamericana S.A., Mexico
Prentice-Hall of India Private Limited, New Delhi
Prentice-Hall of Japan, Inc., Tokyo
Simon & Schuster Asia Pte. Ltd., Singapore
Editora Prentice-Hall do Brasil, Ltda., Rio de Janeiro

CONTENTS

CHAPTER 8   RESERVOIR SYSTEM SIMULATION MODELS   201

PREFACE

This book focuses on the practical application of computer models in determining optimal reservoir storage capacities, storage allocations, and release rules. Reservoir operation practices and associated simulation and optimization modeling methods are outlined. The evaluation strategies and tools covered are applicable to both existing and proposed reservoir projects. The reservoir/river systems addressed range in complexity from a single reservoir serving a single purpose to large-scale multiple-purpose multiple-reservoir systems.

River/reservoir system management and associated computer modeling capabilities are major concerns throughout the world. The book is written based primarily on reservoir operation and model development efforts in the United States. However, the concepts and methods presented are pertinent to any river basin, regardless of geographical location.

A broad array of reservoir system analysis techniques have been reported in the literature over the past several decades. Water management practitioners and university researchers have viewed reservoir system analysis from quite different perspectives. Although various aspects of the extensive academic research literature are noted, the book emphasizes approaches adopted by reservoir management agencies and practitioners.

Chapter 1 provides a general overview of reservoir system modeling and analysis approaches. Reservoirs and their operation are reviewed in Chapters 2 and 3, prior to addressing modeling and analysis topics in Chapters 4 through 12. Chapter 2 presents summary statistics regarding reservoirs in the United States and describes the operation of several major reservoir systems. Considerations are outlined in Chapter 3 regarding storage allocations and release policies for flood control, water supply, hydroelectric power, navigation, recreation, water quality, fish and wildlife, and sedimentation.

Chapter 4 addresses the formulation of objectives, criteria, and strategies for evaluating the performance of alternative storage allocation and operating plans. Hydrologic data required for reservoir system analysis studies are discussed in Chapter 5. Chapter 6 covers the basic water accounting, routing, and hydraulics features of reservoir/river system analysis models. Chapter 7 describes alternative approaches for evaluating the reliability of a reservoir/river system in meeting water use requirements and preventing flood damages. Various types of reservoir system simulation models, including several readily available generalized software packages, are categorized and inventoried in Chapter 8. A series of simulation analyses of a particular 12-reservoir system in the Brazos River Basin is presented in Chapter 9 as an illustrative case study. The fundamentals of linear programming, network flow programming, dynamic programming, and search techniques are outlined in Chapter 10. Chapter 11 categorizes and reviews a number of reservoir system optimization models reported in the literature. Water quality aspects of reservoir management and several generalized water quality simulation models are discussed in Chapter 12.

ACKNOWLEDGEMENTS

I worked in the water resources development program of the Fort Worth District of the U.S. Army Corps of Engineers for nine years prior to joining the Texas A&M University faculty in 1980. Much of my research since 1980 has been sponsored by Corps of Engineers entities including the Waterways Experiment Station, Hydrologic Engineering Center, and Institute for Water Resources. I have also conducted several reservoir system analysis studies, including those noted in Chapter 9, sponsored by the Texas Water Resources Institute, Brazos River Authority, and other state agencies in Texas. The background acquired in working with many colleagues at Texas A&M University and these federal and state agencies contributed greatly to this book. I am grateful for the support of these people and institutions over the years. My associations with Wayne R. Jordan, Director, Texas Water Resources Institute, and Darryl W. Davis, William K. Johnson, Richard J. Hayes, Michael W. Burnham, Vernon R. Bonner, and others at the USACE Hydrologic Engineering Center have been particularly beneficial.

A number of former graduate students who have worked with me on river basin planning and analysis research projects since 1980, have also contributed significantly to this book. Moris Cabezas, Michael Tibbets, Lonnie Roy, Patrick Carriere, Carla Bergman, Brian Walls, Stuart Purvis, Garrett Sullivan, Keith Ganze, Susan Sayger, Ishtiaque Saleh, David Dunn, Gerardo Sanchez-Torres, Awes Karama, and Anil Yerramreddy are acknowledged for their excellent research, much of which has been incorporated into the book.

Finally, thanks are extended to my wife Kerry and our children Jeremy, Amy, and Sarah for their patience in allowing me to devote so much time to this and other projects. The assistance provided by Kerry and Sarah in word processing and preparing figures is also greatly appreciated.

Ralph Wurbs

1

INTRODUCTION

Reservoir storage is necessary to use the highly variable water resources of a river basin for beneficial purposes such as municipal and industrial water supply, irrigation, hydroelectric power generation, and navigation. Dams and appurtenant structures also regulate rivers to reduce damages caused by floods. Public recreation, water quality, erosion and sedimentation, and protection and enhancement of fish, wildlife, and other environmental resources are important considerations in managing reservoir/river systems.

Numerous reservoir projects, located throughout the United States, are operated by the Corps of Engineers, Bureau of Reclamation, Tennessee Valley Authority, other federal agencies, state and regional agencies, local water districts, cities, and private industry. Most of these projects were constructed during the period from 1900 through the 1970s, which has been called the construction era of water resources development. Although additional new reservoir projects are needed and continue to be developed, most of the major reservoir systems required to manage our rivers are in place. Economic, environmental, and institutional considerations constrain construction of water resources development projects. Since the 1970s, water resources management policy and practice have shifted to a greater reliance on improving water use efficiency, managing floodplain land use, and optimizing the operation of existing facilities.

Public needs and objectives and numerous factors affecting reservoir management change over time. Population and economic growth in various regions of the nation are accompanied by increased needs for flood control, water supply, energy, recreation, and the other services provided by water resources development. Depleting groundwater reserves are resulting in an increased reliance on surface water in many areas. Concerns have grown in recent years regarding maintenance of instream flows for preservation of riverine habitat and species, wetlands, and freshwater inflows to bays and estuaries. Each occurrence of a major flood or drought in a region motivates reevaluation of water management practices. With an aging inventory of numerous dams and reservoirs being operated in an environment of change and intensifying demands on limited resources, operational improvements are being considered increasingly more frequently.

1.1 RESERVOIR OPERATIONS

Reservoir system operations can be categorized as:

- Operations during relatively normal hydrologic conditions from the perspective of optimizing the present day-to-day, seasonal, or year-to-year use of the reservoir system
- Operations during normal hydrologic conditions from the perspective of maintaining capabilities for responding to infrequent hydrologic extremes expected to occur at unknown times in the future
  — Maintaining empty flood control storage capacity
  — Maintaining reliable supplies of water
- Operations during hydrologic extremes
  — Operations during flood events
  — Operations during low-flow or drought conditions

An operating plan or release policy is a set of rules for determining the quantities of water to be stored and to be released or withdrawn from a reservoir or system of several reservoirs under various conditions. Typically, a regulation plan includes a set of quantitative criteria within which significant flexibility exists for operator judgment. The operating rules provide guidance to the water managers who make the actual release decisions. In modeling exercises, the reservoir system analysis model contains some mechanism for making release decisions within the framework of user-specified operating rules and/or criteria functions.

Reservoir operating rules and the operating decisions made within the framework of these rules involve:

- Allocating storage capacity and streamflow between multiple water users and types of use
- Minimizing the risks and consequences of water shortages and flooding

- Optimizing the beneficial use of water, energy, and land resources
- Managing environmental resources

1.2 RESERVOIR SYSTEM ANALYSIS APPLICATIONS

A broad array of computer modeling and analysis techniques is available for developing quantitative information for use in evaluating storage capacities, water allocations, and release policies. Reservoir system analysis models support:

- Sizing of reservoir storage capacities and establishing operating policies during preconstruction planning of new projects
- Re evaluation of the operating plans of existing reservoir systems
- Administration of water allocation systems involving water rights and agreements between water suppliers and users
- Operational planning for developing management strategies for the next year or season
- Real-time operations

In many countries such as the United States, the earlier construction era of water resources development has evolved into the current focus on better management of existing facilities and water resource allocations. Thus, in developing and applying reservoir system analysis models, the emphasis has shifted to evaluating the operations of existing reservoirs, but the planning and design of proposed new projects are also still important.

Reservoir/river system analysis models are used for various purposes in a variety of settings. Models are used in feasibility studies to aid in the formulation and evaluation of alternative plans for responding to water-related problems and needs. Feasibility studies range from broad, comprehensive river basin planning to detailed project implementation studies. Feasibility studies may involve proposed construction projects as well as reallocations of storage capacity or other operational modifications at completed projects. Other modeling applications involve studies performed specifically to reevaluate operating policies for existing reservoir systems. Periodic evaluations may be made to ensure system responsiveness to current conditions and objectives. Modeling studies are often accomplished in response to a particular perceived problem or need. Studies may be motivated by drought conditions, a major flood event, water-quality problems, or environmental losses such as fish kills. Development of drought management plans, in preparation for future droughts, is receiving increasing attention. Water availability models support the administration of water rights permit programs. Operational planning studies are conducted annually for some reservoir systems to establish operating strategies for the next year. Execution of models during actual reservoir operations in support of real-time release decisions represents another major area of application.

1.3 COMPUTER MODELING SYSTEMS

Microcomputers, workstations, minicomputers, mainframes, and supercomputers have all been used in reservoir system analysis applications. IBM-compatible microcomputers, operating under the Microsoft Disk Operating System (MS-DOS), became extremely popular during the 1980s in water resources planning and management as well as in many other fields. The majority of the reservoir system analysis models cited in this book have been executed on MS-DOS-based microcomputer systems, with versions being run on other types of computer systems as well. Windows, first introduced by Microsoft in 1990 and expanded as Windows95 in 1995, is an operating system extension that provides a menu-driven user interface and multitasking. With Windows, an IBM-compatible desktop computer provides a user environment somewhat similar to both workstations and the popular Apple Macintosh family of microcomputers. The more expensive workstations (supermicrocomputers), often operating under UNIX, have also established a significant role in reservoir system modeling since the mid-1980s. Mainframe and minicomputers also continue to be widely used in water resources planning and management. The large-scale systems provide certain advantages for multiple users in large organizations.

A computer modeling system includes the following software components:

- Computational engine for modeling the real-world system
- User interface
- Preprocessor programs for acquiring, preparing, checking, manipulating, and storing model input data
- Post-processor programs for managing, analyzing, interpreting, summarizing, displaying, and communicating modeling results

The reservoir/river system is modeled by a set of simulation and/or optimization algorithms incorporated in the basic computational code. A user interface allows the human user to enter information and instructions, control the software and data files, and access modeling results. Data management programs facilitate the handling of input and output. Model development in the 1990s has been characterized by an emphasis on interactive user interfaces and data management systems oriented toward using advances in computer technology to make models more efficient to use. Graphical user interfaces are popular. A graphical user interface is particularly important for applications involving the production or processing of graphic images such as maps, diagrams, drawings, and charts. A variety of database management programs have been used with reservoir system analysis models to store, manipulate, and analyze various types of data, including time series, parametric, and spatial data.

A reservoir system analysis application often involves integration of several different types of models. For example, optimization and simulation models dealing with water quantities might be used along with one or more other models addressing water-quality aspects of reservoir operations. Other models are used to establish diversion and instream flow requirements to be met by reservoir releases. A watershed (precipitation-runoff) model is used to develop runoff hydrographs and pollutant loadings for input to the reservoir operation models, which in turn determine discharges and contaminant concentrations at pertinent locations in the river/reservoir system. The example modeling system could also include a river hydraulics model to compute flow depths and velocities. A geographic information system, database program, spreadsheet, graphics package, and other data management programs are included in the modeling system to (1) develop and manage voluminous input data, (2) perform statistical and graphical analyses of simulation output, and (3) display and communicate results.

The concept of decision support systems became popular during the 1980s in the water management community as well as in business, engineering, and other professional fields in general. A decision support system is a user-oriented computer system that supports decision makers in addressing unstructured problems. The general concept emphasizes:

- Solving unstructured problems which require combining the judgment of manager-level decision makers with quantitative information
- Capabilities to answer "what if" questions quickly and conveniently by making multiple runs of one or more models
- Use of enhanced user-machine interfaces
- Graphical displays

Decision support systems include a collection of software packages and hardware. For example, decision support systems are used for real-time flood control operations. Making release decisions during a flood event is a highly unstructured problem because reservoir operations are highly dependent on operator judgment as well as prespecified operating rules and current and forecasted streamflows, reservoir storage levels, and other available data. The decision support system includes user interface software; data management programs; watershed runoff, stream hydraulics, and reservoir/river system operation models; a computer with various peripheral hardware devices; and an automated real-time hydrologic data collection system.

The reservoir system analysis models cited in this book are often used as components of decision support systems. The models are also often applied in other planning, design, and resource management situations that do not exhibit all the characteristics attributed to decision support systems. For example, models are frequently used to develop firm yield versus storage capacity relationships for feasibility studies, within a setting which may not necessarily qualify as a decision support system. The relationships between decision-making processes and modeling systems vary depending on the particular water management application.

1.4 BASIN/RESERVOIR/USE SYSTEM REPRESENTATION

Reservoir/river system analysis models typically include features for representing:

- The spatial configuration of the river basin system
- Basin hydrology
- Physical characteristics of reservoirs, spillways and outlet works, hydroelectric power plants, and other water control facilities
- System operating rules
- Water use requirements
- Effects of basin-wide water management on the reservoir/river/use system of concern
- Measures of system performance

The spatial configuration of a river/reservoir/use system is typically represented by a set of index locations called nodes, stations, or control points. The location of reservoirs, diversions, return flows, instream flow requirements, streamflows, and other system features are specified by node.

Basin hydrology is represented by sequences of streamflows and reservoir evaporation rates, which are provided as input to the reservoir operation model. Meteorological and water-quality parameters are also included in some models.

Historical gaged streamflows are adjusted to represent flow conditions at pertinent locations for a specified past, present, or future condition of river basin development. In some cases, particularly in flood control studies, watershed modeling is used to compute streamflows from precipitation data. In applying sequential reservoir operation models, the input streamflow sequences consist of either adjusted historical period-of-record streamflows; adjusted historical streamflows during a critical drought period, flood event, or other selected subperiod of the period of record; synthetically generated flows which preserve selected statistical characteristics of the adjusted historical streamflow record; or flows computed with a watershed model from either gaged precipitation or synthetic storms.

Sequential models, based on sequences of streamflow inflows, are used in most analyses performed by water agencies and practitioners. However, in some cases, particularly in the academic research literature, streamflows are represented by probability distributions, assuming negligible autocorrelation between time periods, or by stochastic processes such as Markov chains that condition streamflow probabilities on flow in the previous period.

Reservoir evaporation volumes are typically computed within a model by combining input evaporation rates with water surface areas determined as a function of storage. The evaporation rates may be provided for each individual time interval of the overall period of analysis, or averages may be provided for each month or season of the year. Net evaporation less rainfall rates are often used.

Reservoir characteristics are specified in a model by: (1) storage capacities defining the pertinent pools, (2) storage volume versus water surface area relationships for use in the evaporation computations, and (3) storage volume versus elevation relationships for use in hydropower computations. Reservoir storage capacity is diminished over time due to sediment deposition. Reservoir storage characteristics are reflected in a model based on a specified condition of sedimentation. Discharge capacities of outlet structures and conveyance facilities are also specified in the model. Hydroelectric power plant characteristics must also be furnished as appropriate.

Models contain a variety of mechanisms for defining operating rules which range from very simple to relatively complex. Operating rules may involve seasonal storage allocations; minimum and maximum streamflow targets; coordination of multiple-purpose operations; multiple-reservoir release decisions; flood control operations; and priorities for allocating shortages to competing demands during periods of insufficient supply.

Water use is reflected in a model by a set of diversion and/or instream flow requirements representing either actual historical or projected future water use for a specified past, present, or future point in time; water amounts committed to users by water rights, agreements, or other allocation systems; or hypothetical yields. Water use is often expressed in terms of a constant annual amount which varies monthly or seasonally. Water use may be a function of water availability. For example, certain water uses may be curtailed or reduced in accordance with some triggering mechanism such as reservoir storage falling below specified levels. Diversion return flow specifications are also included in the water use scenario. Hydroelectric energy requirements are specified in a similar manner.

The water resources of a river basin are typically shared by many water users and regulated by complex systems of reservoirs and other facilities. Diversions and return flows of each water user affect water available for other users. Likewise, reservoir storage and evaporation affects downstream flows. Capabilities for representing interactions between the particular storage/use system of concern in a study and other reservoirs and water use activities may be quite important. Removal of the historical effects of basinwide water management may be significant in developing homogeneous sets of unregulated flows provided as input to a model. Streamflows may be adjusted to reflect certain regulation structures or basin activities prior to providing them as input to the reservoir operation model. Alternatively, these other water management facilities may be included in the model along with the reservoir/use system of primary concern.

The output of a reservoir/river system analysis model is typically voluminous. Model results can be summarized in various formats such as tabulations and plots of (1) time series of storage, releases, streamflows, diversions, diversion shortages, energy generated, water quality parameters, or other variables, and (2) frequency or duration relationships for these variables. Measures of system performance in meeting water and energy demands, reducing flooding, and otherwise fulfilling water management objectives are formulated to fit the purposes of a particular study. These measures often include expressions of reliability in meeting demands and/or the economic consequences of alternative plans.

1.5 PERIOD OF ANALYSIS AND TIME STEP

Reservoir system analysis models are generally formulated in terms of a reservoir system being operated, in the model, during a specified hydrologic period of analysis. Unregulated streamflows, reservoir evaporation rates, and possibly water-quality parameters and other data representing basin hydrology are input for each time step or discrete interval of the overall analysis period. Reservoir releases, storage, regulated streamflows, and other variables are computed by the model for each time step.

Selection of a hydrologic period of analysis and computational time interval is a key consideration for a particular modeling application. The choice of analysis period and time step depends on the scope of the application, data availability, and time-variability characteristics of streamflow, water demands, and other factors. The validity of modeling assumptions and the choice of time interval are closely related. For example, flow attenuation in a river reach with a travel time of several days is appropriately neglected in a monthly time step model, but storage routing computations may be necessary in a daily time step model.

A several-decade historical period-of-record hydrologic simulation period and one-month computational interval are typically adopted for planning studies involving water supply, hydropower, and other conservation storage purposes. However, weekly or daily computational time steps during the many-year analysis period are also common. In some cases, a several-month time interval representing wet and dry seasons may be adequate for a particular study.

A model for scheduling irrigation or hydroelectric power releases over the next year might involve a daily or weekly time interval and one-year simulation period, with simulations being repeated for alternative annual sequences of daily or weekly streamflows. Each simulation starts with present storage levels. The scheduling problem may combine multiple analyses performed at different levels of detail, using a monthly interval to allocate water over the next year, a daily time step for a particular month, and an hourly interval for analyzing daily operations.

Simulation of flood control operations involves short time intervals since flows are rapidly changing. A major flood event with a duration of several days or weeks is typically simulated with a one-hour, several-hour, or one-day time step.

Some models use a variable computational time step, with a daily or shorter interval specified during flood events and a longer interval, such as a month, adopted during extended periods of more normal hydrologic conditions.

1.6 MODEL CATEGORIZATION

Reservoir system analysis models can be categorized using a variety of classification schemes. Different types of models are outlined here within the framework of the following comparisons:

* Reservoir operation models versus other related models
* Flood control versus conservation purposes
* Generalized versus site specific
* Simulation versus optimization
* Descriptive versus prescriptive

1.6.1 Reservoir Operation Models versus Other Related Models

The terms *reservoir system analysis model* and *reservoir operation model* are used synonymously to refer to computer programs that simulate and/or optimize storage, release, and diversion of water in a reservoir or system of multiple reservoirs and associated river reaches. This type of model is used specifically to evaluate storage allocations and release policies. For purposes of organizing this book, reservoir/river system analysis models are grouped into the categories of simulation (Chapters 8 and 9), optimization (Chapters 10 and 11), and water quality (Chapter 12). Basic concepts reflected in these reservoir operation models are covered in Chapters 3 through 7. Although the emphasis is on reservoir operation models, other related types of models are noted as pertinent throughout the book.

Water-quality management aspects of reservoir operations involve both flow rates and quality parameters. Maintaining minimum instream flows is a primary water-quality operating objective of many reservoir systems. Thus, the quantities (flow rates) of reservoir releases and streamflows may be the primary consideration in meeting water-quality objectives. In other cases, the quality of the water released may be the primary concern. At many reservoir projects, the quality of the releases may be controlled through multiple-level selective withdrawals, since thermal stratification can significantly affect water-quality characteristics. Many reservoir system analysis models deal only with water quantities, but may still address instream flow quantity requirements for managing water quality. Other models, such as those noted in Chapter 12, incorporate temperature, dissolved oxygen, salinity, and other more complex physical, chemical, and biological parameters.

This book focuses on reservoir/river system simulation and optimization models. These models are referred to as *reservoir system analysis models* or *reservoir operation models*. As noted throughout the book, an array of other types of computer models and evaluation procedures also play important roles in analyzing reservoir/river systems, including:

- Water requirements models for estimating present and future needs for ecosystem, agricultural, municipal, and industrial water uses
- Watershed (precipitation-runoff) models for synthesizing streamflows from rainfall and snow data
- Stochastic hydrology models for synthesizing streamflows which preserve statistical characteristics of the historical streamflows
- Statistical analysis methods for analyzing model input and output and other data
- Methods for analyzing the hydraulics of water control structures
- River hydraulics models for predicting flow rates, stages, and velocities
- Water-quality models for predicting the physical, chemical, and biological characteristics of a river/reservoir system
- Groundwater models and methods for analyzing interactions between surface and groundwater
- Comprehensive water budget models for accounting for water uses and surface and groundwater supplies
- Economic evaluation methods
- Environmental impact analysis methods
- Data management software

Models and methods from several or all of the categories listed above may be applied together in evaluating the operations of a particular reservoir system. All of the modeling and analysis categories listed have been incorporated into reservoir operation models and/or applied in combination with reservoir operation models. These modeling and analysis methods are also applied in many other water resources planning and management situations which do not necessarily involve reservoirs.

1.6.2 Flood Control versus Conservation Purposes

Reservoir purposes represent a key consideration in formulating a modeling and analysis approach. The distinction is particularly significant between flood control and conservation storage purposes such as municipal, industrial, and agricultural water supply, hydroelectric power, navigation, recreation, and maintenance of instream flows for fish and wildlife or water-quality purposes. Reservoir operating rules are different for flood control and conservation purposes. Computational algorithms and data requirements are also different. Although capabilities for analyzing both flood control and conservation operations are combined in some models, other models focus on one type of operation or the other.

Hydrologic analyses of floods are probabilistic-event-oriented, and droughts are long-term stochastic-time-series-oriented. Major flood events have durations of several hours to several weeks, with discharges changing greatly over periods of hours or days. Flood analyses are typically performed using a computational time step in the range of one hour to one day. Modeling flood wave attenuation effects is important.

Conservation storage reservoirs supply water during dry seasons with durations of several months and during extreme droughts with durations of several years. Evaporation is important. Although conservation analyses are often based on daily or weekly streamflow and evaporation rates, a monthly interval is more typical for planning studies. Conservation storage models also vary in their capabilities of analyzing for the various conservation purposes. For example, a model may be categorized by whether or not capabilities are provided for analyzing hydroelectric power operations.

The U.S. Army Corps of Engineers has played a particularly notable role in developing models for analyzing flood control operations as well as operations of conservation storage. The Tennessee Valley Authority also has an extensive background in modeling both flood control and conservation operations. Modeling studies in the United States other than Corps and TVA studies have tended most often to be primarily concerned with conservation purposes, with flood control being a secondary consideration.

1.6.3 Generalized versus Site Specific

A computer model may be developed for a specific reservoir/river system or generalized for application to essentially any river basin. With site-specific models, unique features of the particular reservoir system are built into the computer code. Numerous site-specific models are routinely applied in evaluating the operation of various systems throughout the nation. Models developed for the Tennessee Valley Authority system, Corps of Engineers Missouri River System, Bureau of Reclamation Colorado River Basin system, and California Central Valley Project are among those noted later in the book. The concepts and methods adopted in these models, and the lessons learned in their development and application are certainly pertinent to other river basins. However, the computer programs are not designed to be applied to other reservoir systems.

A few examples of the many generalized computer programs cited in later chapters include the Hydrologic Engineering Center's HEC-5 Simulation of Flood Control and Conservation Systems and HEC-PRM Prescriptive Reservoir Model; the Interactive River-Aquifer Simulation model (IRAS), developed at Cornell University; and MODSIM, developed at Colorado State University. Generalized models are designed for application to a range of problems dealing with systems of various configurations and locations, rather than to analyze one specific reservoir system. With a generalized model, the unique features and information reflecting the particular reservoir system of concern are provided in the input data. Thus, the precise meaning of the generic term *model* varies with the context. For example, the HEC-5 *model* is a generalized package of computer software available from the Hydrologic Engineering Center of the Corps of Engineers. A *model* of the Allegheny River Basin consists of the HEC-5 computer programs combined with an input data file developed for that particular reservoir/river system.

Models may also fall somewhere between the extremes of being site-specific and perfectly generalized. A computer program may be developed specifically for a particular reservoir system, with features included in the code to facilitate future adaptation to other reservoir systems.

1.6.4 Simulation versus Optimization

Reservoir system analysis models have traditionally been categorized as simulation, optimization, and hybrid combinations of simulation and optimization. Development and application of decision support tools within the water resources development agencies that construct and operate reservoir projects have traditionally focused on simulation models. Simulation has been the workhorse of practical applications. Interestingly, optimization has dominated the published literature. Since the 1960s the academic community and research literature have emphasized optimization techniques, particularly linear and dynamic programming, but other nonlinear programming methods as well.

A simulation model is a representation of a system used to predict its behavior under a given set of conditions. Computer models for simulating reservoir operations reproduce the performance of a reservoir/stream system for given hydrologic inputs and operating rules. Simulation is the process of experimenting with a simulation model to analyze the performance of the system under varying conditions. Alternative runs of a reservoir simulation model are made to evaluate alternative storage capacities and operating plans.

In a broad sense, optimization includes human judgment, use of either simulation and/or optimization models, and use of other decision support tools. However, following common usage in the literature, the term *optimization* is used here synonymously with *mathematical programming*, to refer to a mathematical formulation in which a formal algorithm is used to compute a set of decision variable values which minimize or maximize an objective function subject to constraints. Optimization methods include linear, network flow, and dynamic programming, search techniques, and various other nonlinear programming techniques. Optimization models automatically search for an *optimal* decision policy.

Although optimization and simulation are two alternative modeling approaches with different characteristics, the distinction is somewhat obscured by the fact that many models, to varying degrees, contain elements of both approaches. All *optimization* models also *simulate* the system. An *optimization* approach may involve numerous iterative executions of a *simulation* model, possibly with the iterations being automated to various degrees. Optimization algorithms are embedded within many major reservoir system simulation models to perform certain computations. Alternative simulation models may be applied in essentially the same manner, even though one incorporates an optimization algorithm to perform the internal computations and the other does not.

Simulation has the advantage of generally permitting a more detailed and realistic representation of the complex characteristics of a reservoir/river system. Since the computational algorithms are not restricted to a particular mathematical format, simulation provides greater modeling flexibility. The advantages of mathematical programming are related to providing more systematic and efficient computational algorithms, providing capabilities for searching through an infinite or extremely large number of possible combinations of values for decision variables, and facilitating a more prescriptive analysis. Many different models, representing diverse applications in engineering, science, and business, can be developed based on the same standard linear and nonlinear programming algorithms.

Since simulation models are limited to predicting the system performance for a given decision policy, optimization models have the distinct advantage of being able to search through an infinite number of feasible decision policies to find the optimal policy. Optimization (mathematical programming) algorithms systematically and automatically search through all feasible decision policies (sets of values for the decision variables) to find the decision policy which minimizes or maximizes a defined objective function. Mathematical programming methods provide useful capabilities for analyzing problems characterized by a need to consider an extremely large number of combinations of values for decision variables. For example, consider the problem of temporally distributing limited available water over a year. The problem might be formulated in terms of determining daily water supply and hydroelectric power releases from each of 10 reservoirs, for each of the 365 days in a year, which optimizes a specified objective function. Thus, the problem involves 7300 decision variables (365 water supply releases and 365 hydroelectric power releases from 10 reservoirs), which can each take on a range of values. The extremely large number of decision variables and infinite number of possible decision policies illustrate the need for a mathematical programming algorithm.

Mathematical programming requires adherence to the proper mathematical formulation. Methods for dealing with restrictions on mathematical form are available, such as schemes for linear approximations of nonlinear relationships. However, representing the objectives, operating rules, and physical and hydrologic characteristics of the system in the required format, without unrealistic simplifications, is a particularly difficult aspect of the modeling process which limits the application of optimization techniques.

Significant complexities in developing optimization models are related to model capabilities to satisfy needs for defining and evaluating operating rules and the fact that, in actual operations, future streamflows are unknown at the time a release is made. Optimization techniques tend to naturally fit the format of computing releases, which minimize or maximize a specified objective function for either given sequences of streamflows or stochastic representations of statistical character-istics of streamflows. Many optimization models compute the releases that optimize an objective function without directly using operating rules. Simulation models generally provide mechanisms for the user to define and evaluate the operating rules in greater detail. Also, simulation models perform computations period by pe-riod in such a way that future streamflows are not reflected in release decisions, except for some models which include features for limited short-term forecasts. Certain descriptive deterministic simulation models, which incorporate mathemati-cal programming algorithms, make period-by-period release decisions just like other, more conventional simulation models. Many optimization models make all release decisions simultaneously, considering all streamflows covering the entire period of analysis.

Simulation and optimization models can be used in combination. For exam-ple, a typical reservoir system analysis problem consists of establishing operating rules which best achieve certain water management objectives. An optimization model may be used to determine reservoir release decisions that maximize or mini-mize a criterion function that provides a measure of performance in meeting the water management objectives. Professional judgment and various analyses are then used to develop operating rules that appear to be consistent with the sequences of release decisions reflected in the optimization model results. These rules are then tested using a simulation model. In various other types of applications as well, pre-liminary screening of numerous alternatives using an optimization model may be followed by a more detailed evaluation of selected plans using a simulation model.

1.6.5 Descriptive versus Prescriptive Orientation

System analysis models are also categorized as being either descriptive or prescriptive. Descriptive models demonstrate what will happen if a specified plan is adopted. Prescriptive models determine the plan that should be adopted to satisfy specified decision criteria. Although it is desirable for models to be as prescriptive as possible, the real-world complexities of reservoir system operations often neces-sitate model orientation toward the more descriptive end of the descriptive/prescrip-tive spectrum. In general, models should be as prescriptive as the scope of the study demands and the complexities of the application allow.

Simulation models are generally descriptive. In many situations, mathemati-cal programming techniques enhance capabilities to develop models that are more prescriptive. Optimization models automatically determine decision variable values which optimize an objective function, which is consistent with the concept of pre-scriptive modeling. However, capabilities for formulating objectives and assessing performance in meeting these objectives are the driving considerations. Simulation and optimization models should not be rigidly characterized as being descriptive and prescriptive, respectively. Both optimization and simulation models can be more or less oriented toward being either descriptive or prescriptive. The following examples of descriptive and more prescriptive models are presented to illustrate this point.

Assume a multiple-reservoir system is operated for flood control and hydroelectric power. The total storage capacity in each reservoir of this existing system is fixed. The problem is to determine the optimal seasonal allocation of storage capacity between flood control and hydroelectric power in each reservoir.

Examples of descriptive simulation and optimization models are noted first. A linear programming model is formulated for computing firm energy for a given storage allocation plan. Linearization techniques are adopted for approximating the nonlinear evaporation and hydropower computations. Likewise, a simulation model incorporating iterative algorithms can be used to compute firm energy for a given storage allocation plan. Alternative runs of either the linear programming or simulation models can be made to determine firm hydroelectric energy for alternative storage allocations. Firm energy as well as firm water supply yield analyses are quite common. The information provided is very useful in evaluating alternative storage allocation plans. However, the modeling exercise does not directly result in determining the optimal plan, assuming that our objectives entail more than just maximizing firm energy. Thus, both the optimization and simulation models can be categorized as being descriptive.

Optimization and simulation models are now formulated that have a more prescriptive orientation. The objective of maximizing economic benefits in dollars is adopted. The objective is expressed in terms of maximizing average annual hydropower revenues minus average annual flood damages. A linear programming model is developed that determines releases from each reservoir for each time interval which maximizes the total dollar value of hydropower revenues minus flood damages over the period of analysis. The optimization model is executed for a given alternative storage allocation plan, assuming all past and future streamflows are known at the time each release is made. The complexities involved in computing flood damages necessitate a fairly rough approximation in the linear programming model.

A simulation model is also developed which computes both hydropower revenues and flood damages for a given storage allocation plan. The simulation model provides more detailed computational procedures for estimating expected annual flood damages. Other aspects of multiple-reservoir flood control and hydropower release rules are specified in more detail than in the linear programming model.

Alternative runs of either the optimization model or the simulation model result in hydropower and flood control benefits for alternative seasonal storage allocation plans. Since the two alternative models reflect several different basic premises, the results are not equivalent but are meaningful from different perspectives. Multiple alternative runs of these models, combined with adoption of the economic efficiency criterion, represents a more prescriptive evaluation approach than the firm yield analysis of the previous paragraph.

Finding the storage allocation plan that maximizes the specified objective function is not necessarily guaranteed with a finite number of executions of either model. None of the models are absolutely prescriptive, since capabilities are not provided for automatically determining the economically optimum storage realloca-tion plan in a single run of a model.

1.7 MODEL DEVELOPMENT SOFTWARE ENVIRONMENTS

Water-resources-related computer programs are available from federal and non-federal water agencies, universities, and private firms [Wurbs, 1995]. Most of the generalized programs for reservoir system analysis and other water management applications are in the public domain. Additionally, much of the proliferation of proprietary software on the commercial market, which is widely used in various business, scientific, and engineering fields, is also pertinent to reservoir system analysis. Software products are dynamic, with new versions being released peri-odically. Water management professionals will continue to discover new uses for currently available software as well as for the new products continually being mar-keted.

A large array of software products are pertinent to reservoir system analysis applications in regard to providing capabilities for database management, develop-ing geographic information systems, computer-aided drafting and design, develop-ing expert systems, graphical displays, and performing statistical analyses [Wurbs, 1995]. However, the present discussion focuses on developing the actual reservoir operation simulation and/or optimization model.

A model for a particular reservoir system analysis application can be con-structed using either of the following sets of software tools:

- Programming languages
  — Traditional procedural languages such as FORTRAN, BASIC, or C
  — Object-oriented languages such as C++
- General-purpose commercial software
  — Spreadsheet packages
  — Object-oriented simulation packages
  — Mathematical programming packages
- Generalized operational reservoir system analysis models
- Some variation or combination of the above

Use of each type of computer software represents a different model-building approach or environment in which to work. In developing a model for a particular application, a key question is which modeling environment or set of software tools should be adopted. In some situations, one model-building approach may be clearly advantageous over the others. In other modeling situations, the relative merits and trade-offs between the sets of tools will be more balanced. The background and personal preferences of the model builders are typically a major consideration in selecting a software environment. Considerations in comparing the relative advantages and disadvantages of the alternative approaches include:

- Flexibility in using mathematical equations, computational algorithms, and data to realistically represent real-world systems and concerns
- Capabilities which can be incorporated into the model for analyzing, displaying, and communicating results
- The expertise, time, and effort required to build a model, and then to apply and maintain the model, and later to modify the model in response to changing analysis requirements

1.7.1 Programming Languages

Numerous computer language translation software packages are marketed for developing application programs. Although water management models have been written in a variety of high-level languages, FORTRAN dominates. A majority of the reservoir system analysis and other water management models currently in use are coded in FORTRAN. FORTRAN (FORmula TRANslator) is the oldest high-level programming language, dating back to the 1950s, and continues to be widely used in engineering and science. FORTRAN77 and FORTRAN90 are standardized versions of FORTRAN approved by the American National Standards Institute in 1978 and 1994, respectively. FORTRAN compilers are available for all categories of computers.

BASIC (Beginner's All-Purpose Symbolic Instruction Code) and Pascal (named after Blaise Pascal) are examples of other programming languages that have been used in water resources planning and management. These two relatively simple languages are popular for use on microcomputers. BASIC is particularly popular as an easy-to-learn general-purpose language for nonprofessional programmers. The disadvantages of BASIC are a tendency toward long unstructured programs that are difficult to modify, and cumbersome input of text and graphics. Pascal is a highly structured language suitable for scientific and technical applications.

C and the newer object-oriented C++ are multilevel (assembly and compiler levels) programming languages which are popular with system programmers who develop system software and commercial application packages. C++ is an extension of C which retains all of the C language while adding special features for object-oriented programming. C and C++ provide excellent graphics and optimize computational efficiency. These languages have been used fairly extensively in recent years to develop water management models. In some cases, complete models are coded in these languages. In other cases, C (or C++) and FORTRAN are used in combination, with C providing graphical user interface development capabilities and FORTRAN being used for developing computational routines.

FORTRAN and C are based on a traditional structured approach to programming in which programs are organized into algorithmic procedures and control structures. The alternative approach of object-oriented programming is experiencing increasing popularity in recent years. C++ is one of the more popular object-oriented programming languages. Instructions and information are coded and stored as objects or modules. A program is treated as a collection of objects. Objects can be reused in different programs and subprograms, and programs are easier to modify.

Three main properties characterize an object-oriented programming language: encapsulation, inheritance, and polymorphism. Encapsulation involves combining code and data together into a single class-type object. Inheritance is the creation of a hierarchy of classes. New derived classes inherit data and functions from one or more previously defined classes, while possibly redefining or adding new data and actions. Polymorphism gives an action one name or symbol that is shared throughout a class hierarchy, with each class in the hierarchy implementing the action in a way appropriate to itself.

1.7.2 General-Purpose Commercial Software

Reservoir system analysis models can be constructed using commercially available general-purpose software which is widely applied in many other areas of engineering, science, business, and education. These modeling environments include spreadsheet programs, object-oriented simulation systems, and mathematical programming packages [Wurbs, 1995].

Numerous spreadsheet, graphics, and database programs have been introduced to the market since the early 1980s. The more popular programs include Lotus 1-2-3, Quattro Pro, and Excel, marketed by Lotus Development Corporation, Borland International, and Microsoft Corporation, respectively. Excel, Lotus 1-2-3, and Quattro Pro have similar capabilities. All are available for desktop computers operating under the various popular operating systems. These spreadsheet programs are used extensively in many fields. Water management professionals recognized the potential of electronic spreadsheets soon after they were first marketed. The software packages are routinely used in a variety of water resources planning and management applications. Spreadsheet programs have the advantage of applying the same familiar software to many different types of problems. A reservoir system analysis problem can be addressed using software which is already being used in the office for other purposes as well. For relatively simple applications, spreadsheets provide capabilities for developing complete reservoir system analysis models. Spreadsheets are also used to manage input and output data for other reservoir operation models.

Object-oriented simulation environments do not have as vast a market as spreadsheet programs, but are used in a broad range of applications in education, business, science, engineering, and other professional fields. They provide another efficient approach for constructing reservoir operation models. STELLA [High Performance Systems, Inc., 1990] is an example of this type of software. STELLA is a general-purpose, object-oriented modeling package designed to simulate dynamic (time-varying or otherwise changing) systems characterized by interrelated components. The user builds a model for a particular application, using the operations and functions provided, and designs the tabular and/or graphical presentation of simulation results.

As previously discussed, optimization or mathematical programming provides the advantage of applying standard computational algorithms to many different types of problems. Spreadsheet programs, such as the previously noted Lotus 1-2-3™, Quattro Pro™, and Excel™, include linear programming capabilities, but are not designed for solving problems with extremely large numbers of variables and constraints. Other mathematical programming packages are available which are designed specifically for solving linear and, in some cases, nonlinear programming problems, including very large problems. The user inputs values for the coefficients in the objective function and constraint equations for the problem formulation of concern. The optimizer program computes values for the decision variables. The General Algebraic Modeling System (GAMS) is a notable example of a general-purpose optimization package designed for developing large linear, nonlinear, and mixed integer programming models [Brooke et al., 1992]. GAMS is a high-level language that provides data management and model formulation capabilities as well as a set of mathematical programming optimizers.

Water management optimization models are often written from scratch, in FORTRAN or other languages, for a particular application. Already-written FORTRAN subroutines for performing linear or nonlinear programming computations are often incorporated into the coding of various models. The same code for the optimizer routines may be used in any number of different models.

1.7.3 Generalized Operational Reservoir System Analysis Models

A number of generalized operational reservoir system analysis models are described in Chapters 8–12. Most are public-domain software packages available from federal water agencies or universities. The majority are coded in FORTRAN77, but some are written in the other languages previously discussed.

Generalized means that the model is designed for various types of analyses of essentially any reservoir/river system. The user provides input data for a particular application without being concerned with formulating mathematical algorithms or writing code. Most of the generalized packages provide flexible, optional capabilities which the user selects through input data entries.

Operational means that a model is reasonably well documented and tested and is designed to be used by professional practitioners other than the original model developers. Generalized models should be convenient to obtain, understand, and use, and should work correctly, completely, and efficiently. Documentation, user support, and user friendliness of the software are key factors in selecting a software package. The extent to which a model has been tested and applied in actual studies is also an important consideration.

1.7.4 Comparison of Alternative Model-Building Approaches

Given unlimited time, funds, and computer programming expertise, developing a program from scratch using FORTRAN, C, C++, or another similar language will provide the greatest flexibility in developing a model to fit the particular needs of the water management application. This approach also requires the greatest programming expertise and time resources. Specific analysis needs may warrant coding new software specifically for a particular reservoir system and/or type of analysis. Many engineers and scientists naturally prefer the flexibility of working with programs they have coded themselves. However, for complex models, formulating algorithms, devising data management schemes, writing and debugging code, and testing new programs are extremely time-consuming and expensive. Thus, the time and personnel resources required for detailed model construction may reduce the resources available for other, more crucial aspects of the modeling study. Application of available general-purpose software or generalized reservoir system analysis models is often the optimal use of available funding and time resources.

Lotus 1-2-3, Quattro Pro, Excel, STELLA, GAMS, and other similar general-purpose modeling systems are polished products which reflect attention to enhanced user interfaces and graphics capabilities. They provide programming capabilities for developing computational algorithms which are simpler than the actual programming languages such as FORTRAN, C, and C++. These are flexible modeling environments that can be used for a broad range of applications. They are perhaps most pertinent for simpler applications, with the generalized reservoir systems analysis models becoming particularly advantageous for more complex problems.

Generalized reservoir systems analysis models have the advantage of already being written. The user provides input data without being concerned with formulating mathematical algorithms and writing code. Most of the packages provide flexible sets of user-selected optional modeling capabilities. Generalized modeling packages also play an important role in transferring knowledge. State-of-the-art concepts and methods are organized into the format of generalized reservoir system modeling packages.

2

RESERVOIRS IN THE
UNITED STATES

Examples of the numerous reservoir projects in operation throughout the United States are noted in this chapter. The chapter begins with summary statistics regarding dams worldwide and in the United States. The inventories of reservoirs operated by the Corps of Engineers and Bureau of Reclamation are briefly discussed. These federal agencies are the largest reservoir managers in the nation. Large multiple-reservoir systems operated by the Tennessee Valley Authority, Corps of Engineers, Bureau of Reclamation, State of California, and other agencies in the Tennessee, Missouri, Columbia, Colorado, and Sacramento/San Joaquin River Basins are described. These major systems have a long history of computer modeling, including significant studies during the 1990s. Models developed for these systems are cited in later chapters. The statewide inventory of reservoirs in Texas is used to illustrate the full range from small to large reservoirs operated by a variety of entities. A reservoir system in the Brazos River Basin of Texas is described in this chapter and adopted in Chapter 9 as a case study to illustrate modeling and analysis methods.

2.1 WORLDWIDE INVENTORY OF DAMS
AND RESERVOIRS

History does not record exactly when dams were first constructed; however, reservoir projects have served people for at least 5000 years, beginning in the cradles of civilization in Babylonia, Egypt, India, Persia, and the Far East. The history of dams closely follows the rise and decline of civilizations, especially in cultures highly dependent on irrigation [Smith, 1971; Schnitter, 1994].

TABLE 2–1   Distribution of Dams by Type and Height

Dam Height (m)	Type of Dam					
	Embankment	Gravity	Arch	Buttress	Multarch	Total
10-30	24,567	2222	775	175	74	27,813
30-60	3657	1294	428	110	48	5537
60-100	477	361	204	40	13	1095
100-150	116	65	83	12	—	276
150-200	21	8	24	—	—	53
Over 200	6	4	13	—	1	24
Total	28,844	3954	1527	337	136	34,798

Source: 1984 ICOLD World Register of Dams

The International Commission on Large Dams (ICOLD), which was founded in France in 1928, is a primary source of published data covering the worldwide inventory of dams. The ICOLD developed and maintains the *World Register of Dams*, which is a listing of large dams with information on type, dimensions, and ownership. The register is a compilation of data developed by national committees in the participating countries. For purposes of inclusion in the register, the ICOLD has defined a large dam as either greater than 15 m in height or between 10 and 15 m in height if at least one of several other criteria are met. The number of large dams in the world increased from 5196 in 1950 to over 35,000 in 1982. Data constraints caused several hundred dams to be omitted from the register, resulting in 34,798 dams being included in the 1984 register [ICOLD, 1984].

Table 2–1 shows the distribution of large dams by type and height. Most dams are of the embankment type, most earthfill but some rockfill. For heights above 60 m, there are more concrete gravity and arch dams than embankments. The number of dams decrease with height. Eighty percent of the large dams have heights less than 30 m. About 1% have heights greater than 100 m.

The distribution of large dams amongst continents is shown in Table 2–2 The 23 countries which had more than 100 large dams in 1982 are listed in Table 2–3. These 23 countries had 95% of the world's large dams in 1982 and 94% in 1950.

TABLE 2–2   Distribution of Dams by Continent

Continent	Number of Dams			
	In 1950		In 1982	
Asia	1541	29.7%	22,701	65.2%
Americas	2090	40.2%	7,241	20.8%
Europe	1293	24.9%	3,800	10.9%
Africa	123	2.4%	610	1.8%
Australia	150	2.9%	446	1.3%
Total	5196	100.0%	34,798	100.0%

Source: 1984 ICOLD World Register of Dams

TABLE 2–3   Countries With More
Than 100 Dams

Country	Number of Dams	
	In 1950	In 1982
China	8	18,595
United States	1543	5338
Japan	1173	2142
India	202	1085
Spain	205	690
Korea	116	628
Canada	189	580
Great Britain	378	529
Brazil	142	489
Mexico	109	487
France	164	432
Italy	199	408
Australia	122	374
South Africa	79	342
Norway	48	219
Germany	46	184
Czechoslovakia	44	142
Sweden	32	134
Switzerland	35	130
Yugoslavia	12	114
Austria	20	112
Bulgaria	4	108
Romania	6	106
Total	4,876	33,368

Source: 1984 ICOLD World Register of
Dams

Over half the dams in the world are located in China. Practically all the Chinese dams are of the embankment type, and 80% are less than 30 m in height. Most were constructed since 1950.

The extremely large dams and reservoirs are not necessarily located in the countries with the most numerous projects. In terms of storage capacity as of 1982, only six of the 25 largest reservoirs are located in the 23 countries listed in Table 2–3. Of the world's 25 highest dams, 14 are located in the 23 countries with more than 100 dams. Russia is particularly notable for constructing extremely large projects. As of 1982, the two highest dams in the world and eight of the 50 highest dams were located in the former USSR. In terms of storage capacity, the USSR had 14 of the 50 largest reservoirs in the world.

Turkey's development of its portion of the Tigris-Euphrates River Basin is a notable example of recent construction of major reservoir systems. The Southeast Anatolia Development Project (called GAP) is a comprehensive large-scale water resources management plan implemented during the 1980s and 1990s that includes 15 dams, 14 hydroelectric power stations, and 19 irrigation projects [Kolars and

Mitchell, 1991]. The Ataturk Dam on the Euphrates River is the largest dam in the GAP. Construction of the 179 m high rockfill dam and appurtenant structures was accomplished during the period from 1981 through the early 1990s. The Ataturk storage capacity of 49,000 million m^3 is the largest of any reservoir in Turkey, fifth largest in the world, and much larger than any reservoir in the United States.

2.2 RIVERS AND RESERVOIRS IN THE UNITED STATES

Numerous dam and reservoir projects regulate streams throughout the United States ranging from small creeks to major rivers. These projects range in size from small farm ponds owned by individual farmers, with storage capacities of several hundred m^3, to reservoirs on the Colorado and Missouri Rivers operated by the Bureau of Reclamation and Corps of Engineers, which have storage capacities exceeding 30 billion m^3. As indicated in Table 2–4, as of the early 1980s, an estimated 2654 reservoirs, including controlled natural lakes, in the United States and Puerto Rico contain storage capacities of 5000 acre-feet (6,167,000 m^3) or greater [U.S. Geological Survey 1985]. In addition, there are at least 50,000 reservoirs with capacities ranging from 50 to 5000 ac-ft (61,670 to 6,167,000 m^3) and millions of smaller farm ponds. The reservoir storage capacities shown in Table 2–4 include active and inactive conservation storage but exclude flood control and surcharge storage. Many thousands of flood-retarding dams with ungated outlet structures, constructed by the Soil Conservation Service (renamed the Natural Resource Conservation Service in 1994) and numerous other entities, store water only during floods and are not included in Table 2–4.

In terms of total storage capacity, the largest reservoirs in the United States are Lake Mead and Lake Powell on the Colorado River, owned by the Bureau of Reclamation, and Sakakawea, Oahe, and Fort Peck Reservoirs on the Missouri River, owned by the Corps of Engineers. The locations of these reservoirs are shown in Figure 2–1. Hoover Dam and Lake Mead are shown in Figure 2–2. Information provided in Table 2–5 includes both total storage capacity below the highest controlled water surface and the active portion of the storage capacity which is

TABLE 2–4   Reservoirs in the United States by Ranges of Storage Capacity

Storage Capacity Range (acre-feet)	Number of Reservoirs	Storage	
		(acre-feet)	(10^9 m^3)
Greater than 10,000,000	5	107,655,000	133
100,000–10,000,000	569	322,852,000	398
50,000–100,000	295	20,557,000	25
25,000–50,000	374	13,092,000	16
5,000–25,000	1411	15,632,000	19
Total	2654	479,788,000	592

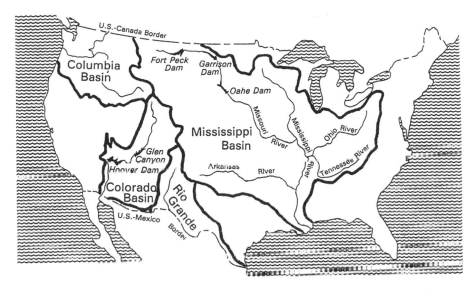

Figure 2–1   The four largest river basins and five largest reservoirs in the conterminous United States.

Figure 2–2   Lake Mead, impounded by Hoover Dam on the Colorado River, is the largest reservoir in the United States. (Courtesy U.S. Bureau of Reclamation)

TABLE 2–5   Five Largest Reservoirs in the United States

Reservoir	Mead	Powell	Sakakawea	Oahe	Fort Peck
Dam	Hoover	Glen Canyon	Garrison	Oahe	Fort Peck
River	Colorado	Colorado	Missouri	Missouri	Missouri
State	Nev & Ariz	Ariz & Utah	N Dakota	S Dakota	Montana
Owner/Operator	USBR	USBR	USACE	USACE	USACE
Year Completed	1936	1966	1956	1962	1940
Storage Capacity (million cubic meters, 10^6 m^3)					
Total	36,702	33,304	29,508	28,787	23,324
Active	21,405	25,071	23,354	22,063	18,041
Area (hectares)	65,840	65,070	154,600	150,900	100,800
Storage Capacity (thousand acre-feet)					
Total	29,755	27,000	23,923	23,338	18,909
Active	17,353	20,325	18,934	17,886	14,626
Area (acres)	162,700	160,800	381,900	372,800	249,000

Source: U.S. Bureau of Reclamation (1992)

Figure 2–3   Oroville Dam on the Feather River in California is the highest dam in the United States. (Courtesy California Department of Water Resources)

above the lowest outlet [Bureau of Reclamation, 1992]. The water surface area noted in Table 2–5 is at the top of conservation pool. These five largest reservoirs in the United States rank 27th, 29th, 34th, 36th, and 44th in storage capacity worldwide [Tilsley, 1983].

Oroville Dam, shown in Figure 2–3, is the highest dam in the United States and 14th highest in the world. Oroville Dam is a 235-m-high earth embankment on the Feather River, which was constructed by the State of California in the 1960s as the key unit of the State Water Project. The 221-m-high concrete gravity and arch Hoover Dam is the second highest dam in the United States and 18th highest in the world [Tilsley, 1983].

In terms of installed power capacity, Grand Coulee, John Day, and Chief Joseph Dams on the Columbia River, with installed capacities of 6180, 2160, and 2069 megawatts, respectively, are the largest hydroelectric power projects in the nation. The Grand Coulee project, shown in Figure 2–4, is owned by the Bureau of Reclamation. John Day and Chief Joseph Dams are owned by the Corps of Engineers.

The Mississippi River is the largest river in the United States in terms of discharge, watershed area, or length. With a length of 5970 km (3710 miles), the Mississippi is the fourth longest river in the world. The Nile, Amazon, and Yangtze

Figure 2–4   Grand Coulee Dam on the Columbia River is the most massive concrete structure and largest hydroelectric project in the United States. Franklin D. Roosevelt Reservoir extends 240 km upstream to the Canadian border. (Courtesy U.S. Bureau of Reclamation)

Rivers, with lengths of 6650, 6400, and 6300 km, respectively, are longer. The Mississippi River is the seventh largest river in the world ranked in terms of the mean discharge at its mouth of 17,300 m³/s (611,000 ft³/s). The Mississippi River has a drainage area, shown in Figure 2–1, of 3,230,000 km² (1,250,000 mi²), which includes 41% of the conterminous United States and a small portion of Canada. The largest subwatersheds of the Mississippi River Basin are the Missouri River Basin (1,370,000 km²), Ohio River Basin (528,000 km²), Arkansas River Basin (416,000 km²), and Red River Basin (93,200 km²).

Outside of the Mississippi and its subbasins, the next largest river basins in the United States are the Yukon (848,000 km²), St. Lawrence (782,000 km²), Columbia (668,000 km²), Colorado (629,000 km²), and Rio Grande (444,000 km²) [Iseri and Langbein, 1974]. The Yukon River Basin encompasses more than half of Alaska and a small portion of Canada. The St. Lawrence River Basin lies primarily in Canada. The Mississippi, Columbia, Colorado, and Rio Grande Basins, which are the largest river basins in the conterminous United States, encompass about 62% of the land area of the 48 states. These four largest basins are delineated in Figure 2–1.

The largest rivers, in terms of 1931–1970 mean flow rates, in the conterminous United States are shown in Figure 2–5 along with a graphic comparison of their mean flows [Iseri and Langbein 1974]. The rivers shown are those with mean

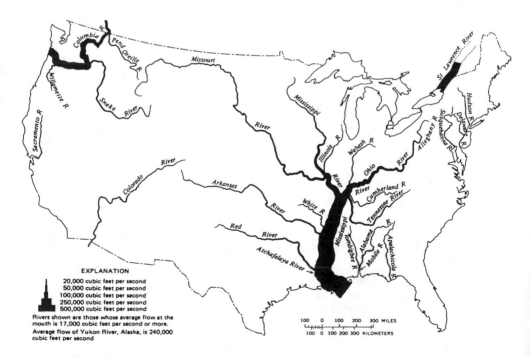

Figure 2–5 Mean flows of the largest rivers in the conterminous United States [Iseri and Langbein, 1974].

flows at their mouths that equal or exceed 17,000 ft^3/s (480 m^3/s). The outflow of all streams from the conterminous United States into the oceans or across national borders is about 2,000,000 ft^3/s (56,600 m^3/s). The rivers shown in Figure 2–1 account for more than 75% of this total flow. The remaining flows to the oceans are discharged by numerous smaller rivers and streams. Several rivers in the humid east (such as the Hudson, Delaware, and Susquehanna) and Southeast (such as the Appalachicola and Alabama Rivers) have relatively small drainage areas but much larger flows than rivers in the arid southwest (such as the Rio Grande) with much larger watershed areas.

2.3 CORPS OF ENGINEERS RESERVOIRS

The U.S. Army Corps of Engineers (USACE) is the nation's oldest and largest water resources development and management agency. The military history of the USACE dates back to the founding of the United States. Its mission, to improve the nation's rivers and harbors for navigation and flood control, originated in the 1800s. The present nationwide civil works program of the Corps of Engineers includes planning, design, construction, operation, and maintenance of facilities for multiple-purpose water resources development as well as other resource management and regulatory functions.

Information for each individual Corps of Engineers reservoir is available from a computer database [Hydrologic Engineering Center, 1994]. Summary data are presented by the Hydrologic Engineering Center [1990]. Storage capacities by purpose for 516 USACE reservoirs, including 114 navigation locks and dams, are summarized in Table 2–6. The 516 reservoirs have a total controlled storage capacity of 272,100 million m^3. About 117,100 million m^3, or 43%, of this capacity is designated for use exclusively for flood control. At least a portion of the storage capacity in each of 330 reservoirs is designated for use exclusively for flood control. An additional 1.1% and 0.2% of the total capacity is used exclusively for navigation and hydroelectric power, respectively. The remaining 151,600 million m^3, or 56%, of

TABLE 2–6   Storage Capacity of Corps of Engineers Reservoirs

Storage Allocation	Number of Reservoirs	Storage Capacity	
		(million m^3)	(acre-feet)
Exclusive flood control	330	117,100	94,950,000
Exclusive navigation	135	2900	2,354,000
Exclusive hydropower	5	475	385,000
Multiple-purpose use	385	151,600	122,926,000
Total storage in 516 reservoirs		272,100	220,615,000

the storage capacity provides multiple-purpose use in 385 reservoirs. The multiple-use storage capacity in each individual reservoir serves two or more purposes, which may include municipal and industrial water supply, irrigation, recreation, fish and wildlife enhancement, low-flow augmentation, flood control, navigation, or hydropower.

A map showing the geographic distribution of the Corps of Engineers reservoirs is presented in Figure 2–6. The reservoirs are operated by USACE district and regional offices. As shown in Figure 2–7, the continental United States is divided into eleven divisions which are further subdivided into 36 districts. Division and district boundaries generally coincide with river basin boundaries. These field offices are responsible for planning, design, construction, maintenance, and operation of the reservoir projects. Centralized reservoir control centers in division offices and water management sections in district offices develop and maintain operating plans and make release decisions for the reservoirs located within the respective districts [USACE, 1987]. Personnel at each individual reservoir project are responsible for the physical operation of the water control structures.

Table 2–7 shows the number of reservoir projects for which construction was completed in each decade. Construction of 149 of the reservoirs was completed during the 1960s. Construction of 237 reservoirs, or 46% of the 516 reservoirs, which account for about 69% of the total storage capacity, was completed during the 1950s and 1960s. Construction of reservoir projects has decreased dramatically since the 1960s.

Figure 2–6   Corps of Engineers reservoirs [Hydrologic Engineering Center, 1990].

Figure 2–7   Corps of Engineers divisions and districts.

The 516 reservoirs are owned and operated by the USACE. There are about 88 other reservoirs owned by other agencies for which the Corps of Engineers shares operational responsibilities [USACE, 1987]. Most of these projects are included in the Bureau of Reclamation database, described next. The Corps of Engineers is responsible for flood control operations at a number of projects constructed by the Bureau of Reclamation. Also, the Corps shares regulation responsibilities for several projects constructed under international agreements with Canada and Mexico. The USACE has also constructed or rehabilitated a number of additional projects which are operated by other agencies such as the Bureau of Reclamation.

TABLE 2–7   Number of Reservoirs Completed by Decade

	USACE Database		USBR Database	
Decade	Reservoirs	Percent	Reservoirs	Percent
Before 1900	9	1.7%	1	0.3%
1900–1909	3	0.6%	6	1.8%
1910–1919	3	0.6%	30	9.0%
1920–1929	8	1.6%	20	6.0%
1930–1939	67	13.0%	35	10.5%
1940–1949	58	11.2%	31	9.3%
1950–1959	88	17.1%	82	24.7%
1960–1969	149	28.9%	74	22.3%
1970–1979	96	18.6%	32	9.6%
1980–1989	35	6.8%	18	5.4%
After 1989	—	—	3	0.9%
Total	516	100.0%	332	100.0%

2.4 BUREAU OF RECLAMATION RESERVOIRS

The water resources development activities of the U.S. Bureau of Reclamation (USBR) of the Department of the Interior date back to the Reclamation Act of 1902. The USBR was created to plan and implement water projects needed to support population and economic growth in the arid west. USBR projects are located in the 17 western states, which are divided into the five regions delineated in Figure 2–8. The regional offices are responsible for the planning, construction, and operation of projects. Headquarters elements in Denver provide technical support and administrative coordination for the regional offices. USBR irrigation and multiple-purpose water resources development projects are major multiple-component systems. Structural components include storage dams, dikes, diversion dams, pumping plants, canals, pipelines, tunnels, and hydroelectric power plants.

Information regarding the 343 storage reservoirs in Bureau of Reclamation systems is provided in Table 2–8 [Bureau of Reclamation, 1992]. Of the 343 reservoirs, 231 were constructed by the USBR. The others are either operated by the USBR, were rehabilitated by the USBR, were financed through the USBR, or serve as non-USBR components of systems developed jointly by the USBR and other agencies, such as local water districts or the Corps of Engineers. The Bureau operates 130 reservoirs. Many reservoirs constructed by the Bureau are turned over to local irrigation or water districts or other entities for operation.

Figure 2–8 Bureau of Reclamation regions.

TABLE 2–8   Storage Reservoirs in Bureau of Reclamation Projects

	Number of Reservoirs	Active Capacity		Total Capacity	
		$(10^6 m^3)$	$(10^3 ac\text{-}ft)$	$(10^6 m^3)$	$(10^3 ac\text{-}ft)$
Constructed and operated by USBR	112	102,723	83,279	145,945	118,319
Rehabilitated and operated by USBR	9	235	191	238	193
Constructed by others and operated by USBR	9	4672	3788	4973	4032
Under construction by USBR	4	1011	819	1303	1057
Constructed by USBR and operated by others	115	32,571	26,406	39,138	31,729
Rehabilitated by USBR and operated by others	14	824	668	824	668
Constructed and operated by others	68	76,840	62,295	109,619	87,248
Constructed or rehabilitated under loan program	12	148	120	187	151
Total	343	219,025	177,566	300,227	243,397
Subtotal—constructed by USBR	231	136,305	110,504	186,386	151,105
Subtotal—operated by USBR	130	107,630	87,258	151,156	122,544

Source: U.S. Bureau of Reclamation (1992)

Almost all of the 343 reservoirs include irrigation as a primary purpose. The majority of the reservoirs serve multiple purposes which include irrigation. About 28% are single purpose irrigation reservoirs. Eighty-seven reservoirs include municipal and industrial water supply storage; 85 reservoirs are operated for hydroelectric power, including a few single-purpose hydropower reservoirs. Eleven of the multiple-purpose reservoirs are operated for navigation. Other purposes include fish and wildlife, recreation, river regulation, and sediment control.

The active storage capacity above the lowest outlet and total capacity below the highest controlled water surface are summarized in Table 2–8. The total storage capacity of individual reservoirs range from less than 100,000 m^3 at several reservoirs to 36,702,000,000 m^3 in Lake Mead behind Hoover Dam. The five largest storage reservoirs in the nation, listed in Table 2–5, are included in the USBR inventory summarized in Table 2–8. These five reservoirs account for half of the total storage capacity of the 343 reservoirs. Lake Mead and Lake Powell on the Colorado River are owned and operated by the Bureau of Reclamation. Garrison, Oahe, and Fort Peck Dams on the Missouri River are owned and operated by the Corps of Engineers. They are included in the USBR statistical compilation, as well as the USACE data cited earlier, because the overall development of the Missouri River Basin has been a joint effort of the two agencies. The Corps and Bureau partnership dates back to their formulation of a comprehensive basin-wide plan in the 1930s and 1940s called the Pick-Sloan Plan.

The number of reservoirs for which construction was completed in each decade is shown in Table 2–7. The numbers total 332 rather than 343 reservoirs, because the completion date is missing for several of the reservoirs in the database. The decades of 1950–1959 and 1960–1969 were most active, with construction of 82 and 74 reservoirs, respectively, being completed. The completion dates for the five largest reservoirs, listed in Table 2–5, range from 1936 for Hoover Dam to 1966 for Glen Canyon Dam.

2.5 TENNESSEE VALLEY AUTHORITY SYSTEM

As shown in Figures 2–1 and 2–5, the Tennessee River flows into the Ohio River just above its confluence with the Mississippi River. The Tennessee River Basin encompasses about 106,000 km^2 of seven southeastern states. The mean annual rainfall over the basin is about 132 cm. Annual snowfall averages about 20 cm, but does not create a snowpack. The mean flow of the Tennessee River at its mouth is about 1870 m^3/s. Terrain in the basin ranges from mountains and forests in the east to hills and open fields in the west.

The Tennessee Valley Authority (TVA) was created by the federal government in 1933 as a regional agency to promote the economic development and social betterment of a depressed area of the nation. In 1933, land in the Tennessee River Valley was underdeveloped and neglected, heavy rainfall had eroded the soil, and the forests had been cut over and burned. Almost every community along the major streams was subject to flood damage. Heavy flows from the Tennessee River also contributed to flooding on the Ohio and Mississippi Rivers. The TVA was charged with planning for the proper use, conservation, and development of the natural resources of the Tennessee River Basin. This was to be accomplished through flood control, power production, navigation, reduction of soil erosion, afforestation, elimination of agricultural use of marginal lands, industrial development and diversification, and community development.

The TVA system now includes 50 dams with a total storage capacity of about 16.9 million m^3, which is about 30% of the mean annual runoff. Norris Dam, shown in Figure 2–9, was the first dam constructed by the TVA. Seven reservoirs located on the main-stem Tennessee River and several of the larger storage reservoirs on tributaries are listed in Table 2–9. The primary TVA operating objectives are flood control, navigation, and hydroelectric power. Recreation, fish and wildlife, water quality, water supply, and vector control are also important. Streamflows are highest during the winter months and at minimum levels during the summer and fall. The reservoir system is operated to provide flood control, particularly in the winter months, and to augment streamflows in the summer and fall months for navigation, power generation, and other purposes.

The TVA schedules releases from 36 hydroelectric power projects, including 22 multiple-purpose projects and 14 single-purpose power dams. The TVA constructed 16 of the projects, acquired six from others, and operates 14 under contract

Figure 2-9   Norris Dam on the Clinch River in Tennesee, completed in 1936, was the first dam constructed by the TVA. (Courtesy Tennessee Valley Authority)

TABLE 2-9   Selected Tennessee Valley Authority Reservoirs

Project	River	Project Purposes[1]	Active Storage (10^6m^3)	Installed Capacity (megawatts)
Kentucky	Tennessee	PNFC	4944	175
Pickwick Landing	Tennessee	PNFC	515	220
Wilson	Tennessee	PN	66	630
Wheeler	Tennessee	PNFC	430	359
Guntersville	Tennessee	PNCA	200	102
Watts Bar	Tennessee	PNFC	467	157
Fort Loudon	Tennessee	PNFC	137	136
Tims Ford	Elk	FC	349	—
Hiwassee	Hiwassee	PCP	377	117
Norris	Clinch	PFC	2371	101
Fontana	Little Tennessee	PFC	1167	238
Douglas	French Brood	PFC	1543	121
Cherokee	Holston	PFC	1416	133
S. Holston	S. Fork Holston	PFC	435	35

[1]Project purposes are flood control (F), navigation (N), hydroelectric power (P), and conservation (C).

with others. Several other multiple-purpose reservoirs that do not include hydro-electric power are also operated by the TVA.

The TVA hydroelectric facilities were originally designed to provide most of the system power demand. However, TVA electrical energy generation has evolved into a thermal-based power system, with the hydroelectric plants being used primarily for carrying intermediate and peak loads. A major objective in developing operating plans is to minimize system fuel costs. A set of guide curves is developed each year based on the expected value of power generation. Decisions to draft storage at any point in time are based on the amount of reservoir storage available and the cost of the thermal generation that would otherwise be required. Hydropower release decisions are also influenced by minimum flow requirements for nonpower purposes and the desirability of maintaining reservoirs as high as possible during the summer months in the interest of reservoir recreation.

The operating plans include seasonal rule curves for allocating storage capacity between flood control and conservation purposes. The major flood control season extends from January through March. The total flood control storage capacity in each reservoir is kept empty during the winter flood season, except when regulating floods. Much of the flood control storage capacity is converted to conservation purposes during the spring months. The operating plan refill curves are based on balancing the diminished flood control requirements with a reasonable probability of refilling the conservation storage.

The 10 dams located on the main stem of the Tennessee River have navigation locks permitting barge traffic to be maintained on the 1000-km (625-mile) reach from the Ohio River to the city of Knoxville. In addition to the reservoirs listed in Table 2–9, the Raccoon Mountain off-stream pumped-storage hydropower project uses flows from the main river. Hydropower is generated at each of the main-stem projects. The multiple-purpose Kentucky Reservoir, located on the Tennessee River near its confluence with the Ohio River, is the largest reservoir in the system.

2.6 MISSOURI RIVER SYSTEM

The Missouri River Basin, above its confluence with the Mississippi River, encompasses 1,350,000 km^2 of 10 midwestern states and about 26,000 km^2 of Canada. The basin covers about one-sixth of the contiguous United States. Average annual precipitation over the basin ranges from 20 cm just east of the Rocky Mountains to about 100 cm in the southeastern portion of the basin. April through June is typically the wettest season of the year. Snowfall in the northern and central portions of the basin ranges from 50 cm in the lower basin to more than 250 cm in high-elevation Rocky Mountain locations. High streamflows on the Missouri River are caused by snowmelt and rainfall on the plains during March and April and by mountain snowmelt and rainfall during the period from May through July.

Numerous reservoirs located on tributaries throughout the basin are operated by various entities. The Corps of Engineers and Bureau of Reclamation have constructed 44 and 22 reservoirs, respectively, on tributaries. Six other Corps of Engineers reservoirs on the main-stem Missouri River in Montana, North Dakota, South Dakota, and Nebraska are the cornerstone for water management in the basin. The USACE system also includes navigation and bank stabilization improvements along the Missouri River from Sioux City, Iowa, to the mouth at St. Louis, Missouri.

The six main-stem reservoirs are listed in Table 2–10, and the three largest reservoirs are also included in Table 2–5 and Figure 2–1. Construction of the system was completed in 1964, and the reservoirs first filled to their normal operating level in 1967. The storage capacity of these six reservoirs is more than three times the average annual flow of the Missouri River. The system is operated for flood control, hydropower, water supply, water quality, irrigation, navigation, recreation, and preservation of environmental resources. Operating criteria involve storage allocations and release rules for navigation and non-navigation seasons of the year.

Controlled storage capacity in the six reservoirs totals 91.5 billion m^3. The storage capacity in the individual reservoirs is allocated to the following zones: inactive, exclusive conservation, joint conservation and flood control, and exclusive flood control. Immediately above the permanent inactive pools of each reservoir are the conservation storage zones, which account for about 70% of the total active capacity of the system. The uppermost zone in the reservoirs used exclusively for flood control is allocated 9% of the active storage capacity. The remaining 21% of the storage is in joint-use (flood control and conservation) zones below the exclusive flood control and above the conservation zones. The joint-use zone is normally empty in March, which is the beginning of the high runoff season; fills during spring and early summer runoff; and is emptied again during fall and winter. Water is carried over in the conservation pools from year to year in anticipation of infrequent extreme drought conditions [USACE, 1985].

Navigation releases are made to maintain depth requirements in the Missouri River between Sioux City and the confluence with the Mississippi River from early

TABLE 2–10   USACE Reservoirs on the Missouri River

Dam	Project Purposes[1]	Storage Capacity Active ($10^6 m^3$)	Storage Capacity Total ($10^6 m^3$)	Surface Area (hectares)	Installed Capacity (megawatts)
Fort Peck	FINPRW	18,040	23,320	100,800	165
Garrison	FINPRW	23,350	29,510	154,600	400
Oahe	FINPRW	22,060	28,790	150,900	595
Big Bend	IPRW	2,320	2,320	25,900	468
Fort Randall	FINPRW	4,980	6,910	41,300	320
Gavins Point	FINPRW	192	622	13,000	100
Total		70,942	91,472	486,500	2,048

[1] Project purposes are flood control (F), irrigation (I), water supply (S), navigation (N), hydropower (P), recreation (R), and fish and wildlife (W).

April to early December. Ice forms on the river during winter, restricting navigation. Ice can also create local flooding conditions if flows are maintained at the levels required for safe navigation. However, a minimum discharge is normally maintained throughout the winter months for water quality and power production.

The hydroelectric power plants at the six reservoirs have a total installed capacity of 2048 megawatts. The firm power output of the projects is used to meet a portion of the power requirements of the Western Area Power Administration's (WAPA) preference customers. This power is a mix of base load, intermediate, and peaking power. When secondary energy is available, it is marketed by WAPA to the area utilities at large, at a cost based on the value of the thermal plant fuel saved. The peak power demand occurs between mid-December and mid-February in the north due to home heating, and between mid-June and mid-August in the south due to air-conditioning loads.

Reservoir releases are minimal during the winter months. The river freezes over and is closed to navigation in the winter. Navigation resumes about April 1. The reservoirs are typically at their highest levels in July, following the spring snowmelt and early summer rains. Reservoir releases are made as necessary to evacuate the exclusive flood control zones. Releases are then made from conservation storage to maintain navigation flows on the mainstream Missouri through the remainder of the navigation season and to meet power requirements. Additional releases are scheduled in the early spring if the runoff forecast indicates that additional flood control storage is required. The refill period can extend from about the first of March through July or early August, depending on the runoff pattern.

In years of high runoff, the exclusive flood control zone may be used. Because of the danger of floods resulting from summer rainstorms, this storage is evacuated as rapidly as possible within downstream channel capacity constraints, spilling if necessary. The annual joint-use or multiple-purpose storage zone is regulated on an annual cycle. In good water years, this zone is full at the end of the refill season. Sufficient storage is available in this zone to meet full-service navigation requirements and average annual energy production through the drawdown season. The annual storage does not always refill, but in most years it is almost completely drafted.

Impacts on the environment are a major consideration in managing the Missouri River system. For example, the endangered least tern and piping plover have been a particular concern in recent years. These bird species are dependent on a nesting habitat that consists of sparsely vegetated sandbars on the Missouri River. Reservoir releases can cause major portions of the available habitat to be inundated.

2.7 COLUMBIA RIVER BASIN

From its source in Columbia Lake in Canada's Selkirk Mountains, the Columbia River flows northwestward and then southward through British Columbia to the United States. It crosses the international border north of Spokane, Washington, then flows southward across central Washington, where it is joined by a major tributary, the Snake River. The Columbia then turns westward to form the border between Washington and Oregon and enters the Pacific Ocean at Astoria, Oregon.

The climate of the region ranges from arid in parts of the interior to very wet in coastal areas. In parts of eastern Washington and Oregon and the Snake River basin, the mean annual precipitation is less than 20 cm. In contrast, some of the coastal mountain rain forests receive more than 510 cm annual precipitation.

2.7.1 Basin Development

The first navigation locks in the region were constructed during the 1870s on the Willamette River at Oregon City and at the cascades area near the present location of Bonneville Dam [Columbia River Water Management Group, 1994]. Hydroelectric development began in the late 1880s when electric "dynamos," now called generators, were installed on the Spokane River in Spokane, on the Willamette River at Oregon City, and on the Snoqualmie River at Snoqualmie Falls, east of Seattle. Harnessing the energy of the mainstem Columbia River began with the construction of Rock Island Dam near Wenatchee in 1932 and initiation of construction on Grand Coulee (Figure 2–4) and Bonneville Dams in the late 1930s. Bonneville Dam was constructed with state-of-the-art fish ladders to facilitate the passage of returning salmon. There were two other main periods of federal dam construction in the basin: in the 1950s, when the Hungry Horse, Chief Joseph, The Dalles, McNary, and Albeni Falls projects were built; and in the mid 1970s, with construction of the Columbia River Treaty projects, Dworshak Dam, and the lower Snake projects. Most construction of privately owned dams also took place during these periods.

The Columbia River Treaty between the United States and Canada, adopted in 1964, provided for the construction and operation of Mica, Arrow, and Duncan Dams in Canada, and Libby Dam in the United States. Under the terms of the treaty, each nation has designated an operating entity. The Canadian entity is the British Columbia Hydro and Power Authority. The Administrator of the Bonneville Power Administration and the Division Engineer of the USACE North Pacific Division represent the United States. These entities, in turn, appoint representatives to two committees, the Operating Committee and Hydrometeorological Committee, which are charged with carrying out the operating arrangements necessary to implement the treaty.

2.7.2 Reservoir Operation

Over 250 major reservoir projects on the Columbia River and its tributaries are operated for flood control, hydroelectric power, irrigation, navigation, fish and wildlife enhancement, recreation, low-flow augmentation, and municipal and industrial water supply. The 12 reservoirs with active storage capacities exceeding a million acre-feet (1.234 billion m³) are listed in Table 2–11. The larger storage projects are generally filled and emptied in an annual cycle. Some storage reservoirs may not necessarily refill each year if all the active storage is withdrawn. Many of the reservoirs are run-of-river or pondage hydroelectric power projects with little or no active storage capacity relative to mean annual streamflow. On-site power-generating facilities are located at some of the reservoirs. Other storage reservoirs make releases that pass through many powerhouses located different distances downstream. Some projects are designed specifically for daily reregulation of outflows from an upstream reservoir.

Certain seasonal guidelines or rule curves are permanently established, without change from year to year. Other rule curves are redeveloped annually or seasonally to reflect varying conditions. The operating guidelines or rule curves are designed to provide adequate space for flood control and the efficient use of storage and natural flows in meeting electric power needs reliably and efficiently and to reasonably ensure refilling of the reservoirs.

Operation of a number of hydroelectric power projects owned by a variety of public and private entities are coordinated in accordance with the Pacific Northwest Coordinating Agreement [Columbia River Water Management Group, 1994]. The seasonal operation of the system reservoirs are coordinated for the optimal use of their collective storage capacity. Each reservoir has several sets of rule curves.

The Coordination Agreement provides that prior to the start of each operating year (13 months from July 1 through August 31), a reservoir operating and storage schedule is developed to provide the optimum firm energy load carrying capacity for each reservoir in the coordinated system. System regulation studies are performed to define reservoir elevations as critical rule curves on a monthly basis to ensure that adequate firm energy will be available from the coordinated system if there is a recurrence of any critical flow conditions. Ensured refill curves, consisting of monthly reservoir elevations, are also determined to limit reservoir drafts and guide the refill of reservoirs. These curves provide a high degree of assurance that a reservoir will refill by the end of the operating year. In some cases, refill target elevations are recomputed each month during the refill season based on the latest snowpack and precipitation measurements. These are called variable energy content curves.

Two different regulation cycles for storage reservoirs in the Northwest are associated with the different climatic regimes found in the coastal and interior regions. In the interior Columbia Basin, water is released from reservoirs during the cold winter period to generate electricity during this period of relatively high loads. In years with high snow accumulation, flood control requirements may require additional drafting by spring. The reservoirs are then filled during the snowmelt period of April through August, regaining stored water for the next cycle and achieving flood control objectives in the process. After filling, reservoirs are generally maintained as full as possible during the summer, then lowered again in the fall to repeat the cycle.

West of the Cascade Mountains, where much of the winter precipitation falls as rain, the regulation is quite different. During the late summer and fall, reservoirs are lowered to provide flood control space for possible winter floods. Further drafting for power generation during the winter may lower the reservoir below the flood control rule curve. If flood control operation is required, any stored water is released immediately after the flood to regain storage space for future flood control. Most of the reservoirs west of the Cascades begin seasonal filling during February in proportion to the decreasing magnitude and possibility of flooding. This operation continues until the reservoirs finally reach their maximum level, which is normally during May. The reservoirs are usually held as full as possible during the summer for recreational use, although some downstream water uses may require some reservoir drafting. Drawdown must begin in the fall so that by November or early December there is sufficient storage space for winter flood control.

Fishery considerations are important in the Columbia River Basin, particularly in regard to migration, habitat, and production of salmon. Operations include moving fish past the dams with minimal losses and augmenting instream flows to better carry juveniles downstream to the ocean. The stocks of salmon that migrate up the river to reproduce each year have continually declined over the decades as dams and hydroelectric projects have been constructed. Certain types of salmon were listed as endangered in 1992, resulting in an intensification of ongoing efforts to incorporate fishery considerations in reservoir system operations. Several reservoir projects include fish ladders and releases to augment instream flows for protection of fisheries. The Corps of Engineers has operated a program since 1977 of transporting salmon by barge in their annual migration. Studies and discussions during the 1990s continue to address modifications in reservoir release policies and implementation of other measures for reversing the decline in salmon stocks.

TABLE 2–11   Selected Reservoirs in the Columbia River Basin

Dam	River	Owner or Operator[1]	Project Purposes[2]	Active Storage ($10^6 m^3$)	Installed Capacity (megawatts)
Mica	Columbia	B C Hydro	FP	14,860	1740
Revelstoke	Columbia	B C Hydro	FP	1570	1800
Keenleyside	Columbia	B C Hydro	FRPNI	8950	—
Libby	Kootenai	USACE	FPRC	6140	525
Duncan	Duncan	B C Hydro	FPI	1730	—
Hungry Horse	Flathead	USBR	FPIR	3900	328
Kerr	Flathead	Montana	FPR	1500	185
Albeni Falls	Pend Oreille	USACE	FPN	1430	49
Grand Coulee	Columbia	USBR	FIPR	6400	6480
Palisades	Snake	USBR	IFP	1480	130
American Falls	Snake	USBR	IFPM	2060	92
Dworshak	Clearwater	USACE	FPN	2490	460

[1]Owner or operator is British Columbia Hydro & Power Authority, Montana Power Company, U.S. Army Corps of Engineers, and U.S. Bureau of Reclamation.

[2]Project purposes are flood control (F), irrigation (I), municipal and industrial water supply (M), navigation (N), hydroelectric power (P), and recreation (R).

2.8 COLORADO RIVER BASIN

The Colorado River and its major tributaries originate as snowmelt-fed streams high in the Rocky Mountains. The river drains about 629,000 km^2 in seven states as it meanders southward to Mexico and the Gulf of California. For purposes of reservoir operations and other water management activities, the Colorado River Basin is divided into the Upper and Lower Basins. The dividing point is at Lee Ferry, Arizona, which is located below Glen Canyon Dam near the Arizona-Utah border.

Flows vary greatly from year to year as well as seasonally. The natural flow, assuming unregulated conditions without human impacts, of the Colorado River at Lees Ferry, Arizona, varies from 6.1 billion to 30 billion m^3/year for the period of record beginning in the early 1900s, with a mean of about 18 billion m^3/year. Snowmelt in the upper basin results in flows generally being highest from April to July. Mean annual precipitation in the arid lower basin is about 130 mm. Water is in extremely short supply in the Colorado River Basin. In an average water year, all available flow is diverted for consumptive use, and the river is dry before reaching the Gulf of California.

2.8.1 Law of the River

Management and use of the waters of the Colorado River Basin are governed by a series of international and interstate negotiations, legislative acts, U.S. Supreme Court decisions, compacts, and treaties, which were developed historically over several decades and are collectively called the Law of the River. Several of the many legal acts are noted as follows. The first action making up the Law of the River was approval in 1922 of the Colorado River Compact by representatives of the states in the Colorado River Basin. The compact appropriated the waters of the Colorado River between the upper and lower basins but did not divide the water among the states. The Boulder Canyon Act of 1928 approved the compact and authorized construction of Hoover Dam and the All-American Canal System. The Mexican Treaty of 1944 obligated the United States to deliver 1,500,000 acre-feet (1.85 billion m^3) of Colorado River water annually to Mexico. The Upper Basin Colorado River Compact of 1948 divided the Upper Basin Colorado River Compact apportionments among the Upper Basin States. This compact led to the Colorado River Storage Act of 1956, which established an Upper Basin Development Fund and authorized the initial phase of the comprehensive upper basin plan of development. The Colorado River Basin Project, authorized by Congress in 1968, includes several major storage and river regulation facilities. Guidelines were established in the authorizing legislation for the investigation and augmentation of the Colorado River, protection of areas of potential export, Mexican Water Treaty obligations, a Lower Basin shortage formula, and criteria for coordinated operation of Lake Powell and Lake Mead.

From the perspective of legal agreements, the Colorado is perhaps the most regulated river in the world. The water allocation system is complex. Water management is complicated by the fact that the amount of water legally allocated among the two nations and several states is greater than the estimated mean annual flow of the river.

2.8.2 Salinity

Salinity is a major problem in the Colorado River Basin, as well as in other western river basins. Naturally high salt loads are further increased by river basin development projects and irrigated agriculture. Numerous studies and investigations of salinity in the Colorado River Basin have been conducted since the early 1900s by the Bureau of Reclamation, U.S. Geological Survey, Environmental Protection Agency, several state agencies, and universities.

Under Minute No. 242, approved by the United States and Mexico in 1973, the United States agreed to adopt measures to ensure that the 1.36 million acre-feet of water delivered annually to Mexico upstream of Dolores Dam shall have an average salinity of no more than 115 parts per million over the annual average salinity of Colorado River water arriving at Imperial Dam, which is located just north of the international boundary. The agreement further provides that the salinity of the approximately 140,000 acre-feet of water delivered annually at the boundary at San Luis, Arizona, and downstream from Moreles Dam be substantially the same as that of water normally delivered there. Salinity control measures that have been implemented in the basin include irrigation management programs, desalting plants, and construction of facilities such as wells, dikes, pumps, desalters, and evaporation ponds to collect and dispose of saline water.

2.8.3 Reservoirs

The nine Bureau of Reclamation Reservoirs listed in Table 2–12 have a total active storage capacity of about four times the mean annual runoff. They are located in the deep canyons of the Colorado River and its principal tributaries in the upper basin. Lake Mead, impounded by Hoover Dam, and Lake Powell behind Glen Canyon Dam are the two largest reservoirs in the United States. These two reser-

TABLE 2–12 Major USBR Reservoirs in the Colorado River Basin

Dam	River	Project Purposes[1]	Storage Capacity Active ($10^6 m^3$)	Storage Capacity Total ($10^6 m^3$)	Surface Area (hectares)
Blue Mesa	Gunnison	FISP	923	1160	3710
Morrow Point	Gunnison	P	52	145	331
Crystal	Gunnison	P	16	32	122
Flaming Gorge	Green	FISPR	4340	4670	17,000
Navajo	San Juan	FIPS	1280	2110	6320
Glen Canyon	Colorado	FISPR	25,070	33,300	65,070
Hoover	Colorado	FIPS	21,400	36,700	65,800
Davis	Colorado	FIPRWS	1960	2240	11,400
Parker	Colorado	PS	222	797	8260

[1]Project purposes are flood control (F), irrigation (I), water supply (S), hydropower (P), recreation (R), and fish and wildlife (W).

Figure 2–10   Flaming Gorge Dam and Lake on the Green River in Utah is a component
of the Colorado River Storage Project. (Courtesy U.S. Bureau of Reclamation)

voirs are included in Table 2–5 and Figure 2–1 as well as Table 2–12. Flaming
Gorge Dam is shown in Figure 2–10. System operations are based on maintaining
flows to Mexico in accordance with the Water Treaty of 1944 and other agreements,
and on meeting obligations to agricultural, municipal, and industrial water users in
the several states within the United States as required by compacts and other allocation commitments. The reservoirs are also operated for water-quality control, flood
control, recreation, and hydroelectric power. Hydroelectric power generation is essentially limited to water supply releases for downstream users.

2.9 CALIFORNIA CENTRAL VALLEY PROJECT

The Sacramento River and San Joaquin River watersheds make up the Central Valley Basin which lies entirely within the northern two-thirds of the State of California. The Sacramento River and its tributaries flow southward, draining the northern
part of the basin. The San Joaquin River and its tributaries flow northward, draining
the southern portion. The two river systems join at the Sacramento-San Joaquin
Delta, flow through Suisun Bay and Carquinez Straits into San Francisco Bay, and
thence out the Golden Gate to the Pacific Ocean.

Water resources development and management in the Central Valley have been driven by California's pattern of precipitation and water demands. Most snow and rain occur in the north during winter, while water demands are concentrated in the south during the summer. The two largest water systems in California are the interrelated State Water Project, developed by the State of California, and the federal Central Valley Project. Major features of the State Water Project include Oroville Dam and Reservoir (Figure 2–3) on the Feather River; Harvey O. Banks Pumping Plant in the Sacramento-San Joaquin Delta; California Aqueduct delivery to the Southern San Joaquin Valley and Southern California; South Bay Aqueduct delivery to the San Francisco Bay area; and San Luis Reservoir, which stores water on the west side of the San Joaquin Valley for later release into the California Aqueduct. The following discussion focuses on the Central Valley Project.

The Bureau of Reclamation's Central Valley Project consists of the six major storage reservoirs listed in Table 2–13, a dozen smaller reservoirs for regulation and power generation, 39 pumping plants, and more than 800 km of canals. Most of the reservoirs and other project elements were constructed by the Bureau of Reclamation. The Folsom and New Melones Reservoirs were constructed by the Corps of Engineers, but the Bureau is responsible for their operation. San Luis Reservoir and associated facilities were constructed and are operated jointly by the Bureau of Reclamation and State of California, as components of both the State Water Project and Central Valley Project.

The primary purpose of the Central Valley Project is to provide a reliable water supply for the rich agricultural lands of the semi-arid Sacramento and San Joaquin Valleys. Some of the most productive agricultural land in the nation is in this region. Precipitation in the area is almost exclusively in the form of rainfall and ranges from 76 cm annually in the northern sections of the valley to 13 cm in the south. Most of this rainfall occurs in the period December through March, during the nonirrigation season. Agriculture is highly dependent on irrigation, which is supplied from both surface and groundwater. Water is stored in the high-runoff winter and spring months to meet irrigation requirements, which are greatest during the

TABLE 2–13   Major Central Valley Project Storage Reservoirs

Dam	River	Project Purposes[1]	Storage Capacity Active ($10^6 m^3$)	Storage Capacity Total ($10^6 m^3$)	Surface Area (hectares)
Trinity	Trinity	FIPR	2640	3020	6680
Whiskeytown	Clear	IP	264	297	1300
Shasta	Sacramento	FIPNRWS	4900	5610	12,000
Folson	American	FIPRWS	1140	1250	4610
New Melones	Stanislaus	FIPRWS	2990	2990	5060
San Luis	San Luis	IPRS	2420	2520	5260

[1]Project purposes are flood control (F), irrigation (I), water supply (S), navigation (N), hydropower (P), recreation (R), and fish and wildlife (W).

summer months. The extensive system of canals and pumping plants is used to transfer water from the water-rich Sacramento River basin in the north to the water-poor but intensively cultivated San Joaquin Valley in the south.

Flood control and hydroelectric power generation are also important functions. A substantial amount of power generation is required to meet pumping requirements for the water supply conveyance components of the system, and revenues from generation above these requirements help repay the cost of reservoirs and other facilities. Reservoir recreation, navigation on the Sacramento River, municipal and industrial water supply, fish and wildlife, and control of salinity intrusion in the Sacramento-San Joaquin River delta are secondary functions, but they have an important influence on how the system is operated.

Clair Engle (Trinity Dam, shown in Figure 2–11), Shasta, Folsom, San Luis, and New Melones Reservoirs are key storage projects. About 30% of the usable storage space in the major reservoirs is allocated to seasonal joint use storage, and the remaining 70% is allocated to carryover storage used to meet water and energy needs during extended droughts. The reservoirs are at their lowest elevations about the first of October, following the end of the irrigation season. Refill takes place in the winter and spring months. Flood control requirements are also most pronounced during the winter and spring. Water is required for irrigation the year around, but the bulk of demand occurs from May through August.

Figure 2–11   Trinity Dam and Clair Engle Lake on the Trinity River is a component of the California Central Valley Project. (Courtesy U.S. Bureau of Reclamation)

Drafting of the individual reservoirs follows somewhat different operating schedules, due largely to differences in runoff patterns in different areas. Much of the runoff in the basin comes from rainfall. Shasta Reservoir, shown in Figure 2–12, for example, is regulated almost exclusively to control rainfall runoff. A fixed flood control requirement is maintained through the first of January. Filling of the joint-use storage begins on that date, following statistically derived rule curves which are designed to ensure as great a probability of refill as possible while still maintaining flood control requirements through early February. Refill of Shasta is usually completed about the first of May. In contrast, the drainage area above New Melones is at a higher elevation, and most of the runoff is from snowmelt. Winter and early spring drafts are based on snowpack forecasts, thus permitting deeper drafts and greater power generation in high-runoff years. Refill of New Melones is usually not complete until mid-July. For the other storage projects, runoff comes from both rainfall and snowmelt, and provision of flood control space and scheduling of refill are based on a combination of statistically derived refill curves and snowmelt forecasts.

A large part of the irrigated land in the San Joaquin Valley is served by the Delta-Mendota Canal. The canal originates in the Sacramento River delta area and extends in a southeasterly direction, generally parallel to the San Joaquin River, for about 185 km, terminating about 50 km west of Fresno. Although the irrigation demand occurs primarily in the summer months, water is pumped into the canal from the Sacramento River year-round. Water in excess of irrigation needs is pumped

Figure 2–12 Shasta Dam and Lake on the Sacramento River is a component of the California Central Valley Project. (Courtesy U.S. Bureau of Reclamation)

into the San Luis Reservoir, to be held until the peak irrigation demand season, when it is released back into the Delta-Mendota Canal. A portion of the San Luis storage is also allocated to the state-operated California Water Project, with water being pumped from and discharged back into the California Aqueduct, which runs generally parallel to the Central Valley Project's Delta-Mendota Canal.

San Luis Reservoir, shown in Figure 2–13, is a large seasonal pumped-storage reservoir that is filled during the winter and used to provide irrigation and hydroelectric power requirements during the summer and fall. The San Luis Pumping-Generating Plant, located at San Luis Dam, lifts water by pump-turbines from the O'Neill Forebay, also shown in Figure 2–13, into San Luis Reservoir. During the irrigation season, water is released from San Luis Reservoir back through the pump-turbines to the forebay, and energy is reclaimed. As a pumping station to fill San Luis Reservoir, each of the eight pumping-generating units lifts 38.9 m^3/s at 88 m of head. As a generating plant, each unit passes 46.3 m^3/s at the same head. The O'Neil Forebay is used as a hydraulic junction point for federal and state waters of the Central Valley and State Water Projects. The forebay has a storage capacity of 70 million m^3, with the top 25 million m^3 used as regulation storage for the San Luis Pumping-Generating Plant. This is one of only a few seasonal pumped-storage plants in the United States.

Figure 2–13 San Luis Dam and Reservoir and the O'Neill Forebay are components of both the State Water Project and Central Valley Project. (Courtesy California Department of Water Resources)

The hydropower plants of the Central Valley Project provide a dependable capacity of 880 megawatts to the Pacific Gas and Electric Company (PG&E). Contracts with PG&E specify minimum 12-month, 6-month, and monthly energy delivery and provide benefits for exceeding these levels. The USBR submits a daily generating schedule, which is based upon Central Valley Project reservoir conditions, to PG&E, which uses this energy on an hour-by-hour basis in such a way as to minimize system fuel costs.

The water pumped from the Sacramento River delta is a mix of natural runoff and releases from storage projects such as Clair Engle and Shasta Reservoirs. Minimum flows must be maintained in the delta below the pumping plants to prevent salt water intrusion. Delta flows are controlled by regulating both upstream reservoir releases and the amounts of water pumped for export from the delta by the Central Valley Project and State Water Project.

Recreational use of Shasta and Folsom Reservoirs is much higher than the other projects, so the sequence of draft from the various reservoirs is scheduled to maintain Shasta and Folsom as high as possible through late May. This draft sequence ensures that irrigation requirements are met. However, it is less than optimal from the standpoint of power generation.

2.10 RESERVOIRS IN TEXAS

The author's home state of Texas is illustrative of the full range from small to large reservoirs owned and operated by a variety of entities [Wurbs, 1987]. Conservation and flood control storage capacities totaling 49,450 and 22,880 million m^3, respectively, are provided by 189 major reservoirs with individual capacities of 6,165,000 m^3 (5000 acre-feet) or greater. Texas has about 6000 reservoirs with surface areas greater than 4 hectares (10 acres). However, the 189 major reservoirs represent over 95% of the total storage capacity in all Texas reservoirs. Most of the major reservoirs are located on the rivers shown in Figure 2–14 or their numerous tributaries. Several of the reservoirs are located on smaller streams, not included in the figure, that flow directly into the Gulf of Mexico.

Texas is illustrative of the nation and world in general, in that there is a tremendous range of reservoir sizes; most reservoirs fall toward the small end of the size spectrum, and most of the total storage capacity is contained in the relatively few very large reservoirs. Many thousands of Texas natural lakes, farm ponds, flood-retarding and stormwater detention structures, recreation lakes, and small water supply reservoirs range in size from several hundred m^3 to 6,165,000 m^3 (5000 acre-feet). The 189 major reservoirs range in size from 6,165,000 m^3 to over 6.165 billion m^3 (5 million acre-feet). Table 2–14 shows the distribution of 189 reservoirs between various ranges of storage capacity. The 28 reservoirs with storage capacities exceeding 500,000 acre-feet (6165 million m^3) contain 75% and 88%, respectively, of the total conservation and flood control storage capacity of the 189 major reservoirs.

Figure 2–14   Major rivers in Texas.

In terms of total controlled storage capacity, the two largest reservoirs are Texoma on the Red River, owned by the Corps of Engineers, and Amistad on the Rio Grande, owned by the International Boundary and Water Commission. Texoma Reservoir contains 6640 million m³ of storage capacity, including conservation and flood control pools of 3360 and 3280 million m³, respectively. Amistad Reservoir has a capacity of 6470 million m³, allocated 4320 and 2150 million m³, respectively, for conservation and flood control. Toledo Bend Reservoir, with 5520 million m³ of conservation storage and no flood control, is the largest conservation storage reservoir in the state. In terms of water surface area at normal pool, with 73,500 hectares Toledo Bend is the largest reservoir in Texas and fifth largest in the nation.

As indicated by Table 2–15, over three-fourths of the present reservoir storage capacity in Texas was developed during the period 1940–1969. Although a few

TABLE 2–14   Reservoirs in Texas by Ranges of Storage Capacity

Storage Capacity Range (acre-feet)	Number of Reservoirs
Greater than 5,000,000	2
2,000,000–5,000,000	4
1,000,000–2,000,000	7
500,000–1,000,000	15
100,000–500,000	39
50,000–100,000	12
5000–50,000	110
Total	189

TABLE 2–15   Reservoirs Completed by Decade in Texas

Decade	Number of Reservoirs	Storage Capacity ($10^6 m^3$)
1900–1909	3	36
1910–1919	7	633
1920–1929	21	1040
1930–1939	15	5040
1940–1949	17	8150
1950–1959	39	17,850
1960–1969	49	30,540
1970–1979	22	3440
1980–1989	13	4880
1990–1995	3	702
Total	189	72,330

small reservoirs were developed in Texas for irrigation and power purposes prior to 1900, the oldest of the major reservoirs still in existence is a 12 million m^3 irrigation reservoir completed in 1900. The 35 major reservoirs in operation in 1935 were relatively small projects constructed by various local entities primarily for either irrigation or municipal and industrial water supply. Several of these early projects were also used for generating hydroelectric power. Manstield Dam impounding Lake Travis on the Colorado River was the first of the large multiple-purpose projects constructed in Texas by the federal government. The Bureau of Reclamation constructed the project from 1937 to 1942. Denison Dam impounding Lake Texoma on the Red River was the first Corps of Engineers reservoir project in the state. Construction was initiated and completed in 1939 and 1943, respectively. At that time, Denison Dam was the largest rolled earthfill dam in the United States. Lake Texoma is still the largest reservoir in Texas.

The 159 million m^3 Caddo Lake is the only natural lake in Texas with a storage capacity greater than 6.17 million m^3 (5000 acre-feet). All of the major reservoirs, including Caddo Lake now, are impounded by man-made dams. Caddo Lake is located on Cypress Creek in northeast Texas and northwest Louisiana. The Louisiana-Texas boundary runs through the lake. Although originally a natural lake, the Corps of Engineers completed construction of a dam in 1914 to preserve the lake because removal of logs and debris from the river downstream for navigation purposes had resulted in erosion, threatening to drain the lake. Construction of a new dam was completed in 1971 because the 1914 dam was found to no longer be safe.

2.10.1 Reservoir Managers

The 189 major reservoirs are owned and operated by an international agency, three federal agencies, 43 water districts and river authorities, 39 cities, 2 counties, a state agency, and 22 private companies. A number of other water management entities contract with these reservoir owners for the use of water and storage capacity. The reservoirs are categorized in Table 2–16 by type of owner.

Most of the major reservoirs in Texas were constructed by state and local governmental agencies or private industry. However, two-thirds of the total storage capacity is contained in reservoirs constructed by federal agencies. Most of the federal reservoirs are large multiple-purpose projects. Federal involvement in reservoir development and management in Texas is summarized in Table 2–17. The data in Table 2–17 do not include federal grants and loans, such as those provided by the early Works Progress Administration (WPA) program, which helped finance several of the nonfederal projects. The federal agencies constructed 21 of the 28 projects with storage capacities exceeding 500,000 acre-feet (6.17 million m³). The 42 projects constructed by federal agencies contain 52%, 99.9%, and 67%, respectively, of the conservation, flood control, and total storage capacities of the 189 major reservoirs. Seven of the federally constructed projects have been turned over to nonfederal entities for operation and maintenance.

International Falcon and Amistad Reservoirs on the Rio Grande were constructed under the terms of a 1944 treaty between the United States and Mexico. The reservoirs are owned and operated by the United States and Mexico Sections of the International Boundary and Water Commission. The United States has 58.6% and Mexico 41.4% of the conservation storage capacity. The Bureau of Reclamation designed and supervised construction of the Falcon project. The Corps of Engineers designed and supervised construction of Amistad. The two reservoirs are operated as a system for flood control, water supply (primarily irrigation), hydro-

TABLE 2–16   Reservoir Owners in Texas

Type of Owner	Number of Reservoirs	Storage Capacity (million m³)		
		Conservation	Flood Control	Total
International Boundary and Water Commission	2	7118	3272	10,390
Corps of Engineers	32	14,253	17,094	31,347
U.S. Fish & Wildlife Service	1	22	—	22
U.S. Forest Service	1	10	—	10
Texas Parks and Wildlife Department	1	7	—	7
Water Districts and River Authorities	57	19,827	1633	21,460
Jointly Owned by Cities and Water Districts or River Authorities	5	3274	306	3580
Cities	49	3521	576	4097
Counties	5	66	—	4097
Private Companies	36	1348	—	1348
Totals	189	49,446	22,881	72,327

TABLE 2–17   **Federal Involvement in Reservoir Development and Management**

Federal Involvement	Number of Reservoirs	Storage Capacity (million m³)		
		Conservation	Flood Control	Total
Constructed, owned, and operated by International Boundary and Water Commission	2	7120	3270	10,390
Constructed, owned, and operated by Corps of Engineers	31	14,090	17,090	31,180
Major modification by Corps of Engineers	2	553	306	859
Constructed by Bureau of Reclamation and maintained and operated by nonfederal sponsors	5	3800	2190	5990
Constructed by Soil Conservation Service and maintained and operated by nonfederal sponsors	2	22	—	22
Constructed by Soil Conservation Service and owned and operated by Fish and Wildlife Service	1	22	—	22
Constructed, owned and operated by U.S. Forest Service	1	10	—	10
Total	44	25,620	22,860	48,480

electric power, and recreation. Much of the water use occurs in the Lower Rio Grande Valley agricultural region. The Texas Natural Resource Conservation Commission, through its water master program, administers the United States share of diversions from the Rio Grande. Several thousand irrigators, cities, and other entities hold water rights permits. The International Boundary and Water Commission makes water supply releases from Falcon and Amistad Reservoirs as requested by the water master. Hydroelectric energy is marketed through the Western Area Power Administration of the U.S. Department of Energy.

The U.S. Army Corps of Engineers owns and operates two single-purpose flood control reservoirs and 30 multiple-purpose projects, which all provide flood control, water supply, and recreation; three projects also include hydroelectric power. Although the Corps of Engineers operates and maintains its projects upon completion of construction, water supply withdrawals or releases from conservation storage are made at the discretion of the nonfederal sponsors. All construction, maintenance, and operation costs allocated to water supply are reimbursed by nonfederal sponsors through contractual arrangements with the USACE. Hydroelectric power is marketed through the Southwestern Power Administration to electric cooperatives, municipalities, and utility companies. The Corps is directly responsible for flood control operation. The Galveston and Tulsa Districts each operate three of the Texas reservoirs. The Fort Worth District operates the other 26 Corps of Engi-

neers reservoirs. The Corps of Engineers is also responsible for flood control operations at three other reservoirs that were constructed by the Bureau of Reclamation
and are now maintained by two cities and a water district.

The Bureau of Reclamation has constructed five reservoir projects in Texas.
All five projects contain water supply storage, and three contain flood control. The
Bureau of Reclamation maintains and operates a number of reservoirs in the other
western states. However, the reservoirs in Texas have been turned over to local
sponsors for operation and maintenance. The federal government retains ownership
of the reservoirs until all reimbursable costs have been repaid by the nonfederal
sponsor.

State and local governmental entities operate 117 reservoirs, which contain
54%, 11%, and 40%, respectively, of the conservation, flood control, and total capacities of the 189 major reservoirs. This includes the seven projects constructed by
the Bureau of Reclamation and Soil Conservation Service, which are operated and
maintained by nonfederal sponsors. Nonfederal sponsors also control all the water
supply storage in the Corps of Engineers reservoirs and reimburse all costs allocated to water supply.

River authorities own a number of the nonfederal reservoirs and have contracted for the conservation storage in many of the Corps of Engineers' projects.
River authorities are a special type of water district created to develop and manage
water resources from a basin-wide perspective. The 19 river authorities finance
their activities primarily through operational and service fees. Some river basins in
Texas are served by a single river authority, while other basins are served by several
authorities. The Brazos River Authority, created in 1929, was the first authority
ever set up in the United States to administer the waters of a major river basin.
Thus, Texas created its first river authority four years before the creation of the
Tennessee Valley Authority by the federal government.

Private companies constructed, own, and operate 36 major reservoirs containing no flood control and less than 3% of the total conservation storage of the major
reservoirs. Most of these projects were constructed by electric companies to provide
cooling water for steam-electric generating plants.

2.10.2 Hydrology and Water Use

Texas is a large state with diverse climate, geography, economy, and water resources. For example, the mean annual precipitation varies from 20 cm at El Paso in
arid West Texas to over 140 cm in the humid lower Sabine River Basin. Statewide
mean annual precipitation has ranged from 36 cm in 1917, the driest year on record,
to 108 cm in 1942, the wettest year.

Depleting groundwater reserves are resulting in a greater reliance on surface
water relative to groundwater. Of the total water diverted in 1980, about 60% was
from aquifers and 40% from reservoirs and rivers. In 1990 the relationship was re-
versed, with 61% and 39%, respectively, of the total water supply diversions being
from surface and groundwater supplies. In 1990 surface water diversions were dis-

tributed among types of use as follows: irrigation (44.1%), municipal (27.9%), manufacturing (19.5%), steam electric (4.6%), livestock (2.6%), and mining (1.3%). Groundwater use in 1990 was distributed as follows: irrigation (71.7%), municipal (21.3%), manufacturing (3.0%), livestock (1.9%), mining (1.3%), and steam electric (0.8%). These data are based on diversions, except for steam electric cooling, which is consumptive use only [Texas Water Development Board, 1990].

Surface water management in Texas is influenced more by a long-term threat of drought than seasonal fluctuations in streamflow and/or water use. Water must be stored through many wet years to be available during drought conditions. Although reservoir storage may be significantly depleted within several months, severe drought conditions are characterized as a series of several dry years rather than the dry season of a single year. The present water management system, with current demands and development, has not been tested by a major drought comparable to those of the 1910s, 1930s, and 1950s. The hydrologically most severe drought of record in Texas began in 1950 and ended in April 1957 with one of the largest floods on record. Although severe dry conditions and reservoir drawdowns have occurred periodically, the years since 1957 have been characterized by relatively abundant precipitation and streamflow compared to the droughts of the 1950s and earlier.

2.10.3 Water Rights

With implementation of a unified permit system during the 1970s and '80s, water rights became a key consideration in surface water management. Rights to divert and use streamflow in Texas have been granted historically under Spanish, Mexican, Republic of Texas, and State of Texas laws. Early water rights were based on various versions of the riparian doctrine. A prior appropriation system was later adopted and then modified. An essentially unmanageable system evolved, with various types of water rights existing simultaneously, with many rights being unrecorded. The Water Rights Adjudication Act of 1967 was enacted to remedy this situation. The stated purpose of the act was to require a recording of all claims for water rights which were not already recorded, to limit the exercise of those claims to actual use, and to provide for the adjudication and administration of water rights. The water rights adjudication process required to merge all existing rights into a unified permit system was initiated in 1968 and completed in the late 1980s.

The Texas Natural Resource Conservation Commission (TNRCC) administers the approximately 8000 active permits in effect in 1995 for use of the waters of the 15 major river basins and eight coastal basins of the state. Water rights permits are held by individual citizens, private companies, cities, water districts, river authorities, and other public agencies. Applications for additional water rights or modifications to existing rights can be submitted to the TNRCC at any time. Applications are approved if unappropriated water is available, existing rights are not impaired, water conservation will be practiced, and the water use is not detrimental to the public welfare. After approval of an application, the TNRCC issues a permit giving the applicant the right to use a stated amount of water in a prescribed manner.

The Texas Water Code is based on the prior appropriation doctrine. Water rights permits include specification of priority dates. For permits issued based on adjudication of existing rights pursuant to the Water Rights Adjudication Act of 1967, priority dates were established based on historical legal rights and actual water use. Since completion of the adjudication process, priorities for additional new rights are based on the dates that the permits are issued.

2.10.4 Reservoir Operation

As discussed in Chapter 3, reservoir operation is based on the conflicting objectives of maximizing the amount of water available for conservation purposes and maximizing the amount of empty space available for storing flood water to reduce downstream damages. Each of the major reservoirs in Texas is operated either solely for conservation purposes, solely for flood control, or a combination of conservation and flood control, with pools for each being separated by a designated top of conservation pool elevation. Institutional arrangements for constructing and operating reservoirs are based on having separate pools for flood control and conservation. Planning, design, and operational problems associated with flood control are handled separately from those associated with conservation. Allocation of storage capacity between flood control and conservation pools is illustrated in Figure 3–4. Most of the multiple-purpose flood-control reservoirs in Texas have constant top of conservation pool elevations with no seasonal variation. Two of the Corps of Engineers' projects have seasonal rule curves. The International Boundary and Water Commission reservoir operating policies provide for an optional seasonal rule curve which may or may not be adopted in any particular year.

Whereas conservation operations throughout the state are the responsibility of a multitude of entities, the responsibility for flood control operations is highly centralized. The International Boundary and Water Commission is responsible for flood control operations of the Falcon and Amistad Reservoirs on the Rio Grande. One small flood control reservoir is operated by a city. The Corps of Engineers is responsible for the remaining 32 flood control reservoirs, which include its own projects and those constructed by the Bureau of Reclamation. The conventional flood control operating procedures outlined in Chapter 3 are generally followed in Texas.

Approximately 2000 flood-retarding dams constructed by the Soil Conservation Service in rural watersheds and thousands of small urban stormwater detention basins are not included in the 189 major reservoirs. Inflows are passed through uncontrolled (ungated) outlet structures in these smaller flood-retarding structures.

All but three of the major reservoirs in Texas contain conservation storage capacity. The primary conservation purposes served are municipal, industrial, and agricultural (primarily irrigation) water supply, cooling water for steam-electric plants, hydroelectric power generation, and recreation. Reservoir recreation is extremely popular and a major consideration in reservoir operation. Texas has 22 hydroelectric plants with an installed generating capacity of 546 megawatts. These facilities provide about 0.5% of the electrical energy generated in the state. Hydroelectric power is primarily used to provide for peak loads. At some projects, releases may be made specifically for hydropower. However, more typically, hydroelectric energy is generated by passing spills and releases for downstream water supply through the turbines.

Reservoir operation procedures for water supply purposes are based essentially on meeting water demands subject to institutional constraints related to water rights, project ownership, and contractual agreements. The organizational setting for water supply operations involves a multitude of water users and suppliers working under various contractual arrangements. Although most surface water is used within the river basin in which it originates, significant interbasin transfers do occur. Several interbasin transfers occur in upper watersheds, and extensive transfers occur from the major river basins to the adjoining coastal basins.

Water supply withdrawals are made at many projects through pumping plants with intake structures located in the reservoir. In many other cases, withdrawals from a river are made at locations up to several hundred kilometers below the dam or dams that regulate the flow.

A majority of water supply reservoirs are operated as individual units to supply specific customers. However, several multiple-reservoir systems are operated with various degrees of interaction between the component reservoirs. System operation typically involves maintaining a balance between storage levels in the component reservoirs. Hydroelectric power generation is also of concern in system operation. Releases are coordinated to meet water supply demands while minimizing the amount of water bypassing the turbines.

2.11 BRAZOS RIVER BASIN

As indicated by Figure 2–15, the Brazos River Basin extends from eastern New Mexico southeasterly across the state of Texas to the Gulf of Mexico. The basin drainage area is 118,000 km^2, with about 111,000 km^2 in Texas. The portion of the basin lying in New Mexico essentially does not contribute to streamflow. The basin encompasses about 16% of the land area of Texas. Mean annual precipitation varies from about 41 cm in the western (upstream) end of the basin to 130 cm in the lower basin near the Gulf of Mexico. Mean streamflow at the mouth of the Brazos River is 200 m^3/s.

From its inception at the confluence of the Salt Fork and Double Mountain Fork, the Brazos River flows in a meandering path some 1480 km to the city of Freeport at the Gulf of Mexico. In the middle and lower reaches, the Brazos is a rolling river flanked by levees, cotton fields, and hardwood bottoms. In its upper reaches, the Brazos River is a gypsum-salty intermittent stream. Upon its descent from the high plains and Caprock Escarpment, the Brazos River flows through a

Figure 2–15   Brazos River Basin.

semi-arid region of gypsum and salt-encrusted hills and valleys containing numerous salt springs and seeps. Natural salt pollution from emissions from underlying geologic formations results in high salinity concentrations in the upper reaches of the main-stem Brazos River. Water quality improves greatly in the lower reaches due to inflows from good-quality tributary rivers and creeks.

The Brazos River Basin is representative of several major river basins in the southwestern United States regarding natural salt pollution. Geologic formations underlying portions of the upper watersheds of the Arkansas, Brazos, Canadian, Colorado, Pecos, and Red River Basins in the states of Colorado, Kansas, New Mexico, Oklahoma, and Texas are sources of salt emissions to the rivers. Millions of years ago, this region was covered by a shallow inland sea. The salt-bearing geologic formations were formed by salts precipitated from evaporating sea water. Salt springs and seeps and salt flats in upstream areas of the basins now contribute large loads to the rivers. The natural salt contamination significantly impacts water management and utilization.

A total of about 1200 reservoirs included in the Texas Natural Resource Conservation Commission dam inventory are located in the Brazos River Basin. About 600 of the reservoirs in the basin are included in the water rights permit system. Water rights permits are not required for numerous small ponds, lakes, and flood retarding structures. Forty-one reservoirs in the Brazos River Basin have storage capacities exceeding 5000 acre-feet (6,165,000 m^3). The 41 major reservoirs account for about 95% of the total conservation storage capacity of the 1200 reservoirs. A system of 12 reservoirs owned and operated by the Fort Worth District of the USACE and the Brazos River Authority (BRA) is shown in Figure 2–16 and Table 2–18. The 12

Figure 2–16   Brazos River Authority and Corps of Engineers reservoirs in the Brazos River Basin.

USACE/BRA reservoirs contain all of the controlled (gated) flood control storage capacity and more than two-thirds of the conservation storage in the basin.

Nine of the reservoirs in Figure 2–16 and Table 2–18 were constructed by the Corps of Engineers as components of a comprehensive basin-wide plan of development. The Corps projects contain about half of the conservation capacity and all of

TABLE 2–18   USACE/BRA Reservoirs in the Brazos River Basin

		Storage Capacity (million m³)				Surface Area (hectares)
Reservoir	River	Inactive	Active Conservation	Flood Control	Total	
Possum Kingdom	Brazos	273	430	—	703	5840
Granbury	Brazos	65	124	—	189	530
Whitney	Brazos	467	306	1693	2466	9530
Aquilla	Aquilla	—	65	115	180	1330
Waco	Bosque	—	188	708	896	2940
Proctor	Leon	—	73	389	462	1870
Belton	Leon	—	552	794	1346	4980
Stillhouse	Lampases	—	291	487	778	2600
Georgetown	San Gabriel	—	46	115	161	530
Granger	San Gabriel	—	81	220	301	1780
Somerville	Yequa	—	197	428	626	4640
Limestone	Navasota	—	278	—	278	5750
Total		805	2631	4949	8385	42,320

the flood control capacity of the 41 major reservoirs in the basin. Georgetown, Aquilla, Granger, Proctor, Somerville, Stillhouse Hollow, Waco, Belton, and Whitney Reservoirs are each operated by the Fort Worth District for flood control, water supply, and recreation. Whitney Reservoir serves the additional purpose of hydroelectric power generation. USACE Fort Worth District personnel operate and maintain the nine federal multiple-purpose projects. The USACE is responsible for flood control operations. Conservation releases are made as directed by the local project sponsor, which for most of the conservation capacity is the Brazos River Authority. The BRA has contracted for the water supply capacity in each of the USACE projects, except Fort Hood military base has 3.2% of the conservation storage in Belton Lake and the City of Waco has 12.5% of the conservation storage capacity in Lake Waco. The City of Waco is also the primary customer for the 87.5% of the Lake Waco conservation capacity controlled by the BRA. The Southwestern Power Administration is responsible for marketing hydroelectric power from Whitney Reservoir, which it sells to the Brazos Electric Power Cooperative.

In addition to controlling the conservation storage in the nine USACE projects, the BRA constructed, owns, and operates Granbury, Limestone, and Possum Kingdom Reservoirs. The 12 reservoirs are operated as a system to supply municipal, industrial, and agricultural water users both in the vicinities of the reservoirs and at downstream locations. Much of the water diverted from the Brazos River is actually used in the adjoining San Jacinto-Brazos Coastal Basin, which lies between the cities of Houston and Galveston. Water is conveyed to users through extensive canal systems. The BRA sells water through contracts with cities, water districts, farmers, electric power utilities, petrochemical companies, and other industrial users. Possum Kingdom Reservoir provides hydroelectric power as well as water supply. The BRA sells the power to the Brazos Electric Power Cooperative.

The Brazos River Authority operates another reservoir project, Lake Alan Henry, that is not included in Figure 2–16. It is located on the South Fork of the Double Mountain Fork of the Brazos River, in the extreme upper basin near the city of Lubbock, shown in Figure 2–15, and supplies water for Lubbock.

About 1100 public agencies, cities, private companies, and individual citizens hold water rights permits for storage and use of the waters of the Brazos River and its tributaries. These entities hold about 1240 water rights permits with annual diversion rights totalling 2.87 billion m^3 and storage capacities in 592 reservoirs totalling 5.12 billion m^3. A majority of the water rights are held by private citizens and involve relatively small amounts of water. Cities and other public agencies hold most of the permits with larger diversion and storage amounts. The Brazos River Authority holds the water rights permits associated with 11 reservoirs. The City of Waco holds the water rights for Waco Reservoir, which is the other reservoir of the 12-reservoir BRA/USACE system. The total diversion and storage rights associated with the 12 reservoirs account for 42% of the total permitted annual diversion amount and 56% of the storage of all the water rights in the Brazos River Basin. Most of the conservation pool of Whitney Reservoir, which is the largest reservoir in the basin, is allocated to hydroelectric power and has no water right permits.

The major reservoirs in the basin include 29 reservoirs in addition to the 12-reservoir Brazos River Authority and Corps of Engineers system. The BRA's additional newest reservoir, Lake Alan Henry, is committed entirely to supplying water for the city of Lubbock. It contains 143 million m³, which is about 3% of the total conservation storage capacity of the 41 major reservoirs in the basin. Eleven reservoirs, with roughly 7% of the total conservation storage capacity of the 41 major reservoirs, are owned and operated by cities for municipal and industrial water supply and recreation. Six reservoirs, containing approximately another 10% of the total conservation storage capacity, are owned and operated by municipal water districts which supply water to member cities and other users. Six other reservoirs, with about 6% of the total storage, are owned and operated by electric utility companies to provide cooling water for thermal-electric power plants. The five remaining relatively small reservoirs are owned by a chemical company (two reservoirs), manufacturing company, water company, and ranch.

As indicated in Table 2–18, storage capacity in the nine federal reservoirs is allocated between flood control and conservation pools. Unlike the previously discussed seasonally varying rule curves adopted for the TVA, Missouri, Columbia, Colorado, and Central Valley systems, the designated top of conservation pool elevations are constant throughout the year for the Brazos River Basin reservoirs. This is due largely to the significant probability of flooding during any season of the year. Following the conventional procedures outlined in Chapter 3, the Corps of Engineers operates the nine reservoirs as a system to control flooding below the dams and throughout the lower basin.

3

RESERVOIR SYSTEM OPERATION

An operating plan or release policy is a set of guidelines for determining the quantities of water to be stored and to be released or withdrawn from a reservoir or system of several reservoirs under various conditions. In this book, the terms *operating* (or *release* or *regulation* or *water control*) *procedures, rules, schedule, policy,* or *plan* are used interchangeably. Operating decisions involve allocation of storage capacity and water releases between multiple reservoirs, between project purposes, between water users, and between time periods. Operating plans provide guidance to reservoir management personnel. In the modeling and analysis of a reservoir system, some mechanism for representing operating rules and/or decision criteria must be incorporated into the model.

A wide variety of operating policies are presently in use at reservoir projects throughout the United States and the world. They range from operating rules which specify ideal pool levels, but provide no guidance on what to do when deviations from these pool levels become necessary, to operating rules that define very precisely how much water to release under all possible conditions. For many water supply reservoirs, operations are based simply on making withdrawals or releases as necessary to meet water demands. Flood flows pass through uncontrolled spillways, and no predeveloped plans are in place for responding to supply depletion during infrequent severe droughts. On the other hand, complex regulation plans guide the operation of many reservoirs, including major multiple-purpose, multiple-reservoir systems like those discussed in Chapter 2. Typically, an operating plan involves a framework of quantitative rules within which significant flexibility exists for operator judgment. Day-to-day operating decisions may be influenced by a complex array of factors and are often based largely on judgment and experience. Operating procedures may change over time with experience and changing conditions.

3.1 INSTITUTIONAL SETTING

Institutional considerations are fundamental in establishing and modifying reservoir operating plans. Water resources development and management is accomplished within a complex framework of organizations, traditions, agreements, programs, policies, and political processes. Funding and financial arrangements are key considerations in constructing reservoir projects and establishing operating strategies. Water is a publicly owned resource, and its allocation and use are governed by state water rights systems. Water resources are also allocated based on agreements between reservoir owners and water customers, between states, and between nations. River basin management must be consistent with federal and state laws and policies.

Environmental legislation has had a significant influence on reservoir system operations. For example, the Fish and Wildlife Conservation Act of 1958 (PL 85-624) established the policy that fish and wildlife conservation be coordinated with other project purposes and receive equal consideration. The National Environmental Policy Act of 1970 (PL 91-190) articulated the policy of protecting the environment and established requirements for evaluating the environmental impacts of federal actions. Requirements for conservation of endangered species, pursuant to the 1973 Endangered Species Act (PL 93 205), have had significant impacts on the operations of several major reservoir systems. For example, although salmon migration had been for decades an important consideration in reservoir system management in the Columbia River Basin, the 1992 listing of certain types of salmon as endangered resulted in intensified fish protection efforts [Prendergast, 1994].

The water management community consists of water users, concerned citizens, public officials, professional engineers and scientists, special interest groups, businesses, utilities, cities, and local, state, regional, federal, and international agencies. In addition to the entities that own and operate a reservoir system, numerous other public agencies, project beneficiaries, and interest groups play significant roles in determining operating policies. Within this complex institutional framework, a number of organizations are directly responsible for developing and managing reservoir projects. Most reservoirs in the United States are owned and operated by cities, water districts, state and regional agencies, and private electrical and water utilities. However, the majority of the storage capacity is contained in federal reservoirs. Most, though certainly not all, of the very large reservoir systems in the United States are operated by the federal water agencies. The much more numerous nonfederal reservoirs tend to have much smaller storage capacities than the federal projects.

As noted in Chapter 2, the U.S Army Corps of Engineers (USACE) is the largest reservoir management agency in the nation, with over 500 reservoirs in operation. The USACE is unique in having nationwide responsibilities for the construction and operation of large-scale multiple-purpose reservoir projects. The U.S. Bureau of Reclamation (USBR) operates about 130 reservoirs in the 17 western states and has constructed numerous other projects which have been turned over to local interests for operation. The Tennessee Valley Authority (TVA) operates a system of about 50 reservoirs in the seven-state Tennessee River Basin. The various other federal agencies that own and operate reservoirs include the Natural Resource Conservation Service (formerly the Soil Conservation Service), Forest Service, and National Park Service.

The responsibilities of the various organizations involved in operating reservoir systems are based on project purposes. The USACE has played a clearly dominant role nationwide in constructing and operating major reservoir systems for navigation and flood control. The USBR water resources development program was founded upon facilitating development of the arid west by constructing irrigation projects. The TVA reservoir system is operated in accordance with operating priorities mandated by the 1933 Congressional act that created the TVA. This act specified that the TVA system be used to regulate streamflow primarily for the purposes of promoting navigation and controlling floods and, so far as may be consistent with such purposes, generation of electric energy. The activities of the federal water resources development agencies have evolved over time to emphasize comprehensive multiple-purpose water resources management. Hydroelectric power, recreation, and fish and wildlife are major purposes of USACE, USBR, and TVA projects. Municipal and industrial water supply has been primarily a nonfederal responsibility, though significant municipal and industrial storage capacity has been included in federal reservoirs for the use of nonfederal project sponsors. Numerous cities, municipal water districts, and other local agencies operate their own reservoir projects. Private companies as well as governmental entities play key roles in hydroelectric power generation, thermal-electric cooling water projects, and industrial water supply.

Contractual arrangements and other institutional aspects of reservoir operations vary greatly between purposes. For example, flood control operations for a USACE reservoir are simpler institutionally (through not necessarily otherwise) than water supply and hydroelectric power operations due to the USACE being directly responsible for flood control operations. The USACE is responsible for flood control operations at projects constructed by the USBR as well as at its own projects.

Nonfederal sponsors contract with the USACE and USBR for municipal and industrial water supply storage capacity [Wurbs, 1994]. All costs, including construction and maintenance, allocated to municipal and industrial water supply are reimbursed by nonfederal sponsors in accordance with the Water Supply Act of 1958, as amended by the Water Resources Development Act of 1986 and other acts. Construction costs are reimbursed, with interest, through annual payments over a period not to exceed 50 years. Nonfederal sponsors for federal projects are often regional water authorities who sell water to municipalities, industries, and other water users under various contractual arrangements.

The Reclamation Acts of 1902 and 1939 and other legislation dictate the policy that costs allocated to irrigation in federal projects be reimbursed by the project beneficiaries. The details of repayment requirements for irrigation projects have varied over the years with changes in reclamation law. Congressional acts authorizing specific USBR projects have often included repayment provisions tailored to the circumstances of the individual project. Consequently, local sponsor repayment contracts for water supply for irrigation vary between projects.

Water supply operations are controlled by agency responsibilities, contractual commitments, and legal systems for allocating and administering water rights. Water allocation and use is regulated by state water rights systems and permit programs. Many of the major reservoir systems in the United States are on interstate rivers, and several are on rivers shared with either Mexico or Canada. Operations of some reservoir systems are strictly controlled by agreements between states and/or nations which were negotiated over many years.

Hydroelectric power generated at USACE and USBR reservoirs is marketed to electric utilities by the five regional power marketing administrations of the Department of Energy. The power administrations are required by law to market energy in such a manner as to encourage the most widespread use at the lowest possible rates to customers consistent with sound business principles. The power administrations operate through contracts and agreements with the electric cooperatives, municipalities, and utility companies that buy and distribute the power. Reservoirs are operated in accordance with the agreements. The TVA is directly responsible for marketing, dispatching, and transmitting power generated at its plants. Many private and public electrical power companies operate their own reservoirs and hydropower plants. Several large hydroelectric power systems are composed of multiple storage and generating components owned and operated by federal, state or local, and private entities. Hydroelectric power facilities are typically components of systems that rely primarily on thermal plants for the base load, with hydropower supplying peak loads.

The purposes to be served by a federal reservoir project are established with Congressional authorization of project construction. Later, additional purposes are sometimes added or the original purposes modified by subsequent congressional action. When the original purposes are not seriously affected and structural or operational changes are not major, modifications in operating policies can be made at the discretion of the agency. Wurbs [1990, 1994] reviews legislative authorities and policies regarding modifying operations at completed federal reservoir projects. Johnson *et al.* [1990] survey federal projects for which storage reallocations and operational modifications have been proposed or implemented.

3.2 OUTLET STRUCTURES

Reservoir projects include dams and appurtenant outlet structures, pumping plants, pipelines and canals, channel improvements, hydroelectric power plants and transmission facilities, navigation locks, fish ladders, recreation facilities, and various other structures. The design, maintenance, and physical operation of these facilities are outside the scope of this book. However, the reservoir system release policies addressed by this book are obviously dependent on the capabilities provided by these water control facilities. Releases and withdrawals of water from a reservoir are constrained by the configuration and capacities of the project structures.

Reservoir releases to the river below a dam are made through spillways and outlet works. Spillways provide the capability to release high flow rates during major floods without damage to the dam and appurtenant structures. Spillways are required to allow flood inflows to safely flow over or through the dam, regardless of whether the reservoir contains flood control storage capacity. Spillways may be controlled or uncontrolled. A controlled spillway is provided with crest gates or other facilities that allow the outflow rate to be adjusted. For an uncontrolled spillway, the outflow rate is simply a function of the head or height of the water surface above the spillway crest. Since spillway flows involve extremely high velocities, stilling basins or other types of energy dissipation structures are required to prevent erosion damage to the downstream river channel and dam. For many reservoir projects, a full range of outflow rates are discharged through a single spillway. Some reservoirs have more than one spillway. A service spillway conveys smaller, more frequently occurring release rates, and an emergency spillway is used only rarely, during extreme floods.

The major portion of the storage volume in most reservoirs is located below the spillway crest. Flows over the spillway can occur only when the storage level is above the spillway crest. Outlet works are used for releases from storage both below and above the spillway crest. Discharge capacities for outlet works are typically much smaller than for spillways.

Outlet works are used to release water for various beneficial uses, such as water supply diversions and maintenance of instream flows. Flood control releases may also be made through outlet works. The components of an outlet works facility includes an intake structure in the reservoir, one or more conduits or sluices through the dam, gates located in either the intake structure or conduits, and a stilling basin or other energy dissipation structure at the downstream end.

Water supply diversions may be either lakeside or downstream. Lakeside withdrawals involve intake structures, along with pumps and pipeline or canal conveyance facilities. Downstream releases through an outlet works may be diverted from the river at locations which are great distances below the dam. Downstream releases may be made through hydroelectric power penstocks, navigation locks, or other structures, as well as outlet works and spillways.

Release requirements specified in operating plans are expressed in terms of flow rates or discharges. Rating curves are used by reservoir operators to relate release rates to storage levels and gate openings. The rating curves are developed by hydraulic analyses of the outlet structures, typically in conjunction with preconstruction design of the project.

The various types and configurations of dams, spillways, outlet works, gates, energy dissipators, and related structures are described by the USBR [1987], Golze

[1977], Jansen [1988], and Linsley *et al.* [1992]. Design, construction, maintenance, and operation considerations are addressed by these references.

3.2.1 Examples of Outlet Facilities

The three reservoir projects shown in Figures 3–1, 3–2, and 3–3 illustrate a few of the many types and configurations of water control facilities. Shasta, Folsom, and Somerville are federal multiple-purpose reservoirs with storage capacities of 5610, 1250, and 626 million m^3, respectively. Thus, Shasta Reservoir is extremely large, and Somerville Reservoir is representative of medium-sized projects. The outlet facilities at Somerville are much simpler than those at Shasta and Folsom. Each of the three projects serves the purposes of flood control, water supply, and recreation. Folsom and Shasta Reservoirs also have hydroelectric power plants.

Folsom Dam and Reservoir on the American River in California is shown in Figure 3 1. As discussed in Chapter 2, Folsom was constructed by the USACE and transferred to the USBR for coordinated operation as an integral part of the Central Valley Project. The dam was constructed during the period from 1948 to 1956. The dam has a concrete river section with a height of 104 m and length of 427 m,

Figure 3–1   Folsom Dam and Lake is operated by the Corps of Engineers for flood control and by the Bureau of Reclamation for hydroelectric power, irrigation, municipal and industrial water supply, fish and wildlife, and recreation. (Courtesy U.S. Bureau of Reclamation)

Figure 3–2 Shasta Dam and Lake serves to control floodwater, store surplus winter runoff for irrigation use in the Sacramento and San Joaquin Valleys, maintain flows in the Sacramento river for navigation, fish conservation, and protection of the Sacramento-San Joaquin Delta from intrusion of saline ocean water, to generate hydroelectric power, and to supply water for municipal and industrial use. (Courtesy U.S. Bureau of Reclamation)

flanked by long earthfill wing dams extending from the ends of the concrete section on both abutments. The Folsom Powerplant, constructed and operated by the USBR, is located at the foot of the dam. The plant has a total capacity of 198,720 kilowatts. Water from the reservoir is released through three 4.7-m-diameter penstocks to three generating units. The much smaller Nimbus Dam, located 11 km below Folsom Dam, reregulates hydroelectric power releases from Folsom. Releases from Folsom Reservoir to the river are also made through eight 1.5-m-by-2.7-m outlet conduits controlled by two slide gates. A 2.1-m-diameter conduit conveys water from the reservoir to Folsom Pumping Plant. Flood flows over the two adjacent overflow spillway sections of the dam are controlled by five 13-m-by-15-m and three 13-m-by-16-m radial gates.

Shasta Dam and Reservoir, shown in Figures 2–12 and 3–2, on the Sacramento River, is also a component of the California Central Valley Project. It was constructed by the USBR in the early 1940s. The curved concrete gravity dam is 184 m high with a crest length of 1060 m. The 539,000-kilowatt powerplant is located just below the dam. Water is released from the reservoir through five 4.6-m-diameter penstocks to the five main generating units and two station service units.

Figure 3–3 Somerville Reservoir is operated by the Corps of Engineers for flood control and recreation. The Brazos River Authority has contracted for the conservation storage, which is used to supply the city of Brenham through a pipeline with intake structure in the reservoir and to contribute to downstream demands by releases through the dam

Keswick Dam, located 14 km below Shasta Dam, stabilizes the uneven hydropower releases. The overflow section near the center of Shasta Dam is controlled by three 34-m-by-8.5-m drum gates. The outlet works also include eighteen 2.6-m-diameter conduits through the dam controlled by 14 2.4-m wheel-type gates and four 2.6-m tube valves.

Somerville Dam and Reservoir, located in Texas on Yequa Creek, a tributary of the Brazos River, is a component of the case study reservoir system discussed in Chapter 9. The dam is an earthfill embankment with a height of 24 m and crest length of 6160 m. The conservation pool and flood control pool have storage capacities of 197 and 428 million m³, respectively. If the flood control storage capacity is ever exceeded, excess flows will pass over an uncontrolled ogee-shaped overflow spillway with a length of 381 m. The spillway crest elevation coincides with the top of the flood control pool. The outlet works intake structure shown in Figure 3–3 contains two 1.5-m-by-3-m gates that control releases through a 3-m-diameter conduit through the dam that discharges to a stilling basin and then to the channel shown in Figure 9–7. Releases to Yequa Creek are made through this outlet works from either the conservation or flood control pools. Water is supplied via pipeline to the city of Brenham through another intake structure located in the conservation pool.

3.3 RESERVOIR POOLS

Reservoir operating policies typically involve dividing the total storage capacity into designated pools. A typical reservoir consists of one or more of the vertical zones, or pools, illustrated by Figure 3–4. The allocation of storage capacity between pools may be permanent or may vary with seasons of the year or other factors.

Water releases or withdrawals normally are not made from the inactive pool, except through the natural processes of evaporation and seepage. The top of inactive pool elevation may be fixed by the invert of the lowest outlet or, in the case of hydroelectric power, by conditions of operating efficiency for the turbines. An inactive pool may also be contractually set to facilitate withdrawals from outlet structures which are significantly higher than the invert of the lowest outlet structure at the project. The inactive pool is sometimes called dead storage. It may provide a portion of the sediment reserve, head for hydroelectric power, and water for recreation and fish habitat.

Conservation storage purposes, such as municipal and industrial water supply, irrigation, navigation, hydroelectric power, and instream flow maintenance, involve storing water during periods of high streamflow and/or low demand for later beneficial use as needed. Conservation storage also provides opportunities for recreation. The reservoir water surface is maintained at or as near the designated top of conservation pool elevation as streamflows and water demands allow. Drawdowns are made as required to meet the various needs for water.

The flood control pool remains empty except during and immediately following a flood event. The top of flood control elevation is often set by the crest of an uncontrolled emergency spillway, with releases being made through other outlet structures. Gated spillways allow the top of flood control pool elevation to exceed the spillway crest elevation.

The surcharge pool is essentially uncontrolled storage capacity above the flood control pool (or conservation pool if there is no designated flood control storage capacity) and below the maximum design water surface. Major flood events ex-

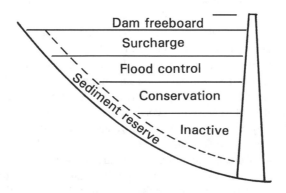

Figure 3–4   Reservoir pools.

ceeding the capacity of the flood control pool encroach into surcharge storage. The maximum design water surface profile, or top of the surcharge storage, is established during project design from the perspective of dam safety. Reservoir design and operation is based on ensuring that the reservoir water surface will never exceed the designated maximum design water surface elevation under any conditions. For most dams, particularly earthfill embankments, the top of dam elevation includes a freeboard allowance above the top of surcharge pool to account for wave action and provide an additional safety factor against overtopping.

3.4 SEDIMENT RESERVE

Reservoir storage capacity is lost over time due to sedimentation. The rate of sediment deposition various tremendously between reservoir sites, depending on flow rates and sediment loads in the rivers flowing into the reservoirs and the trap efficiencies of the reservoirs. Since sediment transport increases greatly during flood events, reservoir sedimentation also varies greatly over time with the random occurrence of floods. As illustrated in Figure 3–4, sediment deposits occur throughout the reservoir in each of the designated pools. As streamflow velocities decrease in the upper reaches of a reservoir, sediments are deposited, forming deltas. Smaller particles will move further into the reservoir before depositing. Reservoir sediment surveys may be performed periodically to determine current bottom topography and resulting storage capacities. However, due to the expense of conducting the bottom elevation measurements, many reservoirs have existed for decades without sediment surveys ever having been performed. Thus, storage capacity estimates are somewhat uncertain.

For many smaller reservoirs constructed by local entities, no special provisions are made to allow for sedimentation. Although it is recognized that the storage capacity of these reservoirs will significantly decrease over time, no attempt is made to estimate the volume and location of the sediment deposits at future points in time. However, for most federal projects and other large reservoirs, sediment reserve storage capacity is provided to accommodate sediment deposition expected to occur over a specified design life, typically 50 to 100 years. The volume and location of the sediment deposits and resulting changes in reservoir topography are predicted using methods outlined by the USBR [1987] and USACE [December, 1989]. Storage capacity reserved for future sediment accumulation is reflected in water supply contracts and other administrative actions.

3.5 RULE CURVES AND WATER CONTROL DIAGRAMS

The terms *rule curve* and *guide curve* are typically used to denote operating rules which define ideal or target storage levels and provide a mechanism for release rules to be specified as a function of storage content. Rule curves are typically expressed as water surface elevation or storage volume versus time of the year. Al-

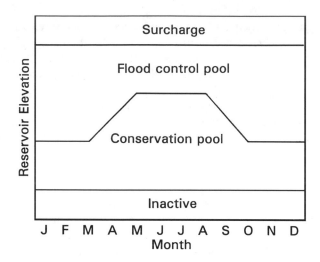

Figure 3–5 Seasonal top of conservation pool.

though the term *rule curve* denotes various other types of storage volume designa-tions as well, the top of conservation pool is a common form of rule curve.

The top of conservation pool is varied seasonally at many reservoirs, particu-larly in regions with distinct flood seasons. The seasonal rule curve illustrated by Figure 3–5 reflects a location in which the summer months are characterized by high water demands, low streamflows, and a low probability of floods. The top of conservation pool could conceivably also be varied as a function of watershed moisture conditions, forecast inflows, floodplain activities, storage in other system reservoirs, or other parameters as well as season of the year. A seasonally or other-wise varying top of conservation pool elevation defines a joint-use pool which is treated as part of the flood control pool at certain times and part of the conservation pool at other times. Figure 3–6 illustrates such an operating plan, where upper and

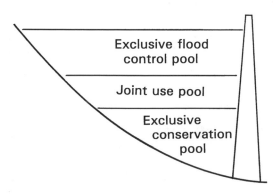

Figure 3–6 Exclusive and joint use pools.

lower zones are used exclusively for flood control and conservation purposes, respectively, and the storage capacity between is used for either purpose depending on the season or other factors. Also, as discussed later, either the flood control pool or conservation pool can be subdivided into any number of vertical zones to facilitate specifying reservoir releases as a function of the amount of water in storage.

Operating plans may be expressed in various formats. A water control diagram represents a compilation of regulating criteria, guidelines, rule curves, and specifications that govern the storage and release functions of a reservoir. A water control diagram or set of rule curves specify release rules as a function of storage levels, season of the year, and related factors. The format and types of rules reflected in water control diagrams vary greatly for different reservoir projects.

An example of a water control diagram for a particular reservoir is presented in Figure 3–7 [U.S. Army Corps of Engineers, 1987]. The Youghiogheny Reservoir on the Youghiogheny River, a tributary of the Monongahela River in Pennsylvania, is operated as a component of a multiple reservoir system in the Ohio River Basin. The Youghiogheny Reservoir is operated for flood control, hydroelectric power, and low-flow augmentation for downstream navigation, water quality, and recreation (whitewater rafting). Releases from the conservation pool are specified in the water control diagram of Figure 3–7 as a function of uncontrolled streamflow at a gaging station located downstream, time of the year, and storage content. Reservoir

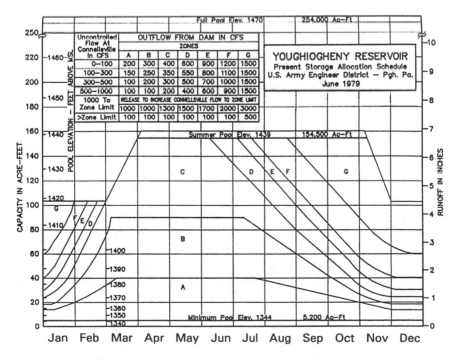

Figure 3–7   Example water control diagram [USACE, 1987].

storage levels are expressed alternatively as volume in acre-feet, volume equivalent in inches of runoff depth over the 434-square-mile (1120 km^2) watershed above the dam, and water surface elevation in feet above mean sea level. Storage capacities at the top of inactive pool and top of flood control pool are 5200 acre-feet (6.4 × 10^6 m^3) and 254,000 acre-feet (3.13 × 10^8 m^3), respectively. The 248,800 ac-ft (3.07 × 10^8 m^3) of active storage capacity is allocated to flood control and conservation purposes by a designated top of conservation pool which varies from 103,500 ac-ft (1.28 × 10^8 m^3) during December through February to 154,500 ac-ft (1.91 × 10^8 m^3) from April to early November.

3.6 OPERATING RULES FOR FLOOD CONTROL STORAGE CAPACITY

Flood control pool operations are based on minimizing the risk and consequences of making releases that contribute to downstream flooding, subject to the constraint of ensuring that the maximum design water surface is never exceeded. Flood control pools must be emptied as quickly as downstream flooding conditions allow to reduce the risk of future highly damaging releases being necessitated by filling of the available storage capacity. Minimizing the risks and consequences of storage backwater effects contributing to flooding upstream of the dam is also an important tradeoff consideration at some reservoir projects.

One type of reservoir system operation problem consists of developing an operating plan, often called a *regulation schedule*. Another related but distinctly different reservoir system operation problem involves making release decisions during real-time flood control operations within the framework of the regulation schedule. The operation plan provides guidance for real-time release decisions, but typically leaves a significant degree of flexibility. Information regarding current storage levels and streamflows is used, in combination with the regulation schedule, to make release decisions. Real-time operations often involve collection of current precipitation and streamflow data and forecasting flows to be expected at pertinent locations during the next several hours or days, to enable more effective release decisions. During normal nonflooding conditions, flood control operations consist simply of passing inflows to maintain empty storage capacity.

The USACE is responsible for operating a majority of the major flood control reservoir systems in the nation. Corps of Engineers procedures are representative of flood control operations in general. Flood control regulation plans are developed to address the particular conditions associated with each individual reservoir and multiple-reservoir system. Peculiarities and exceptions to standard operating procedures occur at various projects. However, the operating schedules for most reservoirs follow the same general strategy, which is outlined as follows.

Release decisions depend upon whether or not the flood control storage capacity is exceeded. Flood control storage capacities in federal reservoirs are typically sized to contain at least a 50-year recurrence interval (2% probability of exceedance in any year) flood and, for many projects, design floods greater than the 100-year flood (1% annual exceedance probability), perhaps much greater. Thus, complete filling of the flood control pool of a federal project is rare.

A specified set of rules, based on downstream flow rates, are followed as long as sufficient storage capacity is available to handle the flood without having to deal with the water surface rising above the top of flood control pool. Operation is switched over to an alternative approach, based on reservoir inflows and storage levels, during extreme flood conditions when the anticipated inflows are expected to deplete the controlled storage capacity remaining in the reservoir. The reservoir release rates necessitated by the flood control storage capacity being exceeded will contribute to downstream flooding. The objective is to ensure that reservoir releases do not contribute to downstream damages as long as the storage capacity is not exceeded. However, for extreme flood events which would exceed the reservoir storage capacity, moderately high damaging discharge rates beginning before the flood control pool is full are considered preferable to waiting until a full reservoir necessitates much higher release rates.

3.6.1 Regulation Based on Downstream Flow Rates

Assuming the flood control storage capacity is not exceeded, flood control operations are based on target allowable flow rates and stages at selected index locations or control points. The allowable flow rates are typically related to bankfull stream capacities, stages at which significant damages occur, environmental considerations, and/or constraints such as inundation of road crossings or other facilities. Stream gaging stations are located at the control points. Releases are made to empty the flood control pool as quickly as possible without contributing to streamflows exceeding the allowable flow rates at each downstream control point. The regulation schedule includes specified maximum allowable flow rates at the designated stream gages above which releases are not made.

When a flood occurs, the spillway and outlet works gates are closed. The gates remain closed until a determination is made that the flood has crested and flows are below the target levels specified for each of the control points. The gates are then operated to empty the flood control pool as quickly as possible without exceeding the allowable flows at the control points.

Normally, no flood control releases are made if the reservoir level is at or below the top of conservation pool. However, in some cases, if flood forecasts indicate that the inflow volume will exceed the available conservation storage, flood control releases from the conservation storage may be made if downstream conditions permit. The idea is to release some water before the stream rises downstream, if practical, to maximize storage capacity available for regulating the forecast flood. Prereleases are particularly important in operating reservoirs with only limited amounts of flood control storage capacity.

For many reservoirs, the allowable flow rate associated with a given control point is constant regardless of the volume of water in storage. At other projects, the

flood control pool is subdivided into two or more zones. The allowable flow rates at
one or more of the control points vary depending on the level of the reservoir water
surface with respect to the specified storage zones. This allows stringently low flow
levels to be maintained at certain locations as long as only a relatively small portion
of the flood control pool is occupied, with the flows increased to a higher level, at
which minor damages could occur, as the reservoir fills. The variation in allowable
flow rates at a control point may also be related to whether the reservoir level is rising or falling.

Flood control reservoirs are typically operated based on maintaining flow
rates at several control points located various distances below the dam. The most
downstream control points may be several hundred kilometers below the dam. Lateral inflows from uncontrolled watershed areas below the dam increase with distance downstream. Thus, the impact of the reservoir on flood flows decreases with
distance downstream. Operating to downstream control points requires streamflow
forecasts. Flood attenuation and travel time from the dam to the control point and
inflows from watershed areas below the dam may be estimated as an integral part of
the reservoir operating procedure.

Most flood control reservoirs are components of basin-wide multiple-
reservoir systems. Two or more reservoirs located in the same river basin will have
common control points. A reservoir may have one or more control points which are
influenced only by that reservoir and several other control points which are influenced by other reservoirs as well. For example, in Figure 3–8, streamflow gage 3 is
used as a control point for both reservoirs A and B; and gage 4 controls releases
from all three reservoirs. Multiple-reservoir release decisions may be based on
maintaining some specified relative balance between the percentage of flood-control storage capacity utilized in each reservoir. For example, if unregulated flows
are below the maximum allowable flow rates at all the control points, the reservoir
with the greatest amount of water in storage, expressed as a percentage of flood
control storage capacity, might be selected to release water. Various balancing crite-

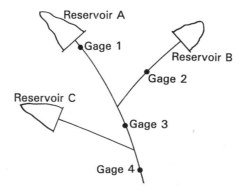

Figure 3–8 Multiple-reservoir flood control operations.

ria may be adopted. Releases from all reservoirs, as well as runoff from uncontrolled watershed areas, must be considered in forecasting flows at control points.

Maximum allowable rate of change of reservoir release rates are also specified. Abrupt gate openings causing a flood wave with rapid changes in stage are dangerous from the perspective of downstream hazards to public safety. Rapid variations in flow rates also contribute to streambank erosion.

3.6.2 Regulation Based on Reservoir Inflows
and Storage Levels

For an extreme flood event, limiting reservoir releases based on allowable downstream flow rates, as discussed above, could result in the storage capacity of the flood control pool being exceeded. If the releases are based on downstream target flows until the flood control pool fills, later uncontrolled spills at high flow rates could result. The higher peak release rate necessitated by this hypothetical release policy would typically be more damaging than a lower release rate with a longer duration beginning before the flood control pool is full. On the other hand, an operator would not want to make releases in excess of allowable downstream flow rates during a storm and then later learn that the flood control pool never filled and the releases unnecessarily contributed to downstream damages. Although streamflows that will occur several hours or days in the future are often forecast during real-time operations, future flows are still highly uncertain.

Consequently, the overall strategy for operating the outlet works and spillway gates of a flood control reservoir typically consists of two component types of regulation procedures. The type of procedure requiring the largest release rate controls for given flooding and storage conditions. The regulation approach discussed previously, based on downstream allowable flow rates, is followed until such time, during a flood, that the release rate indicated by the schedule outlined next is higher than that indicated by the downstream allowable flow rates. The regulation procedure outlined next is based on reservoir inflows and storage levels.

An example regulation schedule is presented in Figure 3.9 [U.S. Army Corps of Engineers, 1987]. This type of schedule controls releases during an extreme flood which would otherwise exceed the capacity of the flood control pool. Downstream flooding conditions are not reflected in the family of curves illustrated in Figure 3–9. The reservoir release rate is read directly from the graphs as a function of current water surface elevation and inflow rate. An alternative version of the schedule provides release rates as a function of the current water surface elevation and rate of rise of water surface. The two forms of the schedule are intended to result in the same release rate. Release rates are typically determined at a reservoir control center which has access to real-time streamflow measurements and can base release rates on inflow rates. If communications between the control center and operator at the project are interrupted during a flood emergency, the operator can determine gate releases based on rate of rise of the water surface without needing measurements of inflow rates.

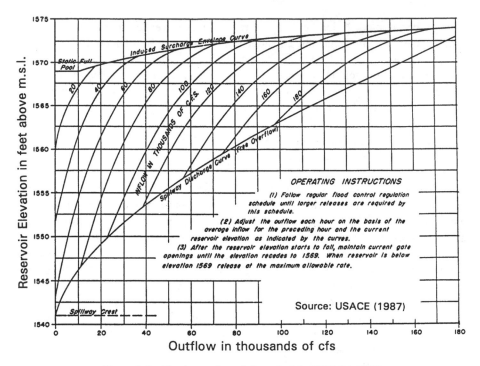

Figure 3–9 Flood control regulation schedule. [USACE, 1987]

The operating plan is prepared during preconstruction project planning. The regulation schedule curves are developed based on estimating the minimum volume of inflow that can be expected in a flood, given the current inflow rate and reservoir elevation. Having estimated the minimum inflow volume to be expected during the remainder of the flood, the outflow required to limit storage to the available capacity is determined by mass balance computations. For a given current inflow rate, the minimum inflow volume for the remainder of the storm is obtained by assuming the inflow hydrograph has just crested and computing the volume under the recession side of the hydrograph. For conservatively low inflow volume estimates, the assumed recession curve is made somewhat steeper than the average observed recession. The complete regulation schedule, which allows the outflow to be adjusted on the basis of the current inflow and empty storage space remaining in the reservoir, is developed by making a series of computations with various assumed values of inflows and amounts of remaining storage available.

The family of curves in Figure 3–9 also illustrates the concept of incorporating induced surcharge into the regulation plan. The release rates are set to allow specified encroachments into surcharge storage above the static full flood control pool. For most of the range of conditions reflected in the regulation schedule the gates are not fully open; thus, additional storage in the surcharge pool is induced over that which results from fully opening the gates sooner. The example regulation schedule of Figure 3–9 is for a gated spillway. However, the same general approach is applicable for reservoirs with uncontrolled spillways combined with outlet works with ample release capacity.

3.7 OPERATING RULES FOR CONSERVATION STORAGE CAPACITY

A multitude of factors and considerations may be important in the operation of reservoirs for water supply, hydroelectric power, and other conservation storage purposes. Each reservoir and multiple-reservoir system has unique aspects, and a variety of mechanisms are used to define operating rules. There is no standard format for specifying operating rules which is applicable to all situations. However, several basic concepts pertinent to a wide range of operating policies are noted in the following paragraphs.

In general, conservation operations can be categorized as being primarily influenced by either seasonal fluctuations in streamflow and/or water use or long term threat of drought. In some regions of the United States and the world, a reservoir will be filled during a distinct season of high rainfall or snowmelt and emptied during a dry season with high water demands. Thus, the reservoir level fluctuates greatly each year in a predictable seasonal cycle. In other cases, surface water management is predominately influenced by a long-term threat of drought. Water must be stored through many wet years to be available during drought conditions. Although reservoir storage may be significantly depleted within several months, severe drought conditions are characterized as a series of several dry years rather than the dry season of a single year. Reservoir operation during infrequent drought periods is significantly different than during normal or wet conditions. Although the relative importance of seasonal fluctuations versus long-term threat of drought varies between reservoir systems, both aspects of reservoir operations will typically be of some concern in any system. The terms *within-year storage* and *carry-over storage* are sometimes used to differentiate between the storage capacity required to handle seasonal variations in streamflows and water demands, and the additional capacity required for variations between years.

Conservation storage capacity serves a variety of project purposes or types of water use. Reservoir operation for municipal and industrial water supply is based on meeting demands subject to institutional constraints related to project ownership, contractual agreements, and water rights. Municipal and industrial water supply operations are typically based on ensuring a high degree of reliability in meeting demands during anticipated infrequent but severe droughts. Supplying water for irrigation often involves the acceptance of greater risks of shortages than municipal and industrial water supply, and is based more on maximizing economic benefits. Irrigation involves consumptive withdrawals and significant fluctuations in reservoir storage levels. Conversely, in steam-electric power plant cooling water reservoirs, most of the water withdrawn is returned to the reservoir, and water surface levels fluctuate very little. Hydroelectric power plants are typically components of complex energy systems which include thermal-electric as well as hydroelectric generation. Reservoir operations are based on both maintaining a high reliability of meeting hydroelectric power and energy commitments, and minimizing total costs, including both thermal and hydro generation. Reservoir storage for navigation purposes involves ensuring sufficient water depths in downstream navigation channels and sufficient water supply for lockages. Instream flow needs also include maintenance of streamflow for water quality, fish and wildlife habitat, livestock water, river recreation, and aesthetics. Reservoir operating policies may include specified flow rates to meet instream needs. Operating considerations for reservoir recreation typically involve maintaining desirable storage levels and minimizing fluctuations in storage levels.

Reservoir operations also address requirements other than the primary project purposes. For example, due to water rights considerations, releases may be required to pass inflows through the reservoir to more senior water users and management entities located downstream which are not directly served by the reservoir. Such requirements may be specified in terms of maintaining minimum stream flows at specified downstream locations, subject to the stipulation that reservoir releases in excess of inflows are not required. Another consideration involves restricting the rate of change in release rates to prevent public safety hazards. Rapid increases in stage and velocity can be dangerous for people recreating in the river downstream of a reservoir. Rapid changes in release rates are also undesirable from the perspective of river bank erosion. Storage-level fluctuations are sometimes made to help control vectors such as mosquitos. Water-quality storage has been included in reservoirs as a primary project purpose to provide releases for low-flow augmentation. Water quality is often an important incidental consideration in operations for other purposes. The quality of downstream flows and water supply diversions is sometimes controlled by selection of the vertical storage levels from which to make the releases. Operation during floods is an important consideration for conservation-only projects without flood control storage capacity.

3.7.1 Multiple-Purpose and Multiple-User Operating Considerations

Multiple-purpose reservoir operation involves various interactions and trade-offs between purposes, which are sometimes complementary but often competitive or conflicting. Reservoir operation may be based on the conflicting objectives of maximizing the amount of water available for conservation purposes and maximizing the amount of empty space available for storing future flood waters to reduce downstream damages. Conservation pools are shared by various purposes that involve both consumptive withdrawals and in-reservoir and instream uses.

Common practice is to operate a reservoir for conservation only, flood control only, or a combination of flood control and conservation with separate pools designated for each. Interactions between flood control and conservation purposes in a multiple-purpose reservoir involve allocation of storage capacity as represented by the designated top of conservation pool elevation, which is a form of rule curve. Modifications to the operations of completed projects may involve either permanent long-term reallocations of storage capacity, or establishing or refining seasonally varying rule curves for joint-use storage. Studies of long-term storage reallocations and designing seasonal rule curves are two important types of reservoir system modeling and analysis applications.

Interactions between flood control and conservation purposes may also involve flood control pool release rates. For example, in some cases flood control pool releases may be passed through hydroelectric power plants and limited to the maximum discharge that can be used to generate power. Also, releases from conservation storage may be made to partially draw the pool down in anticipation of forecast flood inflows. Releases from the conservation pool in anticipation of forecast flood inflows are particularly important for reservoirs with little or no designated flood control storage capacity.

Conservation pools typically serve multiple purposes with at least some complementary characteristics. Water stored for water supply and hydroelectric power provides opportunities for recreation and reservoir fisheries. Hydroelectric power releases contribute to other instream flow uses and can be diverted at downstream locations for water supply. On the other hand, sharing of reservoir storage capacity and limited water resources by multiple users involves conflicting demands.

Conservation operations may include designing triggering mechanisms by which certain demands are curtailed whenever storage falls below prespecified levels. This allows water supply withdrawals, instream flows, and/or hydroelectric energy levels at different levels of reliability to be provided by the same reservoir. Specifying the release or withdrawal rate as a function of storage (or storage plus inflow) is sometimes called a *hedging rule*. The storage designations, or rule curves, used as a triggering mechanism in allocating water between competing users and uses are sometimes called *buffer zones*. Full demands are met as long as the reservoir water surface is above the top of buffer pool, which certain demands being curtailed whenever the water in storage falls below this level. The top of buffer pool elevation may be constant, or may be specified as a function of time of the year or other parameters. A range of different storage levels in one or more reservoirs may be designated as triggering mechanisms for various management decisions.

Certain water users require a high degree of reliability. For other water users, obtaining a relatively large quantity of water with some risk of shortage may be of more value than a supply of greater reliability but smaller quantity. Storage triggering designations may also provide a mechanism for reflecting relative priorities or tradeoffs between purposes. For example, a reservoir operating plan may involve ensuring a high degree of reliability for a municipality and less reliability for agricultural irrigators. All demands are met as long as storage is above a specified level, but the irrigation withdrawals are curtailed whenever storage falls below the specified level. Release requirements for maintaining instream flows for fish and wildlife habitat and/or freshwater inflows to estuaries may be conditioned upon storage being above a specified buffer level. Implementation of drought contingency plans may be triggered by the storage level falling below a specified buffer level. More severe demand management options may be implemented as storage contents fall below various prespecified levels.

3.7.2 Multiple-Reservoir System Operations

Multiple-reservoir release decisions occur in situations in which water needs can alternatively be met by releases from two or more reservoirs. In Figure 3–10, diversions 1 and 3 are from specific reservoirs, but diversion 4 can be met by releases from either of the three reservoirs. Instream flow, as well as diversion, requirements at diversion location 4 can be met by releases from the reservoirs.

One criterion for deciding from which reservoir to release is to minimize spills, since they represent water loss from the system. Spills from an upstream reservoir (such as reservoir A in Figure 3–10) may still be stored in a downstream reservoir (reservoir B) and thus are not lost to the system. The term *spill* refers to discharges through an uncontrolled spillway or controlled releases made simply to prevent the reservoir surface from rising above the designated top of conservation pool. For reservoirs in series, such as reservoirs A and B in Figure 3–10, the downstream reservoir would be depleted before using upstream reservoir water to meet downstream demands. In addition to minimizing spills from the downstream reservoir, this procedure maximizes the amount of water in storage above, and thus accessible to, each diversion location. For example, water stored in reservoir A can be used to meet diversions 1, 2, and 4, but water stored in reservoir B can be used to meet only diversions 2 and 4.

For reservoirs in parallel, such as reservoirs B and C in Figure 3–10, minimizing spills involves balancing storage depletions in the different reservoirs. The sim-

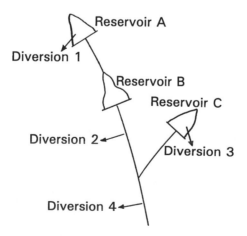

Figure 3–10   Multiple-reservoir conservation operations.

plest approach might be to release from the reservoir with the largest ratio of conservation pool storage content to storage capacity. Thus, release decisions would be based on balancing the percent depletion of the conservation pools. Other more precise and more complex approaches can be adopted to select the reservoir with the highest likelihood of incurring future spills.

Numerous other considerations may be reflected in multiple-reservoir release decisions. If the reservoirs have significantly different evaporation potential, minimization of evaporation may be an objective. The criteria of minimizing spills or evaporation are pertinent to either single-purpose or multiple-purpose systems. Multiple-purpose, multiple-reservoir release decisions can involve a wide variety of interactions and tradeoffs. For example, releases to meet downstream municipal, industrial, or irrigation water supply demands may be passed through hydroelectric power turbines. Thus, multiple-reservoir water supply release decisions may be based on optimizing power generation. Likewise, recreational aspects of the system could motivate release decisions which minimize storage-level fluctuations in certain reservoirs.

As illustrated in Figure 3–11, conservation pools can be subdivided into any number of zones to facilitate the formulation of multiple-reservoir release rules. The multiple-zoning mechanism can be reflected in the operating rules actually followed by reservoir operators. Also, even in cases where operating rules are not actually precisely defined by designation of multiple zones, the multiple-zone mechanism can be used in computer models to approximate the somewhat judgmental decision

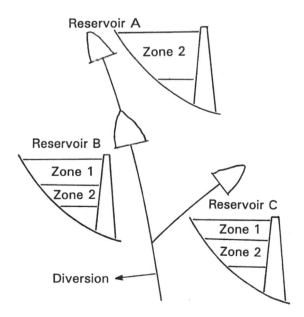

Figure 3–11   Storage zones for defining release rules.

process of actual operators. The zones provide a general mechanism or format for expressing operating rules. Multiple-reservoir release rules are defined based on balancing the storage content such that the reservoirs are each in the same zone at a given time to the extent possible. For example, in meeting the downstream diversion (or instream flow) requirement of Figure 3–11, water is not released from zone 2 of one reservoir until zone 1 has been depleted in all the reservoirs. Since zone 1 in reservoir A is assigned zero storage capacity, no releases are made from reservoir A until zone 1 is empty in the other two reservoirs. With the storage content falling in the same zone of each reservoir, the release is made from the reservoir which is most full in terms of percentage of the storage capacity of the zone. For example, if the storage capacities of zone 2 of reservoirs A, B, and C, respectively, are 55%, 60%, and 68% full, a release is made from reservoir C to meet the downstream diversion requirement. Variations of this general type of multiple-reservoir release rule can be formulated.

3.8 WATER SUPPLY

Water is diverted or withdrawn from rivers and reservoirs for municipal, industrial, agricultural, and other beneficial uses. During normal hydrologic conditions, real-time reservoir operations involve meeting water demands in accordance with the commitments and responsibilities of the water supply agencies. During low-flow or drought conditions, operations may involve allocating limited water resources to competing users within the institutional framework of project ownership and agency responsibilities, contractual agreements, legal systems for allocating and administering water rights, and political negotiations.

Developing and administering water supply contracts and agreements, water rights allocation systems, and reservoir operating plans involve various types of reservoir system operation decision problems, which can be categorized as follows:

- Allocation of a limited amount of water between competing uses and users
- Within-year temporal allocation of a limited amount of water (e.g., distributing available water over the irrigation season)
- Determination of the trade-off between the amount of water to use during the current water year and the amount to be carried over in storage into the next year
- Coordination of water supply operations with demand management strategies and other sources of supply such as groundwater
- Coordination of water supply operations with other project purposes
- Coordination of the releases from each reservoir of a multiple-reservoir system
- Various combinations of the above

Maintaining high reliability for meeting water needs during infrequent drought or low-flow conditions expected to occur at unknown times in the future is a key consideration in water supply management. Municipal and industrial water supply typically requires a particularly high level of reliability. Project planning and design, contractual agreements, and water rights are typically based on ensuring a very dependable supply.

Supplying water for irrigation often involves acceptance of greater risks of shortages than municipal and industrial water supply. Obtaining a relatively large quantity of water with some significant risk of shortage may be of more value than a supply of greater reliability but smaller quantity. An operating plan may involve allocating water to the various users at the beginning of each water year or irrigation season based on current reservoir storage levels and present and forecast future hydrologic conditions.

The amount of water required to meet the demands for growing crops for the entire season is called the *water duty*. This is equal to the amount of water supplied to the land by means of gravity diversions from reservoirs, or pumped from rivers, reservoirs, or groundwater aquifers. *Net duty* is the amount of water delivered to individual farm units, considering losses in canals and laterals and waste from the point of diversion to the point of application to the land. Irrigation water diverted from reservoirs, diversion dams, or natural river channels is controlled in a manner to supply water for the irrigation system as necessary to meet water duty requirements, which vary seasonally. In most irrigated areas of the western United States, the agricultural growing season begins in the spring months of April and May. The diversion requirements gradually increase as the summer progresses, reaching their maximum amounts in July or August. By the end of the growing season, irrigation requirements are terminated. The return flow of water from irrigated lands is collected in drainage channels and flows back into natural creeks and rivers. The return flows may vary from essentially zero to greater than half of the diversion amounts. Increases in salinity concentrations are often associated with irrigation return flows.

Shifting to a greater reliance on demand management has been a major emphasis in recent years in all sectors, including municipal, industrial, and irrigation. Implementation of appropriate demand management strategies is an important consideration in determining water needs to be supplied by reservoirs. Implementation of short-term or emergency demand management measures are dependent on current reservoir storage levels and associated risks of future shortages in supply. Coordination of reservoir operations and demand management programs is important.

Multiple-reservoir system operation involves coordinated releases from two or more reservoirs to supply common diversions or instream flow needs at downstream locations. Under appropriate circumstances, multiple-reservoir system operations can significantly increase reliabilities, as compared to operating each individual reservoir independent of the others. Coordinated releases from two or more reservoirs increase reliabilities by sharing the risks associated with the individual reservoirs not being able to meet their individual demands. Operated independently, one reservoir may be completely empty and unable to supply its users while significant storage remains in the other reservoirs. At other times, the other reservoirs may - be empty. System operation balances storage depletions. Multireservoir system operation can also serve to minimize reservoir spills and evaporation and channel losses due to seepage and evaporation. In some systems, water treatment costs and electrical pumping costs for water conveyance and distribution may vary significantly, depending on which demands are met by releases or withdrawals from which reservoirs.

Another key aspect of system operation involves use of unregulated flows entering the river below the most downstream dams but above the location of water supply diversions. For example, the diversion in Figure 3–11 is partially supplied by surface runoff and baseflow from subsurface sources entering the river below reservoirs B and C. This unregulated streamflow does not flow into any reservoir but flows past pumping plants, where water is diverted from the river for beneficial use. Unregulated river flows are typically highly variable, of significant magnitude much of the time but zero or very low some of the time. Thus, unregulated flows have firm yields of zero or very little. However, when combined with reservoir releases during low-flow periods, the unregulated streamflows may significantly contribute to the overall stream/reservoir system water supply capabilities.

3.9 HYDROELECTRIC POWER

Hydroelectric plants are generally used to complement the other components of an overall electric utility system. Because the demand for power varies seasonally, as well as at different times during the week and during the day, the terms *base load* and *peak load* are commonly used to refer to the constant minimum power demand and the additional variable portion of the demand, respectively. Hydroelectric power is typically used for peak load, while thermal plants supply the base load. Hydroelectric power plants can assume load rapidly and are very efficient for meeting peak-demand power needs. In some regions, hydroelectric power is a primary source of electricity, supplying much or most of the base load as well as the peak load. Availability of water is generally a limiting factor in hydroelectric energy generation.

Hydroelectric plants may be classified as storage, run-of-river, or pumped storage. A storage-type plant has a reservoir with sufficient capacity to permit carry-over storage from the wet season to the dry season or from wet years through a drought. A run-of-river plant essentially has no active storage, except possibly some pondage to permit storing water during off-peak hours for use during peak hours of the same day or week, but may have a significant amount of inactive storage which provides head. Flows through the turbines of run-of-river plants are limited to unregulated streamflows and releases from upstream reservoirs. A pumped-storage plant generates energy for peak load, but during off-peak periods water is pumped from the tailwater pool to the headwater pool for future use. The pumps are powered with secondary energy from some other plant in the system.

At many projects, reservoir releases are made specifically and only to generate hydroelectric power. At other projects, hydroelectric power generation is limited essentially to releases being made anyway for other purposes, such as municipal, industrial, or agriculture water supply. An upstream reservoir may be operated strictly for hydropower, with the releases being reregulated by a downstream reservoir for water supply purposes.

The objective of a electric utility is to meet system demand for energy, capacity (power), and reserve capacity (for unexpected surges in demand or loss of a generating unit) at minimal cost. Power is the rate at which energy is produced. Capacity is the maximum rate of energy production available from the system. The value of hydroelectric energy and power is a function of the reliability with which they can be provided.

Three classes of energy are of interest in hydroelectric power operations: average, firm, and secondary. Average energy is the mean annual amount of energy that could be generated assuming a repetition of historical hydrology. Firm energy, also called *primary* energy, is estimated as the maximum constant annual energy that could be generated continuously during a repetition of historical hydrology. From a marketing perspective, firm energy is electrical energy that is available on an ensured basis to meet a specified increment of load. Secondary energy is energy generated in excess of firm energy. Secondary energy, expressed on an average annual basis, is the difference between average annual energy and firm energy.

Reservoir operating rules for hydroelectric power generation assume many different forms depending on the characteristics of the electric utility system and reservoir system, hydrologic characteristics of the river basin, and institutional constraints. However, designation of a power pool and power rule curve, as illustrated by Figure 3–12, is a key aspect of hydroelectric operations. The power pool is reserved for storage of water to be released through the turbines. Inactive or active storage below the power pool provides additional head. If the reservoir water surface is at the top of power pool, net inflows (inflows less evaporation and withdrawals) are passed through the reservoir. Flows up to the maximum generating ca-

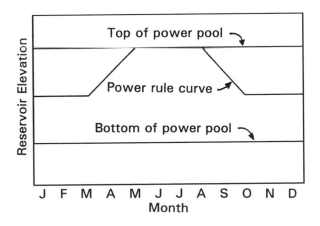

Figure 3–12   Hydroelectric power rule curve.

pacity of the plant may be used to generate energy, and the remainder of the flow is spilled. If the reservoir contains flood control storage, water will be stored in the flood control pool above the top of power pool during flood events. Power generation is curtailed any time the water surface elevation drops below the designated minimum power pool elevation.

Hydroelectric power operations are typically based on two objectives: to ensure firm energy in accordance with contractual agreements or other commitments, and to meet total system energy and power demands at minimum cost. The rule curve is designed to ensure firm energy. Operation is based on meeting firm energy commitments continuously as long as the power pool contains water. Additional secondary energy is generated only if the reservoir storage content is above the rule curve. The seasonal variation of the rule curve over the year is tailored to the hydrologic conditions and power demands of the particular area. For example, the rule curve shown in Figure 3–12 reflects the following considerations. Power storage must be at a maximum during the middle of the calendar year in anticipation of high summer power demands coincident with low inflows. Droughts usually begin during the early summer in this area. A low pool elevation is acceptable in the fall and winter season because demands are lower and inflows higher.

The power rule curve is typically developed based on the historical hydrologic period-of-record streamflows. Droughts more severe than the critical drought of record can result in depleting the power pool and interrupting firm energy generation. Although power rule curves are discussed here from the perspective of a single reservoir, rule curves can also be developed for a multiple-reservoir system on the basis of total system storage or potential energy.

Determining day-to-day and hour-to-hour releases when the storage is above the power rule curve represents a basic real-time decision problem. Only firm energy can be generated if the storage is at or below the rule curve. However, secondary energy can be generated with storage above the rule curve. A variety of approaches can be adopted for utilizing this water. Although in some systems detailed guidelines have been developed to guide secondary energy generation decisions, typically considerable flexibility exists for operator judgment on a day-to-day basis. If opportunities exist for displacing very expensive thermal generation, secondary hydroelectric energy can be very worthwhile. The optimization problem consists of timing secondary energy generation to minimize thermal generation costs or to maximize hydroelectric revenues. However, drawing the storage down to near the rule curve increases the risk of not meeting firm energy requirements if a future inflow sequence is more adverse than the critical period of historical inflows upon which rule curve development was based. Thus, a trade-off also exists between minimizing thermal generation fuel costs or hydroelectric power revenues and maintaining high reliability of firm energy commitments being met in the future. The impacts of secondary energy generation decisions on average annual energy also involves tradeoffs between maintaining a high head and minimizing the probability of spills. In multiple-reservoir systems, the decision problem involves balancing storage and releases between reservoirs as well as timing of releases.

Developing, modifying, and refining reservoir operating policies often involve interactions between hydroelectric power and other project purposes. If the reservoir includes flood control, the top of power pool coincides with the bottom of the flood control pool. The top of power pool may be a seasonally varying rule curve defining a joint-use pool used sometimes for flood control and sometimes for power. Design of the rule curve must reflect both hydroelectric power and flood control objectives. Rule curves can also be established to optimize hydroelectric power operations subject to the constraint of maintaining highly reliable supplies for municipal, industrial, agricultural, and/or low-flow augmentation purposes. Likewise, water supply release decisions may be based on optimizing hydroelectric power operations while meeting water supply demands. Hydropower operations may be constrained by minimum streamflow requirements for fish and wildlife or other instream flow needs. Minimizing the adverse impacts of storage-level fluctuations on recreation may be an important consideration. The rate of change of release rates is often limited to reduce streambank erosion.

3.10 NAVIGATION

The Corps of Engineers is the primary agency in the United States responsible for navigation improvements. During the past century and a half, USACE has been involved in the improvement for navigation of some 35,000 km of inland and coastal waterways. Navigational improvements include canals, locks, dams and reservoirs, maintained channels and estuaries, bank protection, and channel stabilization measures.

Reservoirs provide slack pools for navigation and releases that supplement natural flows in maintaining minimum flow depths in downstream channels. Use of reservoir releases to maintain streamflows for navigation is limited due to the large quantities of water required. Slack-water waterways, such as the Tennessee Valley System, provide required depths by maintaining reservoir storage levels and dredging. Open river waterways like the Missouri and Mississippi rely on channel constriction, dredging, and normal depth of flow to maintain the minimum depth for navigation. When available water is limited, navigation is concerned with depth, width, channel alignment, and length of navigation season at authorized depth. During floods, navigation is affected by flow velocities, cross-currents, bridge clearances, docking and locking difficulties, and shoaling.

Reservoir operations for navigation involve optimizing the use of available water for maintaining storage levels to provide slack pools, releases to augment flows in downstream channels, and providing water for locking operations. Reservoir operations also involve minimizing the adverse impacts of floods on navigation. Typical objectives considered in developing and evaluating reservoir operating plans for navigation include:

- Maximizing the length of the navigation season
- Maximizing the reliability of the dependable minimum depth

- Minimizing fuel and other operating costs
- Minimizing dredging costs
- Minimizing the volume of water released from storage to meet minimum navigation requirements

3.11 RECREATION

The general public uses reservoirs and rivers for boating, swimming, fishing, and other recreational activities. Reservoir operating plans include consideration of recreation in the reservoir, along the shore, in the river just below the dam, and at river locations further downstream.

Recreational aspects of reservoir operations involve maintaining storage levels and minimizing fluctuations in storage levels. Reservoir water surface area, depths, length of shoreline, area and quality of beaches, and usability of facilities such as marinas, docks, and boat ramps are related to storage level. Under most circumstances, the optimal recreation use of reservoirs would require that the water level be maintained at or near the top of conservation pool during the recreation season. This is often infeasible due to other project purposes.

In streams below reservoirs, recreation is impacted by flow rates, variations in flow rates, and water quality. Both high flows and low flows can reduce the recreation potential. Reservoir releases can also cause safety hazards for downstream recreationists. Operating plans often include specification of minimum streamflows and possibly augmented flows during short periods for special activities such as river rafting.

Water quality affects body contact activities such as swimming and water skiing. Temperature, fecal coliform count, dissolved oxygen, and turbidity are important water quality parameters for recreation.

The effects of reservoir regulation on the aesthetics of the riverine environment are closely related to public use. Aesthetic considerations in reservoir operating plans may involve maintaining minimum streamflows, releasing water for special aesthetic purposes, or minimizing the duration of exposure of mud flats or unsightly shoreline resulting from drawdowns.

3.12 WATER QUALITY

Water quality encompasses the physical, chemical, and biological characteristics of water and the abiotic and biotic interrelationships. The quality of the water and the aquatic environment may be significantly affected by reservoir management practices. Water-quality requirements for reservoir releases may involve both flow rates and quality parameters. Low-flow augmentation, or maintenance of minimum streamflow rates at downstream locations, is a primary water-quality operating objective at many reservoir projects. In addition, the quality of the releases is controlled at many projects through multiple-level selective withdrawals.

Most impoundments exhibit some degree of temperature stratification. In general, deeper lakes are more likely to become highly stratified each summer and are not as likely to become mixed by wind or short-term temperature changes. When the surface of the lake begins to receive a greater amount of heat from the sun and air than is lost, it becomes warmer and less dense, while the colder, denser water remains on the bottom. In the layer of colder water near the bottom, little if any oxygen is transferred from the air to replace that depleted by oxidation of organic substances, and eventually anoxia may develop. Under this condition, a reducing environment is created, resulting in elevated levels of parameters such as iron, manganese, ammonia, and hydrogen sulfide. Changes such as these may result in water that is degraded and toxic to aquatic life.

A primary means of managing the water quality of reservoir releases is to control the vertical levels at which water is withdrawn from the reservoir. Many reservoir projects include outlet works intake structures providing multilevel withdrawal capabilities. The reservoir operating decision problem involves establishing the desired temperature, dissolved oxygen, and other water-quality criteria and selecting the elevations at which to make releases to meet the criteria. Water from different levels may have to be mixed to meet the different water-quality criteria. For example, cold water with high dissolved oxygen may be needed to support downstream fisheries. It may be necessary to draw colder water from a lower level, where oxygen is deficient, in combination with water from a higher, warm layer to supply the required oxygen. Because of the uncertainty involved in proportioning flow through two or more intakes in a single wet well, dual wet wells may be provided as a part of the selective withdrawal facilities.

Management of water quality in the reservoir pool may also be a consideration in selective withdrawals from multilevel intake structures. Good and poor quality water can be blended to meet the release criteria with a minimum of good and maximum of poor quality water. This type of release policy will help to prevent a deterioration of quality in the reservoir which could lead to an eventual inability to meet the release criteria.

3.13 FISH AND WILDLIFE

Environmental resources management opportunities and problems associated with reservoir operations vary widely between regions and between reservoirs. Reservoir operations influence fish, wildlife, and ecological systems both in the reservoir pool and in the river downstream.

Reservoir releases contribute to maintenance of instream flows necessary for the support of aquatic habitat and species, protection or enhancement of water quality, preservation of wetlands, and provision of freshwater inflows to bays and estuaries. Reservoir operating plans may include maintenance of specified minimum flow rates at downstream locations. Periodic flooding as well as low-flow augmentation may be important for certain ecosystems. The required flow rates may be specified as a function of season, reservoir storage, reservoir inflows, and other factors.

Reservoir releases for downstream fishery management depend on water-quality characteristics and water control capabilities. Achieving optimal temperatures for either cold water or warm water fisheries through selective multilevel releases may be an operating objective, as may be maintenance of dissolved oxygen. Releases can be beneficial for maintaining gravel beds for certain fish species. Dramatic changes in release rates, typically associated with hydropower and flood control operations, can be detrimental to downstream fisheries.

Migration of anadromous fish, such as salmon in the Pacific Northwest and striped bass in the Northeast, is a concern in some regions. Declines in anadromous fish populations have been attributed to dams due to blockage of migration, alteration of normal streamflow patterns, habitat modification, blockage of access to spawning and rearing areas, and changes in water quality. Regulation for anadromous fish is particularly important during certain seasons of the year.

Project regulation can influence fisheries in the pool as well as downstream. Water-surface-level fluctuation is one of the most apparent influences of reservoir operation. Periodic fluctuations in water levels present both problems and opportunities in regard to reservoir fisheries. The seasonal fluctuations that occur at many flood control projects and daily fluctuations at hydropower projects often eliminate shoreline vegetation and cause subsequent shoreline erosion, water-quality degradation, and loss of habitat. Adverse impacts of water-level fluctuations also include loss of shoreline shelter and physical disruption of spawning and nests. Beneficial fisheries management techniques include pool-level management for weed control; forcing forage fish out of shallow cover areas, making them more susceptible to predation; and maintaining appropriate pool levels during spawning. The lowering of levels during spawning is sometimes also combined with rough fish removal during drawdown to help manage another more desirable fishery. Wave action from slowly draining the water can help maintain clean gravel substrates, which are favorable to some target species.

The effectiveness of fisheries management measures vary from region to region and lake to lake. Predicting the results of changes in reservoir levels and releases for fisheries management is difficult due to variability in physical, biological, chemical, and operational characteristics of the reservoir and uncontrolled environmental influences.

Water-level management in fluctuating warm-water and cool-water reservoirs generally involves raising the water levels during the spring to enhance spawning and survival of young predators. Pool levels are lowered during the summer to permit regrowth of vegetation in the fluctuation zone. Fluctuations may be timed to benefit one or more target species. In the central United States, small increases in pool levels are often recommended during the autumn for waterfowl management.

In addition to the relatively small pool fluctuations discussed above, periodic major drawdowns have been used for fishery management. Drastic lowering of the reservoir over an extended time of at least a growing season, possibly augmented by seeding of plants, permits vegetative regrowth in the dewatered area. Selective removal of various fish species may also occur. Desirable fish are restocked after the reservoir is refilled.

Water-level fluctuations also present both problems and opportunities for wildlife management. The beneficial aspects of periodic drawdowns on wildlife management include permitting natural and artificial revegetation of shallows for waterfowl, installation and maintenance of artificial nesting structures, control of vegetative species composition, and mast tree survival in greentree reservoirs. Wildlife benefits of periodic flooding include inhibiting the growth of undesirable and perennial plants, creating access and foraging opportunities for waterfowl, and ensuring certain water levels in stands of vegetation to encourage waterfowl nesting and reproduction.

Water-level fluctuations can also adversely impact wildlife habitat. Such impacts include the destruction of emergent and terrestrial vegetation, permitting predator access to otherwise inaccessible areas, abandonment of active nest sites, and rendering soils with iron oxides unproductive when dried out.

3.14 EROSION AND SEDIMENT DEPOSITION

Natural stream erosion and deposition processes are significantly altered through the construction and operation of reservoir projects. The impacts of individual projects vary significantly, depending on the streamflow and sediment characteristics of the parent stream, and the specific operating rules of a given project. Interruption of the natural sediment processes of a stream generally results in deposition of sediment in the upstream reservoir area, and corresponding erosion and degradation of the streambed and banks immediately downstream from the project. The location of deposits in the reservoir is a function of the size of the reservoir, the amount and gradation of the sediments being transported, and the pool level at the time of significant inflow. The amount of bank and shoreline erosion is closely related to the rate and magnitude of the pool-level fluctuations.

Large reservoir projects frequently trap and retain essentially all of the suspended sediment and bed material load within the upstream pool, thus releasing sediment-free water. These releases are capable of eroding the bed and banks of the river downstream of the outlet structure. The extent of this erosion is related to the composition of the bed and bank material, the volume of water released on an annual or seasonal basis, release rates and flow velocities, and the manner in which the flow is released. Fluctuating releases often result in an initial loss of the banks. This loss is closely related to the magnitude of the stage fluctuation. The recession of banks due to fluctuating releases usually stabilizes in the first few years of operation, as the underwater slope reaches a state of quasi-equilibrium. Once this equilibrium slope has been achieved, the bank erosion process behaves as in the natural channel. Periodic wetting and drying of the banks through fluctuating releases accelerates this process. Reservoir releases also result in lowering the streambed, with the maximum amount of lowering occurring immediately downstream from the outlet structures, and decreasing in the downstream direction. This degradation process continues until the slope is reduced to its equilibrium value and/or the bed becomes naturally armored by removal of the fines, which exposes the coarser, nonerodible bed materials. After the bed becomes naturally armored, future lowering of the streambed is usually insignificant.

Channels downstream from small and medium-sized reservoir projects often exhibit entirely different characteristics than described above for large reservoirs. Channel capacity below the smaller reservoirs tends to be lost over time. Reservoir projects that make only limited releases may result in extensive deposition and subsequent vegetative encroachment in the downstream channel. With construction of a reservoir, the preconstruction periodic flushing flows that are capable of removing deposits near the mouth of tributaries are often replaced by low nonerosive reservoir releases. This contributes to the loss of channel capacity and reservoir operating flexibility.

Reservoir shorelines are subject to a number of forces contributing to their instability, and frequently undergo major changes during the life of a project. Fluctuating pool levels saturate previously unsaturated material, resulting in massive slides when the pool is drawn down to lower levels. This material accumulates at the base of the slope, and often forms an underwater bench, leaving steep unstable slopes above the water line. Reservoir banks are also subject to attacks by both wind and waves, which tend to remove this material and undercut the banks.

Sediment deposits in the reservoir pool are an important consideration, since storage capacity and many reservoir management activities are adversely impacted. Sediment deposits occur throughout a reservoir, but particularly in the upper reaches where inflow velocities are reduced by the impoundment. The impacts of sediment accumulations over the life of the reservoir should be recognized in project planning and operation.

Much of the erosion and deposition process is beyond the control of reservoir managers. However, the following precautions can significantly minimize problems [U.S. Army Corps of Engineers, 1987].

- Minimize the rate of reservoir pool drawdown.
- Avoid sudden increases in reservoir releases and subsequent downstream stage fluctuations.
- Keep reservoir pool levels as low as possible during known periods of high sediment inflow, thus encouraging sediment to deposit in the lower zones of the reservoir.
- Periodically raise pool levels high enough to inundate existing sediment deposits, thus precluding the establishment of permanent vegetation and subsequent increased sediment deposits in the backwater reaches entering the pool.

- Schedule periodic releases through the outlet works to preclude sediment accumulations in and near the intake structure and in the downstream channel.
- Be aware of conditions that may have an impact on the erosion/deposition process, such as the potential for ice jams, tributary inflow, shifting channels, and local constraints, and adjust regulation criteria to minimize adverse impacts.

4

MEASURES OF SYSTEM PERFORMANCE

Computer models are applied in a variety of river basin management decision-making situations. Each reservoir/river system and each study is unique. A variety of decision variables and decision criteria are incorporated in various simulation and optimization modeling approaches. Measures of system performance in meeting water and energy needs, reducing flooding, and otherwise satisfying water management objectives are formulated to fit the purposes of each particular application.

Defining quantitative measures of system performance is a key element of formulating a modeling and analysis strategy. The reservoir system analysis model must be able to compute values of the performance measures as a function of storage capacities and release policies. Reliability indices, economic efficiency indicators, and other performance measures provide mechanisms for evaluating and comparing alternative reservoir storage allocations and operating plans. The measures of system performance outlined in this chapter are fundamental to the full spectrum of simulation and optimization modeling and analysis methods addressed in subsequent chapters.

4.1 FORMULATING A MODELING AND ANALYSIS APPROACH

Formulating a modeling and analyzing strategy for a particular application requires a thorough understanding of the real-world problems to be addressed and the decision processes to be supported. Models should be designed based on a clear definition of the study objectives, information needs, and intended use of the information to be derived from the modeling exercises. The following questions must be answered in developing a modeling approach and formulating measures of system performance:

- Why is the analysis being conducted? What questions are to be answered by the modeling exercises?
- What information is needed to support planning and management decisions? How can modeling results be most effectively communicated?
- What is to be determined? What variables are to be optimized?
- What factors constrain the decision variables? What are the tradeoffs involved in alternative decision policies?
- What criteria are to be adopted to measure the optimality of the alternative plans to be formulated and evaluated?
- Which models and analysis techniques should be used?
- What are the input data requirements of the alternative modeling approaches being considered? What data are available? What are the limitations regarding data availability?

Simulation and optimization modeling approaches and descriptive versus prescriptive orientations are compared in Chapter 1. Simulation models are addressed in Chapters 8 and 9, and optimization models are covered in Chapters 10 and 11. Developing mechanisms for defining reservoir operating plans and developing criteria for evaluating the performance of alternative plans are fundamental elements of formulating simulation and optimization modeling strategies. In a descriptive simulation model, system performance is evaluated for a specified system design and operating plan. The model is repeatedly executed for each alternative plan being evaluated. An optimization model determines the set of values for releases, storages, and other decision variables that minimize or maximize an objective function. The heart of a prescriptive optimization model is its objective function, which represents a measure of system performance.

4.2 HIERARCHY OF GOALS AND OBJECTIVES

Reservoir projects are constructed and operated to meet a broad spectrum of goals and objectives. Quantitative performance measures are formulated within a hierarchical framework of goals and objectives, which range from general expressions of societal goals to the specific quantitative performance measures adopted for a particular modeling and analysis exercise. Although the concept of a hierarchy of goals and objectives has been expressed in various ways, the following nomenclature is adopted here:

- Societal goals and objectives for water resources development and management
- Project purposes
- Specific planning and operational objectives
- Performance measures incorporated in modeling exercises

4.2.1 Societal Goals and Objectives

Hall and Dracup [1970] suggest that the broad spectrum of goals for water resources systems in meeting the needs of society can be approximated by the following fundamental objectives:

- To control or otherwise manage the freshwater resources of the cognizant geographic or political subdivision so as to provide protection against injurious consequences of excesses or deficiencies in quantity or quality
- To provide or maintain water in such places and times and in adequate quantity and quality for human or animal consumption, wildlife (including native plants), food production and processing, industrial production (including energy), commerce, and for the recreational, aesthetic and conservation purposes considered desirable by the body politic
- To accomplish all of the above with a minimum expenditure of the physical, economic, and human resources available

Broad goals and objectives related to enhancing the welfare of society are established at the policy level. Planning of federal projects follows the Water Resources Council [1983] Principles and Guidelines (P&G), which state:

"The Federal objective of water and related land resources project planning is to contribute to national economic development consistent with protecting the Nation's environment, pursuant to national environmental statutes, applicable executive orders, and other planning requirements."

The P&G outline four accounts to facilitate evaluation and display of effects of alternative plans.

- The national economic development (NED) account displays changes in the economic value of the national output of goods and services.
- The environmental quality (EQ) account displays nonmonetary effects on ecological, cultural, and aesthetic resources.
- The regional economic development (RED) account registers changes in regional economic activity.

- The other social effects (OSE) account registers plan effects from perspectives that are relevant to the planning process, but are not reflected in the other three accounts.

Federal projects involve partnership arrangements with state and local entities to address regional and local problems and needs. State and local objectives are also incorporated into the planning process for federal studies and projects. Many of the major reservoir systems in the United States are on interstate rivers. Objectives articulated by several states may be important in establishing operating policies for a particular reservoir system.

Likewise, managing major reservoir systems may be influenced by the goals and objectives of neighboring nations. Worldwide, some 214 river basins, comprising about 50% of the land area of the world, are shared by two or more countries [United Nations, 1978]. Several major reservoir systems in the United States are on rivers shared with either Mexico or Canada. Reservoir operations are governed by treaties and other agreements between the countries.

4.2.2 Project Purposes

A reservoir typically serves one or more of the following purposes: flood control, hydroelectric power, steam-electric cooling water, industrial water supply, municipal water supply, irrigation, navigation, recreation, fish and wildlife enhancement, and water-quality management. Operating policies are established within the framework of reservoir purposes. Congressional authorization of federal projects includes specification of the purposes to be served. Nonfederal projects are also constructed for specific purposes. Institutional considerations, such as agency responsibilities and cost sharing, are based on project purposes. Modeling and analysis methods vary significantly depending on the purposes being considered.

4.2.3 Planning and Operational Objectives

Defining problems, needs, and opportunities is a major initial step in the planning process for feasibility studies in which either construction projects or operational modifications at existing facilities are being considered. An analysis of problems, needs, and opportunities results in a set of planning objectives for guiding plan formulation and evaluation. In the case of operating reservoirs, planning objectives may be expressed as operational objectives for governing operating plans. Planning and operational objectives are formulated within the framework of the previously discussed general societal goals and are associated with particular project purposes. These objectives are specific to the problems, needs, and opportunities being addressed by a particular study. The effectiveness of alternative plans in meeting the planning objectives should be quantified as much as possible, but may be evaluated qualitatively as well as quantitatively.

A broad range of planning and operational objectives may be pertinent, depending on the circumstances of the particular reservoir system being considered. Examples of objectives include the following:

- Reduce the risk and consequences of flooding of residential and commercial properties along a certain reach of a river through a particular urbanizing area for a full range of flood frequencies.
- Increase the firm yield to that required to meet projected year 2010 municipal and industrial water needs of a particular city.
- Minimize the cost of meeting electrical energy needs in a particular region served by a system of thermal-electric and hydroelectric power plants.
- Reduce the risks and consequences of irrigation water supply shortages for existing farming activities in a region as well as for a projected increase in the amount of irrigated land.
- Maintain specified minimum flows in a particular river reach for preservation of aquatic habitat and species.
- Maximize opportunities for specific reservoir recreation activities consistent with present and projected future needs for recreation in a certain region.

4.2.4 Quantitative Measures of System Performance

Quantitative measures of system performance to be used in modeling and analysis exercises are formulated within the framework of the aforementioned hierarchy of goals and objectives. The purpose of performance measures is to have a mechanism to quantitatively compare, for alternative storage allocations and operating plans, the effectiveness of the reservoir system in meeting specified objectives. Consequently, the performance measures must be a function of the storage and release parameters that define a project design and an operating policy. A performance measure could be a well-defined objective function incorporated in an optimization model. However, performance measures typically include a wide range of information computed by simulation and/or optimization models in addition to values for single-valued well-defined objective functions. An overview of the various performance measures that are used in reservoir system analysis studies follows.

4.3 OPTIMIZATION OBJECTIVES

As discussed in Chapters 10 and 11, optimization (mathematical programming) methods involve determining values for a set of decision variables which will minimize or maximize an objective function subject to a set of constraints. Application of mathematical programming techniques requires development of an objective function consisting of a single mathematical equation expressing the objective as a function of the decision variables. Precisely defined objective functions are also useful for optimization strategies involving multiple comparative runs of a simulation model with alternative operating plans. The requirements placed on the mathematical form of an objective function incorporated in mathematical programming models are stricter than in simulation models.

In an optimization model, the objective function is a single-valued criterion which can be expressed in a format suitable for a mathematical programming algorithm. Simulation results discussed later in this chapter include an array of functional relationships and sets of numbers descriptive of the performance of a reservoir system for a specified operating plan, as well as some performance criteria that may be represented by a single computed value. The objective or criterion functions discussed here provide a single value of a performance measure for a set of values for the specified decision variables.

A wide variety of objective or criterion functions have been incorporated in the different optimization (mathematical programming) models for analyzing reservoir operations that have been reported in the literature. The objective functions may be similar, and are sometimes identical, in various optimization models. However, in general, individual objective functions have been formulated for each specific model and reservoir system operation study. A number of specific reservoir optimization models are discussed in Chapters 10 and 11, along with their objective functions.

4.3.1 Examples of Decision Criteria

The objective function may be a penalty or utility function used to define operating rules based on relative priorities, or may be a mathematical expression of a planning or operational objective such as the examples listed here. The following water management objectives have been reflected in the objective functions of various optimization models:

- Economic benefits and costs
  — Maximize water supply and/or hydroelectric power revenues
  — Minimize the cost of meeting electric power commitments from a combined hydro/thermal system
  — Minimize economic losses due to water shortages
  — Minimize the electrical cost of pumping water in a distribution system
  — Minimize the damages associated with a design flood event
  — Maximize the net benefits of multiple purpose operations
  — Minimize costs associated with multiple purpose operations
- Water availability and reliability
  — Maximize firm yield, yields for specified reliabilities, or reliabilities for specified demands
  — Minimize shortage frequencies and/or volumes

— Minimize shortage indices, such as the sum of the squared deviations between target and actual diversions or instream flows
— Minimize the weighted sum of shortage indices
— Maximize the minimum streamflow
— Maximize reservoir storage at the end of the optimization horizon
— Minimize spills
— Minimize evaporation losses
— Minimize average monthly storage fluctuations
— Maximize the length of the navigation season
— Minimize the total volume of water released for minimum navigation needs
— Minimize the number of days that flood stages are exceeded
• Hydroelectric power generation
— Maximize firm energy
— Maximize average annual energy
— Minimize energy shortages or energy shortage indices
— Maximize the potential energy of water stored in a system

4.3.2 Multiple Decision Criteria

Although several different objectives will typically be of concern in a particular reservoir system analysis study, an optimization model can normally incorporate only one objective function. Multiple objectives can be combined in a single function if they can be expressed in commensurate units, such as dollars. However, the different objectives of concern are typically not quantified in commensurate units. Developing multi-objective programming methods for analyzing trade-offs between objectives has, for many years, been a major field of research in operations research and systems engineering [Cohon, 1978; Goicoechea et al., 1982]. Most of the multi-objective decision techniques proposed in the literature are variations of the following alternative approaches to analyzing tradeoffs between objectives.

One approach is to execute the optimization model with one selected objective reflected in the objective function and the other objectives treated as constraints at fixed user-specified levels. For example, a reservoir operating policy might be developed based on the noncommensurate objectives of maximizing municipal water supply firm yield, firm hydroelectric energy, and average annual energy. The model could be formulated to maximize an objective function representing average annual energy, subject to the constraints that a user-specified water supply firm yield and firm energy be maintained. Alternative runs of the model are made to show how the average annual energy is affected by changes in the user-specified water supply firm yield and firm energy.

Objectives may be prioritized in some applications. Higher-priority objectives are more important than those of lesser priority. A multiple-objective optimization strategy that has been adopted in a number of studies is based on optimizing objectives in priority order. Each objective is optimized subject to the constraint that no higher-priority objectives are adversely affected.

Another alternative strategy for analyzing trade-offs between noncommensurate objectives involves treating each objective as a weighted component of the objective function. The objective function is the sum of each component multiplied by a weighing factor reflecting the relative importance of that objective. The weighing factors can be arbitrary, with no physical significance other than to reflect relative weights assigned to the alternative objectives included in the objective function. The model can be executed iteratively with different sets of weighing factor values to analyze the trade-offs between the objectives with alternative operating plans.

4.3.3 Role of Objective Functions

Two equally valid but somewhat contrasting perspectives on the role of objective functions, and the models into which they are incorporated, in the overall process of evaluating alternative operating plans are presented in the following two paragraphs. Reconciling and balancing these two perspectives is a key consideration in formulating a modeling and analysis approach for a particular reservoir system operation study. This is also a basic philosophical issue in assessing the practical utility of optimization models in general. The question is, how completely and accurately does the objective function incorporated in a model have to reflect actual societal objectives in order to provide meaningful information for use in the decision-making process?

The usefulness of a model depends on how meaningfully the complex real world can be represented by a set of mathematical equations. A key consideration with optimization models is how well the objectives of the decision makers can be represented by a single-equation objective function. Public needs and objectives are complicated and typically ill-defined. The physical and institutional characteristics of multiple-purpose water management facilities, electric utilities, and water allocation and use systems are complex. Likewise, hydrologic and environmental characteristics of a river basin are complex. The risks and consequences of water shortages and flooding and the benefits of water management are difficult to quantify. Functional relationships between the complex characteristics of floods and droughts and the objectives to be served by a reservoir system can only be approximated. Quantitative measures of system performance must meaningfully capture the complexities of the real world. Necessary simplifications and approximations severely limit the utility of models.

On the other hand, objective functions are simply quantitative measures of system performance which can be meaningfully utilized to obtain a better understanding of the system. Even if planning objectives can be precisely articulated, which is typically not the case, it will not likely be possible to incorporate an objective function into an optimization model that captures the total essence of the planning objectives. However, models still provide valuable analysis tools. A model can significantly contribute to the evaluation process even though it can never tell the whole story. The objective function can be a simple index of the relative utility of alternative operating plans, which provides significant information regarding which alternative plan best meets the planning objectives. Modeling exercises with alternative decision criteria help address different aspects of the overall story.

4.4 SIMULATION RESULTS

This discussion of simulation results is generally pertinent to either optimization or simulation models. All optimization models simulate as well as optimize. Modeling studies typically involve numerous executions of one or more models, with each execution generating voluminous output data. For a given reservoir system operating plan and specified set of conditions, an extensive amount of data can be developed. The model output data can be used in a variety of ways, based on the ingenuity and preferences of the analysts, to develop a better understanding of the operation of the reservoir system. Quantitative measures of system performance typically involve an array of information. There is seldom, if ever, a simple single-number answer to a reservoir operation problem.

A wide variety of modeling and analysis approaches are possible. However, a typical model operates a reservoir system during a hydrologic period of analysis, with storages, discharges, and other pertinent quantities being computed at some appropriate time step. Model results may be expressed as tabulations or plots of storage, releases, streamflows, diversions, diversion shortages, energy generated, water-quality parameters, or other variables for each time interval. The primary output of reservoir operation studies is often simply time-series plots of storage levels. Model results are often presented as frequency tabulations and plots of the various variables computed. For example, frequency or duration relationships are commonly displayed of reservoir storage levels, streamflow rates, or salinity concentrations.

Flood control studies typically involve analyses of the effects of alternative storage capacities and operating plans on various characteristics of specified historical or hypothetical design flood hydrographs. Such characteristics include peak stage, peak discharge, duration, volume, velocity, rate of rise or warning time, and sediment transport rates. Discharge and/or stage versus exceedance frequency relationships are typically computed for pertinent streamflow locations. Variations in reservoir storage hydrographs and/or storage versus exceedance frequency relationships, with alternative operating plans, also provide meaningful comparisons.

4.4.1 Flood Control Storage Capacity

Flood control storage capacity has traditionally been evaluated in terms of the annual exceedance probability (P) or recurrence interval ($T = 1/P$) of a flood that can be contained without releases contributing to downstream flooding. For example, a flood control pool may be sized to contain a 1% exceedance probability (100-year recurrence interval) design flood. A flood of the specified magnitude just fills the flood control pool. Thus, a flood control pool may be sized such that a design flood, associated with a probability of 1% of being equalled or exceeded in any year, will just fill the pool without necessitating releases that contribute to flooding at downstream locations. Frequency analyses, using a plotting position formula or probability distribution function, may be performed based on a series of peak annual storage levels computed using a reservoir system simulation model with period-of-record historical streamflow data. Alternatively, design storms developed based on rainfall intensity-duration-frequency relationships are input to a watershed (precipitation-runoff) to develop a reservoir inflow hydrograph associated with a specified annual exceedance probability. This design flood is then routed through the reservoir to determine the peak storage level associated with the specified exceedance probability. Flood control storage capacity in many Corps of Engineers reservoirs was designed based on a standard project flood rather than a design flood based on a specified statistical frequency. The standard project flood is developed following procedures outlined by the USACE [1952]. Although statistical measures of risk are important in sizing flood control pools, dams and spillways are designed based on ensuring the structural safety of the dam from overtopping and related hazards. Emergency spillways are designed to pass the probable maximum flood, which represents the most severe flood that can reasonably be expected to be physically possible. The probable maximum flood is developed with a precipitation-runoff model from a probable maximum storm developed using precipitation data and procedures published by the National Weather Service [National Weather Service, 1982; McCuen, 1989].

4.4.2 Measures of Reliability for Conservation Storage Operations

Conservation storage operations are evaluated in terms of capabilities for meeting water supply and hydroelectric energy requirements. The water supply requirements may involve diversions for municipal, industrial, and agricultural water uses or maintenance of instream flows for preservation of aquatic species, habitat, or other environmental resources. Water use is reflected in a model by a set of diversion and/or instream flow requirements representing either actual historical or projected future water use for a specified past, present, or future point in time; water amounts committed to users by water rights, agreements, or other allocation systems; or hypothetical yields. Water use is often expressed in terms of a constant annual amount combined with monthly or seasonal water use distribution factors. Model output includes shortages or failures in meeting specified target demands. Concise measures of system reliability in meeting water use requirements are useful in analyzing and displaying simulation results.

Models are based on specified conditions or assumptions. Reliability estimates are based on specified scenarios regarding reservoir operating policies and the impacts of other water users and activities in the river basin. The stochastic nature of streamflow and other pertinent variables must be reflected in the analyses.

Since streamflows, reservoir evaporation rates, and other pertinent quantities are highly variable and the future is unknown, water supply capabilities must be viewed from a reliability, likelihood, or percent-of-time perspective. Water availability may be expressed in terms of a firm or dependable yield, the reliability of meeting various demand levels, the percent of time specified quantities of water are available, the risk of shortages, storage frequency relationships, or a tabulation of the amount of water available during each time interval of a simulation.

As discussed in Chapter 7, reservoir reliability is an expression of the likelihood or probability of meeting a given demand or, equivalently, the percent of time the given demand can be met. Reliability (R) is the complement $(R = 1 - F)$ of the risk of failure (F) or probability that the demand will not be met or, equivalently, the percent of time that the demand will not be met. Various reliability indices can be formulated to serve the purposes of the particular application. The objective is to formulate a concise measure of water supply capabilities which is meaningful from the perspective of the particular water management activities being supported by the modeling study.

Variations of the concepts of period and volume reliability are often adopted as indices of capabilities for meeting water use requirements. Period reliability is based on counting the number of time periods during a simulation during which a specified demand target is met. Volume reliability reflects the shortage magnitude as well as frequency.

Volume reliability (R_v) is the ratio of the volume of water supplied (v) to the volume demanded (V):

$$R_v = (v/V)\ 100\% \tag{4-1}$$

or, equivalently, the ratio of the mean actual diversion rate to the mean target diversion rate. The shortage volume is the demand target (V) minus the volume supplied (v) within the constraints of water availability.

Period reliability (R_p) is computed from the results of a historical hydrologic period-of-record simulation as

$$R_p = (n/N)\ 100\% \tag{4-2}$$

where n denotes the number of time periods (such as months) during the simulation for which demands could be met, and N is the total number of time periods in the simulation. Thus, period reliability provides an index representing the proportion of time that the reservoir/stream system is able to meet demands without water use restrictions, based on the water management scenario reflected in the simulation model. Equivalently, the reliability represents the likelihood or probability of the demand being met in any randomly selected period.

Various representations of water supply failure can be incorporated into the computation of period reliability. Reliability can be formulated in terms of meeting all of the demand target or, alternatively, at least a specified portion of the demand target. Alternatively, the demand target can be defined in terms of meeting demands without reservoir storage falling below specified levels. Reliability can be defined as the percentage of periods during the simulation during which a specified storage level is equaled or exceeded. Various definitions of reliability can be formulated for alternative time periods. Monthly periods are typical in simulation modeling studies, but other time intervals are common as well. Period reliability may alternatively be defined in terms of meeting an annual diversion target on a yearly basis during a simulation performed using a monthly or weekly computational time interval.

Firm yield (or safe or dependable yield) is a commonly used measure of water supply capabilities. Yield (sometimes called *draft*) is the amount of water which can be supplied by a stream/reservoir system under specified conditions. Firm yield is the estimated maximum release or diversion rate which can be maintained continuously during a hypothetical repetition of the hydrologic period of record, based on all the premises incorporated in the model, including assumptions regarding operating policies, diversion locations, impacts of other water users and activities in the basin, and other complexities. Firm yield is the draft that will just empty the reservoir or multiple-reservoir system during a hydrologic period-of-record simulation. Firm yield is typically expressed in terms of a mean annual rate with monthly, or some other time interval, distribution factors being incorporated in the model to reflect the within-year seasonal variation in water use. Firm yield and smaller yields have a period and volume reliability of 100%. Yields greater than firm yield have a reliability of less than 100%.

Variations of the resilience and vulnerability criteria are also useful in analyzing reliability [Hashimoto *et al.*, 1982; Moy *et al.*, 1986]. Resilience is a measure of the capability for recovery from failure to some acceptable state within a specified time interval; and resilience may be defined as the maximum number of consecutive periods of shortage during a simulation. Alternatively, resilience may be viewed in terms of the probability of being in a period of no failure given that there was a failure in the previous period. Vulnerability is a measure of the magnitude or significance of failures, and may be defined as the largest deficit during a simulation.

The prior discussion deals with computing reliability from the results of a simulation based on a single set of input streamflows representing the hydrologic period of record. Reliability measures can also be formulated in terms of the likelihood of satisfying specified demands during a long multiple-year period. This type of reliability analysis requires many alternative streamflow sequences in addition to the historical record. Synthetic streamflow generation techniques are used to provide synthesis of the required data [Loucks *et al.*, 1981; Salas, 1993]. Synthetic streamflow generation involves synthesis of equally likely streamflow sequences with a length equal to the time period over which the reservoir system is being analyzed. The synthetic sequences preserve selected statistical characteristics of the historical data. For example, the Markov model is one of the simplest and most widely used approaches to synthesizing streamflows. The Markov model preserves the mean, standard deviation, and lag-1 autocorrelation coefficient of the historical data, at least in the sense that an infinitely long sequence of synthetic flows would have the same values for these parameters as the historical data.

The reliability may be defined as the percentage of the streamflow sequences for which demands can be met without incurring a shortage. For example, a large number (perhaps 100) of monthly streamflow sequences of specified length (say 50 years) could be synthesized. Each sequence represents an equally likely scenario of possible future streamflows. Firm yield could then be computed for each of the 100 streamflow data sets using a reservoir/river system simulation model. The firm yield is the maximum demand level which can be fully met without failure for the given streamflow sequence. The reliability associated with a given yield would be estimated as the percentage of the 100 alternative 50-year streamflow data sets for which the computed firm yield equaled or exceeded the specified yield level.

Reliability is discussed above from the perspective of supplying water for various beneficial uses. The concepts are equally applicable to hydroelectric power. Firm power is the maximum rate of energy production that can be maintained continuously assuming the period-of-record historical inflows are repeated in the future. Firm power and reliability associated with various levels of power production are computed similarly to firm yield and reliability for water supply.

4.5 ENVIRONMENTAL OBJECTIVES

Environmental concerns are often a primary consideration in reservoir/river system analysis studies. Both water quality and quantity are important in meeting environmental objectives. Reservoir releases contribute to meeting instream flow requirements for fish and wildlife, riparian habitat, wetlands, freshwater inflows to bays and estuaries, and low-flow augmentation for water-quality management. Instream flow needs for aquatic and riparian ecosystems involve maintaining minimum flows and optimal fluctuations in flows. Seasonal upper limits may be placed on reservoir releases to prevent adverse impacts on wildlife habitat. Migration of anadromous fish, such as salmon in the Pacific Northwest, may be a major consideration in reservoir operation. Fisheries, wildlife, and other environmental resources are affected by fluctuations in reservoir pool levels. Environmental objectives are translated into demands for flows and storage similarly as are hydroelectric power, water supply, navigation, and other needs. Other evaluation techniques, not covered in this book, are required to establish the streamflow and storage requirements for meeting environmental needs, as well as for meeting human needs for water and electricity. Reservoir/river system analysis studies, using the models described in Chapters 8–11, are conducted to evaluate reliabilities provided by alternative management strategies for meeting these multiple-purpose needs.

Water quality is a primary concern in meeting environmental needs as well as human water supply needs. Temperature, dissolved oxygen concentrations, and other water-quality parameters are of concern in both managing and modeling reservoir/river systems. Reservoir managers may be concerned with meeting state and federal water-quality standards. Reservoir releases may be based on maintaining temperature and dissolved oxygen levels in downstream flows as required for healthy fisheries and ecosystems. In-pool fisheries, habitat, recreation, and aesthetics also depend on reservoir water quality. The simulation models discussed in Chapter 12 provide capabilities for addressing water-quality concerns.

4.6 ECONOMIC EVALUATION

Economic analysis has played an important role in water resources planning and management. The Flood Control Act of 1936 and subsequent statements of policy have required a benefit/cost justification for federal water resources development projects. Economic efficiency criteria are also commonly incorporated in state and local water resources planning. Over the past several decades, procedures have been developed by the federal water agencies for estimating national economic development benefits associated with the various project purposes. The Principles and Guidelines [Water Resources Council, 1983] outline the basic concepts of economic evaluation followed by the federal water agencies. Past federal work in developing and applying economic evaluation methods has been accomplished primarily from the perspective of planning studies involving proposed construction projects rather than reevaluation of completed projects. However, the economic efficiency objective is also applied to reevaluation of operations of existing reservoir systems.

Economic evaluation consists of estimating and comparing the benefits and/or costs, in dollars discounted to a common time base, which would result from implementation of alternative plans of action [James and Lee, 1971; Goodman, 1984]. The economic objective function may be to either maximize net benefits, minimize costs, or maximize benefits. Net benefits are benefits minus costs. If a fixed level of service is provided by all of the alternative plans being considered, the economic objective function may be expressed as minimizing costs, without needing estimates of benefits. Likewise, in operations of an existing facility, the costs may be fixed, and alternative plans are compared based only on benefits. If most of the benefits and costs are fixed for all alternative plans, the evaluation may be based on a specific component of either the benefits or costs.

Cost estimates are usually more straightforward than benefit estimates. Conventional engineering estimates of construction, major replacement, operation and maintenance, and fuel costs are common to many applications.

Benefit estimates are based on the standard of willingness to pay, which is defined as the amount a rational and fully informed buyer should be willing to pay for one more unit of a service or commodity. If a competitive unregulated market exists for a commodity, the market price would be the willingness to pay. However, market prices typically do not reflect the true marginal cost or economic benefit of the output of water projects. Consequently, benefits, representing willingness to pay, are typically estimated using approaches based on alternate cost or cost savings. In the alternate cost approach, benefits are estimated as the cost of the least costly alternative method for providing the service. It is assumed that the service must be provided even if the proposed plan is not implemented. Benefits for municipal and industrial water supply, water-quality management, and hydroelectric power are estimated as the cost of the least costly alternative means for providing these services. Navigation benefit estimates are based on comparing the cost of shipping pertinent goods by barge with the cost of transporting the same goods by the least costly alternative method available, such as by truck or train. Flood control benefits are based on cost savings, including reductions in flood damages.

The concept of shadow price is an example of a less conventional approach that has been used to estimate the relative utility of alternative plans for modifying reservoir operations. Shadow price is the dollar amount associated with one objective given up in order to gain an extra unit of a second objective, with the performance of all other objectives remaining the same. For example, levels of service of other purposes provided by a multiple-purpose reservoir may be shadow-priced to hydroelectric power revenue. The cost of modifying the operating policy to enhance water supply, recreation, or other purposes is related to the corresponding reduction in hydroelectric power revenue.

Benefit estimation methods vary depending on water resources development and management purposes. Procedures have been developed for evaluating benefits for flood control, municipal and industrial water supply, irrigation, hydroelectric power, navigation, recreation, and other project purposes.

4.6.1 Flood Control Benefits

The USACE and other federal agencies have developed detailed procedures [Davis *et al.*, 1988; Hansen, 1987; U.S. Army Corps of Engineers, 1994] for quantifying the economic benefits associated with reducing the risk of flooding, within the framework of the Principles and Guidelines [Water Resources Council, 1983]. As discussed in Chapter 7, estimation of average annual damages is a central component of economic evaluation methodology. Expected or average annual damage is a frequency-weighted sum of damage for the full range of damaging flood events and can be viewed as what might be expected to occur on the average in any year. The expected value of annual damages is computed as the integral of the exceedance frequency versus damage relationship representing a flood plain reach. Exceedance frequency versus peak discharge, stage versus discharge, and stage versus damage relationships are combined to develop the exceedance frequency versus damage relationship.

Economic feasibility, as measured by estimated benefits exceeding costs when discounted to a common time base using a specified discount rate, is a strict requirement for justification of a federal flood control project. Maximizing net economic benefits is a key consideration in sizing flood control storage capacity. The objective in sizing storage capacity has often been to maximize net benefits subject to the constraint of providing at least a specified minimum level of protection, such as containing a 2% or 1% annual exceedance probability (50- or 100-year recurrence interval) flood without releases contributing to downstream flooding.

4.6.2 Water Supply Benefits

Municipal and industrial water supply planning and management has tradition-ally been based on ensuring that firm yield exceeds water needs by some reasonable margin of safety. The major policy emphasis since the 1970s on demand management and achieving more efficient water use has resulted in reservoir planning studies now reflecting demand management strategies in projections of future water needs. Water supply studies typically include two key elements: (1) projections of water needs; and (2) firm yields, or yield versus reliability relationships, for alternative water supply augmentation plans. New water supplies are developed and/or operations of existing facilities modified to ensure that firm yields are maintained in excess of needs.

In economic evaluations of proposed new federal multiple-purpose reservoir projects, municipal and industrial (M&I) water supply benefits are estimated as the cost of the least costly alternative means of providing the same quantity and quality of water assuming the proposed project is not implemented. Separable costs must be less than benefits for inclusion of M&I water supply storage to be economically jus-tified. Estimating municipal and industrial water supply benefits based on a least-costly-alternative analysis does not provide the precision and sensitivity needed to evaluate modifications in reservoir operating plans. Young, Taylor, and Hanks [1972], Dziegielewski and Crews [1986], and Wurbs and Cabezas [1987] suggest economic evaluation procedures in which average annual losses due to water short-ages are computed that may be more pertinent for evaluating alternative reservoir operating policies.

Irrigation benefits are estimated as the increase in net farm income projected to result from an irrigation project. Land areas planted with various crops, produc-tion costs, and income are estimated alternatively for conditions with and without the irrigation project. For some situations, the Principles and Guidelines [Water Re-sources Council, 1983] allow irrigation benefit estimates to be based on the in-creased market value of land in lieu of farm budget analyses.

4.6.3 Hydroelectric Power Benefits

The USACE [1985] outlines economic evaluation procedures for hydroelec-tric power within the framework of the Water Resources Council P&G. The general measurement standard is the willingness of users to pay for a project's output. In most instances, willingness to pay cannot be measured directly. Four alternative techniques are used to estimate hydroelectric power benefits in lieu of direct mea-surement of the consumers' willingness to pay:

- Actual or simulated market price
- Change in net income
- Cost of the most likely alternative
- Administratively established values

The first three approaches stem from the willingness to pay criterion. The fourth approach, administratively set prices, relates to willingness to pay but also may reflect other social objectives and procedures. The first and third approaches are most often adopted in actual practice. In order for any measure of benefits to be valid, a need for the electrical energy or power capacity must be demonstrated.

Actual or simulated market price is the preferred basis for estimating benefits in situations where energy is sold at its marginal cost; and new power plant additions or increases in production are small compared to the system load. However, electric power is not normally priced at its marginal cost. Power prices are typically based on the average cost of generation, which includes older power plants as well as newer additions to the system, rather than the marginal cost.

The most commonly adopted basis for estimating hydroelectric benefits is alternate cost analysis. The alternative costs are estimated as either the cost of constructing and operating an alternative thermal plant or an increment thereof, or the value of generation (primarily fuel costs) from existing thermal plants that would be displaced by the output of the proposed hydroelectric plant. Energy conservation measures and other hydroelectric power projects may also be considered in the evaluation of least costly alternatives to the proposed plan.

4.6.4 Recreation Benefits

Construction of a reservoir project or modification of the operating policy of an existing reservoir may both create and displace recreational opportunities [Vincent *et al.*, 1986]. For example, construction of a reservoir project provides reservoir-related recreation while displacing stream and terrestrial recreation. Net recreation benefits are the difference between the value of recreational opportunities gained and those lost. Total willingness of users to pay is the sum of two components: (1) the actual entrance fees and user charges plus (2) any excess amount they would be willing to pay but do not have to pay.

Several techniques have been developed to estimate recreational demand and value. Three alternative approaches are recommended by the Principles and Guidelines [Water Resources Council, 1983]: unit day method, travel cost method, and contingent value method. Under the P&G, the unit day method is acceptable for only limited situations, since it is more approximate. The unit day method does not attempt to account for the impact of price on visitation to a recreational site. An assigned user day value is applied to the total number of estimated visitors. Both the travel cost and contingent value methods determine the value of a recreational site by attempting to approximate the price-quantity demanded relationship. Use as well as the willingness to pay for that use are estimated simultaneously. A separate use estimation technique is not required by the travel cost and contingent value methods, but is required by the unit cost method [Vincent *et al.*, 1986].

The unit day value method applies a simulated market value, in dollars per unit of use, to projected use. The simulated value is judgmentally derived from a published range of values, agreed to by federal agencies, for generalized and specialized types of recreation [Water Resources Council, 1983]. Alternative methods for estimating person-days of recreation use include regional and site-specific use estimating models, the similar project method, and the capacity method.

The travel cost method uses the variable costs of travel as a proxy for price in determining net willingness to pay for increments or decrements of supply above fees and user charges. Travel costs depend on travel distance and the value of travel time. The method is based on empirical data on the behavior of individuals in visiting recreation facilities.

The contingent valuation method estimates changes in benefits based on the willingness to pay and the level of participation of individual recreationists. The method is based on asking recreationists questions that indicate their willingness to pay entry fees for recreation opportunities, through means such as mail questionnaires or telephone or personal interviews.

4.6.5 Operations of Existing Reservoirs Compared with Proposed Construction Projects

As discussed in Chapter 1, the several-decade construction era of developing numerous major reservoir projects transitioned, during the 1970s and 1980s, to the present focus on operations of existing facilities. The economic feasibility criterion, which requires that benefits exceed costs for a construction project to be justified, has been a major consideration in developing traditional economic evaluation procedures. Economic evaluation of operations of existing reservoir systems differ significantly from evaluating proposed construction projects, in regard to refinements and incremental differences, tradeoffs between project purposes, and data availability.

Economic evaluation of proposed construction projects incorporates a comprehensive array of benefits and costs, but the evaluation procedures do not necessarily have to be sensitive to refinements and relatively small changes in operating policies. Analyses of proposed modifications in the operations of existing reservoir systems are more concerned with incremental differences. Modeling approximations and simplifications may become more important when relatively small incremental changes in storage allocations or refinements in release rules are evaluated. Greater precision may be required; for example, estimating average annual economic flood damages necessarily involves significant approximations and simplifications. Average annual damages with and without a proposed project can be meaningfully estimated and compared to quantify the reduction in flood damages to be derived from construction of the project. However, estimates of the change in average annual flood damages to result from reallocation of a relatively small amount of storage capacity between flood control and conservation purposes, adoption of a seasonal rule curve, or changes in target release rates are much less meaningful because the change in annual damages is too small relative to the inaccuracies caused by modeling approximations and data uncertainties. For example, frequency versus discharge relationships can be most accurately estimated for the more frequent flood events. However, storage reallocations and other types of modifications in reservoir operations typically affect the release only for the extreme, less frequent events, for which data is most uncertain.

The increased need to develop a detailed understanding of the interactions and trade-offs between project purposes may make evaluation of operational modifications more complex. In sizing storage capacities and establishing operating rules during preconstruction planning of proposed multiple-purpose reservoir projects, the flood control and conservation pools are handled largely independent of each other. The various conservation purposes are also separated in many respects. Although site characteristics may somewhat limit total storage capacity, the amount of storage capacity provided for conservation purposes usually does not significantly constrain the amount of storage capacity for flood control, and vice versa. Economies of scale tend to favorably impact the economic justification of the multiple project purposes. On the other hand, operational modifications of completed projects often involve reallocating limited fixed total available storage capacity and/or water resources between project purposes. A reallocation study may involve evaluation of the comparative worth of storage capacity and water used for alternative purposes, identification and quantification of benefits lost by project beneficiaries due to a change in operations, and formulation of operational strategies for enhancing one purpose while minimizing adverse impacts on other purposes. Evaluating tradeoffs between purposes may be the primary focus of modeling and analysis efforts.

In other respects, operational modifications are simpler to evaluate than construction projects. In evaluating modifications to reservoir operating rules, most of the project benefits and costs are fixed, and only selected pertinent benefit and/or cost components are reflected in the analysis. Quantitative dollar-related measures of system performance are often adopted in reservoir system analysis studies which do not necessarily reflect the willingness-to-pay benefit standard. For example, the objective function of optimization models are sometimes based on revenues from selling water or hydroelectric energy. Although in some situations such revenues may accurately reflect willingness to pay, water supply and hydroelectric power revenues do not typically represent marginal cost or the actual price consumers are willing to pay. However, water supply or hydroelectric revenues may still provide a meaningful index or performance measure for comparing alternative operating plans, even though the willingness to pay standard is not accurately represented. For purposes of comparing alternative operating plans, a high degree of accuracy in estimating economic parameters may not necessarily be required as long as sensitivity relative to differences between alternative plans is properly reflected.

Studies of modifications to existing projects typically have the advantage of a much better data base than was available when the project was originally planned and designed. For example, there is a longer record of streamflow and other hydrologic data. Information regarding revenues from marketing water, energy, and other services provided by the project will also be available. In addition to greater data availability, experiences and lessons learned from years of actual operations can be reflected in operational modifications.

5

HYDROLOGIC DATA

Reservoir system modeling and analysis applications require a variety of hydrologic and related types of data. Long records representative of historical hydrology are used in planning studies. Current observations are required for real-time operations. Much of this chapter is devoted to considerations in developing streamflow data sets. Reservoir evaporation rates, water quality data, hydrometeorologic data, and reservoir elevation versus storage and area relationships are also addressed.

An array of computational tasks are involved in converting records of field measurements of hydrologic processes to the input data required for reservoir system analysis models. For example, historical records of gaged streamflows are adjusted to represent flows at pertinent locations for a specified condition of river basin development. Streamflows may be synthesized using watershed (precipitation-runoff) models combined with rainfall and/or snowpack observations, meteorologic data affecting snow melt, and watershed parameters required to compute runoff hydrographs. Sedimentation predictions are necessary in developing reservoir elevation/storage/area relationships for future points in time.

5.1 DATA ACQUISITION AND MANAGEMENT

5.1.1 Hydrologic Data Collection and Dissemination Agencies

Numerous federal, state, regional, and local agencies conduct water-related data observation and management activities for various reasons. For example, reservoir management entities routinely collect a variety of data in conjunction with operating their reservoir projects. However, the agencies noted in the following para-graphs play particularly key roles in collecting hydrologic, water quality, and clima-tologic data and making these data available to the water management community.

The U.S. Geological Survey (USGS) has primary federal responsibility for the collection and dissemination of measurements of stream discharge and stage, reservoir and lake stage and storage, groundwater levels, well and spring discharge, and the quality of surface and groundwater. The USGS is also responsible for the coordination of all the federal agencies in the acquisition of certain water data. The USGS works closely with other federal, state, and local agencies, including reser-voir management entities, in its cooperative data collection programs. The USGS maintains hydrologic data collection stations located throughout the United States and a database system called WATSTORE (WATer data STOrage and REtrieval). All data collected by the USGS are stored in WATSTORE. The data are available to other agencies and the public in machine-readable form or as computer-printed tables or graphs, statistical analyses, and digital plots. The data are also published by water year for each state in a publication series entitled *U.S. Geological Survey Water-Data Reports.* Commercial firms also market USGS data on CD-ROM (com-pact disk-read only memory).

The National Water Data Exchange (NAWDEX) is an interagency program, coordinated by the USGS, to facilitate the exchange of water data and to promote the improvement of data-handling procedures. The NAWDEX program includes maintenance of an index of water data available from both federal and nonfederal sources.

The USGS, created in 1879, is an agency of the Department of the Interior. The hydrologic data collection and dissemination mission is the responsibility of the Water Resources Division (WRD) of the USGS. Water Resources Division headquarters are located at the USGS National Center in Reston, Virginia. The pro-grams of the WRD are administered through four regional centers located in Reston, Atlanta, Georgia, Denver, Colorado, and Menlo Park, California. Field op-erations are conducted through 48 district offices, generally located in state capitals, with jurisdictional boundaries corresponding to state boundaries. These offices pro-vide assistance to agencies and the public in acquisition of products or information from WATSTORE and NAWDEX.

The National Oceanic and Atmospheric Administration (NOAA) National Climatic Data Center (NCDC), located in Asheville, North Carolina, is the collec-tion center and custodian of all United States weather records. The National Climatic Data Center stores and distributes data collected by the National Weather Service, the weather services of the Air Force, Army, Navy, and Marine Corps, the Federal Aviation Administration, the Coast Guard, and cooperative observers. The numerous databases maintained by the NCDC include precipitation, temperature, solar radiation, dewpoint, relative humidity, pressure, wind speed and direction, evaporation, Palmer drought index, and various other types of meteorological data in a variety of formats for daily and other time intervals. Data are available on com-puter-readable media as well as in published reports. Numerous weather and climate-related reports and serial publications are available from the National Climatic Data Center. The printed report series *Climatological Data* is distributed to subscribers monthly along with an annual summary issue. This series dates back to the 1880s, is published for each state or (in a few cases) combinations of states, and presents basic daily and monthly climatological data.

The Natural Resource Conservation Service (NRCS), formerly the Soil Conservation Service, of the U.S. Department of Agriculture administers a cooperative federal-state-private system for conducting snow surveys in the western states. From January through May of each year, the NRCS publishes a monthly report entitled *Water Supply Outlook,* which provides snowpack and streamflow forecast data for each state and region. The NRCS also publishes snowpack and water content data in the monthly *Basin Outlook Reports.* Snow data are also collected by various other entities. The California Department of Water Resources maintains a snow survey system. In the eastern United States, several federal, state, and private agencies perform snow surveys. Snow cover data are also available from the NOAA through NCDC.

5.1.2 Hydrologic Data Observations

Streamflow discharge measurement normally involves: (1) establishing the relationship (rating curve) between water elevation or level (stage) and discharge (flow rate) at a gaging station; (2) continuously or periodically measuring stage at the gaging station; and (3) transforming the record of stage into a record of discharge by applying the rating curve. Likewise, reservoir storage content records are developed by combining measurements of water surface elevation with a relationship between elevation and storage volume.

Reservoir inflows are often estimated based on water budget computations. Inflows are calculated considering gaged releases and storage levels along with evaporation volumes estimated by combining evaporation rates with water surface areas determined from an elevation versus area relation.

Stage or water surface levels in rivers and reservoirs are measured by various types of mechanical, electromechanical, and electronic sensors [Singh, 1992; Latkovich and Leavesley, 1993]. The three most commonly used types are float-driven sensors, pressure sensors, and ultrasonic sensors. In a typical installation of a float-driven water-level sensor, the vertical movement of a float in a stilling well, resulting from fluctuations in water level, are translated by a mechanical movement or an electronic signal. Pressure sensors are based on the relationship of water depth to pressure. Ultrasonic sensors use acoustic pulses to sense water levels by either contact or noncontact methods. Development of streamflow rating curves requires measurement of flow velocities. Mechanical, electromagnetic, ultrasonic, and thermal-pulse sensors are used to measure velocity.

Water quality data collection involves field sampling and analysis, laboratory analysis, and in-situ monitoring. Temperature, conductivity, dissolved oxygen, pH, turbidity, and other parameters with very short sample holding times are measured in the field. Many of the methods of measuring inorganic, organic, and radioactive substances in water use analytical procedures that can be performed only in a laboratory. In-situ monitors are used at reservoir projects requiring frequent water quality data for making real-time operating decisions or to monitor their effects in terms of meeting operating objectives. Typical applications involve thermistor strings for measuring temperature stratification in a reservoir, an inflow monitor to detect acid slugs entering a reservoir, and a discharge monitor for gaging results of operations [U.S. Army Corps of Engineers, 1987].

Hydrometeorological observations also include pan evaporation, used to estimate reservoir evaporation; precipitation, as measured at ground stations or estimated by radar, satellites, or other sensors; depth and water equivalent of snow accumulation, as determined from ground measurements or remote sensors; snow-covered area, as determined from aerial or ground reconnaissance, or remote sensors from satellites or aircraft; air temperature, as measured at ground stations or by upper air atmospheric soundings; and conditions of river ice. Other meteorologic data include observations of humidity, wind speed and direction, and solar radiation. Other pertinent measurements may include soil moisture and temperature and groundwater levels.

5.1.3 Automated Data Acquisition Systems

Historically, hydrometeorological data collection systems have relied largely on manual observations. Even with more advanced systems, manual observations serve as backup and are necessary for observing data that cannot feasibly be collected by automated means. However, recent advances in electronics and computer technology have enabled development of automated data observation, transmission, and storage systems. Real-time reservoir system operations represent a major application area for these data collection technologies. Data systems for supporting reservoir operations perform the following tasks: data observation and collection at field stations, transmission of data, and data management and processing in a database [U.S. Army Corps of Engineers, 1987; Latkovich and Leavesley, 1993].

The types of sensors previously noted are used to make field observations of various types of parameters. Gaging stations for measuring precipitation, streamflow, and other quantities may also include a data collection platform with telephone equipment and/or a transmitter for communication via satellite. A structure houses the monitoring and transmission equipment.

A variety of technologies are available for transmitting data from a remote site to a central station for storage and processing. Automatic data transmission media include ground-based VHF radio, environmental or general-purpose communication satellites, meteor-burst-based communication systems, land line equipment utilizing hard wire or switched commercial telephone circuits, and microwave communication systems [U.S. Army Corps of Engineers, 1987; Latkovich and Leavesley, 1993]. The choice of one technology over another depends on several factors, including the minimum allowable time between data collection and reporting, the spatial scale of the application, and the available funds. Operations of relatively simple reservoir systems require much less complicated communication and data-handling capabilities than real-time operations of larger, more complex multiple-purpose, multiple-reservoir systems. Automation of field station reporting has replaced manually observed and transmitted data in many areas.

Geostationary Operational Environmental Satellites (GOES) have been widely used to transmit streamflow, precipitation, and other data in support of real-time reservoir system operations. The Corps of Engineers, in particular, uses this type of technology in conjunction with reservoir system operations. The USACE is the largest single user of the GOES Data Collection System [U.S. Army Corps of Engineers, 1987]. GOES satellites are operated by the National Earth Satellite Data and Information Service of the NOAA. The major components of a GOES-based data collection system are the field gaging sites, the satellite, and direct readout groundstations located at central receiving sites for data retrieval. The equipment located at the gaging station site necessary for the use of a GOES-based data collection system is referred to as a Data Collection Platform (DCP). DCPs are of the following types: interrogable at any time; periodically reporting at specified times; or reporting whenever specified thresholds are exceeded.

5.1.4 Data Management Software for Modeling Applications

Streamflow, water quality, reservoir storage, climatic, and other water related data are voluminous. Likewise, reservoir system analysis model input and output files are voluminous. Data management is important for both real-time operations and planning studies. Many of the reservoir system analysis models cited in later chapters have pre- and/or post-processor programs, developed for the individual models, that are used to store, transport, organize, manipulate, analyze, summarize, and display model input and output data. Other data management software is not associated with one particular model, but rather can serve as a preprocessor for input data and/or post-processor for output data for a variety of different models. Commercially available database management systems are widely applied in business, engineering, and other fields, and can also be used for water management purposes. Several public domain data management systems have been developed by federal agencies and other entities specifically for water resources applications.

The Hydrologic Engineering Center Data Storage System (HECDSS) is available from the USACE Hydrologic Engineering Center [Hydrologic Engineering Center, 1990]. HECDSS is used routinely with several HEC simulation models, including HEC-1, HEC-5, HEC-5Q, HECPRM, and WQRRS, discussed in subsequent chapters. HECDSS can be used with other non-HEC programs as well. The database can include any type of data but would typically be hydrologic, climatic, or water quality data. HECDSS database management capabilities are oriented particularly toward very voluminous sets of time series data. HECDSS uses a block of sequential data as the basic unit of storage. The basic concept underlying the data storage system is the organization of data into records of continuous, applications-related elements, as opposed to individually addressable data items. This approach is more efficient for water resources applications than that of a conventional database system because it avoids the processing and storage overhead required to assemble an equivalent record from a conventional system.

HECDSS is available for desktop computers running the MS-DOS operating system. The software package has also been compiled and executed on various minicomputer and mainframe systems. The programs are written in FORTRAN77.

HECDSS provides capabilities to store and maintain data in a centralized location, provide input to and store output from application programs, transfer data between application programs, mathematically manipulate data, and display the data in graphs and tables. The user may interact with the database through utilities that allow entry, editing, and display of information, application programs that read from and write to the data base, and library routines that can be incorporated in any program to access database information.

A variety of utility programs are included in HECDSS for entering data into a database file. Some are designed for entering data from other databases, such as the USGS WATSTORE system and files provided by the NCDC. Several HEC application programs have been interfaced with HECDSS, allowing users to retrieve data for analysis or store results in a HECDSS file. This provides the user the capability of displaying and analyzing application program results by using the HECDSS utility programs. A set of FORTRAN subroutines are available which can be used to link application programs with HECDSS.

HECDSS also provides means for mathematically manipulating data in a variety of ways. Normal arithmetic operations and many mathematical functions are provided. Various statistical analyses can be performed. Missing data can be synthesized. Hydrologic routing of streamflows can be performed.

5.2 ADJUSTED HISTORICAL STREAMFLOWS

The future, rather than the past, is of concern in conducting modeling studies to support water management decision processes. Since the future is unknown, in planning studies historical streamflow or precipitation data must be used as a representation of the hydrologic characteristics of a river basin to be expected during the future planning horizon. Planning studies may involve analyses of the operations of existing reservoirs and/or evaluation of proposed new projects. Evaluation of proposed long-term improvements in operating policies may be of concern, or the objective may be to formulate operating strategies for the next season or year.

Modeling studies are commonly based on historical measured streamflow data, adjusted to represent flow conditions at pertinent locations for a specified condition of river basin development. The gaged streamflow data are typically nonhomogeneous due to reservoir construction and other activities of man in the river basin. The historical streamflow measurements must be adjusted to reflect conditions of river basin development at a specified past, present, or future point in time. The data must also be transferred from the locations of the stream gages to other pertinent locations required by the study. Gaged streamflow records also typically contain periods of missing data, or various pertinent gages have different periods of record. Missing data may be reconstituted using regression analyses. Watershed (precipitation-runoff) modeling provides an alternative approach for developing streamflows in situations in which gaged streamflow data are unavailable or inadequate.

Basin hydrology is typically represented in a reservoir system analysis model as an input sequence of streamflows at each pertinent location for each time interval of the period of analysis. Input streamflow sequences could consist of any of the following:

- Historical period-of-record streamflows adjusted to represent the specified locations for a specified condition of river basin development
- Adjusted historical streamflows during a critical drought period, flood, or other selected subperiod of the period of record
- Synthetically generated streamflow sequences which preserve selected statistical characteristics of the adjusted historical data
- Flood hydrographs for a hypothetical storm event generated using a precipitation-runoff model
- Long-term streamflow sequences generated using a continuous precipitation-runoff model

Alternatively, some reservoir system modeling and analysis approaches are based on representing the streamflow inflows as probability distributions or stochastic processes in various formats which capture the probabilistic characteristics of the adjusted historical data. For example, as discussed in later chapters, the reservoir inflows in an explicit stochastic model may be represented by a transition probability matrix which describes the discrete probability of a certain inflow conditioned on the previous period inflow. Other deterministic models require that streamflow inflows be simplified to one average or otherwise characteristic flow for each season or subperiod of a representative year.

Reservoir system analysis studies conducted by the federal water agencies, involving conservation storage operations, are typically based on adjusted historical period-of-record or critical-period streamflows. Studies involving flood control operations typically include flood hydrographs computed using watershed (precipitation-runoff) models as well as adjusted historical period-of-record streamflows. The other approaches for developing streamflow input data cited above have all been used in various studies reported in the literature, most typically in university research projects.

5.2.1 Time Intervals

Streamflows are commonly recorded as mean rates over a specified time interval. Mean flows may be expressed in units such as m^3/s or ft^3/s over a time interval such as a month, week, day, several hours, or one hour. Equivalently, flows may be expressed as volumes, in units such as m^3 or acre-feet, over a daily, monthly, or other time interval. Flood hydrographs are often expressed in terms of instantaneous flow rates (flows in m^3/s or ft^3/s at an instant in time). Selection of a computational time interval is a key consideration for a particular modeling application. The choice of time step depends on the scope of the study, data availability, and time-variability characteristics of streamflow. Planning studies involving water supply, hydropower, and other conservation storage purposes are often based on monthly streamflows covering a several-decade historical period-of-record analysis period. Some applications may require weekly or daily computational time steps during the several-decade analysis period. Modeling of flood control operations involves short time intervals because flows are rapidly changing. Floods may be represented by instantaneous or mean flow rates at time intervals of one hour, several hours, or one day. Some modeling studies adopt a variable time step, with a daily or other short interval being used during flood events and a longer interval, such as a month, being used during extended periods of more normal hydrologic conditions.

5.2.2 Homogeneous Streamflow Sequences at Pertinent Locations

The streamflow input for a reservoir system analysis model should represent the stochastic characteristics of the hydrologic processes of the river basin for a specified constant condition of watershed development. Significant nonhomogeneities in the historical record may be caused by human activities such as reservoir construction and water use. For a relatively undeveloped and unchanged watershed, the gaged streamflows may be quite homogeneous, stationary, and consistent. However, more typically, construction of river regulation structures, water use, and changes in watershed land use and runoff characteristics during the period of record significantly affect gaged streamflows. For example, modeling studies for the Brazos River Basin, discussed in Chapter 9, are based on monthly flows during a 1900–1984 hydrologic period of analysis. Flow measurements prior to the 1940s reflect essentially no effects of human activities, but flows became increasingly more regulated as numerous major reservoir projects were constructed during the 1940s through 1970s and water use increased. Adjustments to the measured streamflows are required to develop a homogeneous data set representing a specified condition of river basin development.

Unregulated, natural, or virgin flows are often used as streamflow input data for reservoir/river system modeling studies. The gaged streamflows are adjusted to represent natural undeveloped basin conditions with flows unregulated by structures or human activities. Alternatively, the gaged streamflows may be adjusted to represent some other specified condition of watershed development, such as conditions at a specified past, present, or future point in time. The adjusted flows may reflect the effects of selected water control activities. For example, for a model to analyze alternative operating plans for a particular reservoir system, the streamflow input may reflect the impacts of other reservoirs in the basin that are not included in the reservoir system operation model.

Development of unregulated (naturalized or virgin) streamflow data sets may represent a major portion of the effort required for the overall reservoir system analysis study. Adjustments to gaged flows are typically performed on an ad hoc study-by-study basis, rather than using specific well-established methods and generalized computer programs. Depending on the scope of the study and characteristics of the river basin, judgments are required regarding the level of detail required. An undeveloped river basin may require no adjustments at all to the gaged flows. The more typical basin, with a history of intensive water resources development, may require extensive adjustments. The streamflow adjustments may reflect removal of the impacts of only major reservoirs and diversions or, alternatively, may include numerous smaller reservoirs, diversions, return flows, channel losses, and changes in watershed runoff characteristics caused by urbanization, agricultural activities, or other land use changes.

The unregulated flow (Q_{unreg}) for a given time period is the streamflow that would have occurred if existing reservoirs had not existed and actual historical diversions had not been made. The gaged regulated streamflow (Q_{reg}) for the time period is adjusted for the effects of historical diversions and return flows at pertinent locations as follows:

$$Q_{unreg} = Q_{reg} + \text{Diversions} - \text{Return Flows} \qquad (5\text{-}1)$$

The gaged streamflow is adjusted for the storage and evaporation impacts of an upstream reservoir as follows:

$$Q_{unreg} = Q_{reg} + (S_{t + \Delta t} - S_t) + R_{evap}((A_t + A_{t + \Delta t})/2) \qquad (5\text{-}2)$$

where subscripts t and $t + \Delta t$ denote the beginning and end, respectively, of the computational time interval Δt. Thus, the unregulated flow (Q_{unreg}) during the time interval is computed by adding the storage change and evaporation to the gaged regulated flow (Q_{reg}). The reservoir storage volume (S_t and $S_{t + \Delta t}$) at the end of the previous and current time intervals are used to compute the change in storage during the interval. Thus, historical records of gaged storages are required. Reservoir evaporation volumes are estimated as an evaporation rate (R_{evap}), expressed as a depth, multiplied by the mean water surface area (($A_t + A_{t + \Delta t}$)/2) during the interval, with area being determined as a function of storage volume. Reservoir evaporation rates and storage-area relationships are discussed later in this chapter.

Computations based on equations 5-1 and 5-2 are sufficient for a location near the dam or for a relatively long time interval such as a month. However, flood control studies in particular, and conservation storage operations in some cases, are complicated by attenuation effects in a river reach. The reservoir release is attenuated or changed by the storage effects of the river prior to reaching a downstream location. Thus, flood routing, outlined in Chapter 6, may be incorporated in the computations to develop unregulated flows from gaged flows.

In modeling reservoir systems, streamflow data for locations other than gage sites are commonly required. Various methods are used to estimate streamflows at pertinent ungaged locations based on data available for gage sites on the same river, at either upstream or downstream locations. Flows at an ungaged site may also be related to flows at one or more gages at locations on other nearby streams based on watershed areas and other watershed characteristics.

A simple approach sometimes used to transfer flows from a gaged to an ungaged location on a stream is to ratio flows in proportion to watershed areas as follows:

$$Q_t = Q_{\text{gage},t} \, (A/A_{\text{gage}}) \qquad\qquad (5\text{-}3)$$

where Q_t and $Q_{\text{gage},t}$ denote the flow for time period t at the ungaged location of concern and at the gage, respectively, and A and A_{gage} are the corresponding watershed areas. A better estimate is made when records at two gaging sites are available, preferably one upstream and one downstream of the ungaged site of concern. The flows at the ungaged site can be interpolated or extrapolated, in proportion to drainage area, from the data at the gaged sites. Alternatively, in some cases, interpolation of flows might be more appropriately performed on the basis of distance between the gaging sites.

5.2.3 Filling in Missing Flows and Extending Records

Gages are installed at different times; thus, periods of record have different lengths. For example, one streamflow gage may have records extending back to 1925, another gage has been recording flows since 1937, while other gages in the basin were not installed until the 1960s. Regression analyses are used to extend the shorter records based on flows during preceding months at the location and flows during the current and preceding periods at nearby locations. Likewise, gaps or periods of missing data can be filled in using regression analyses.

Least-squares multiple linear regression and associated correlation analysis techniques are described in most statistics and numerical methods textbooks and incorporated in numerous statistical analysis software packages. The regression equation is

$$y = a_0 + a_1 x_1 + a_2 x_2 + \ldots + a_n x_n \qquad\qquad (5\text{-}4)$$

where y is the independent variable; $x_1, x_2, \ldots, x_n$ are the n independent variables; and $a_0, a_1, a_2, \ldots, a_n$ are regression coefficients computed from the data. Thus, y represents the missing streamflow and the xs are flows at other gages in the same and/or preceding time periods and possibly the same gage for the preceding time period. Nonlinear relationships are developed by applying linear regression to various transformations of the dependent and/or independent variables. For example, application of the linear regression equation 5-4 to the logarithms of the variables is equivalent to the following commonly used power form of regression equation:

$$y = b_0 x_1^{a1} x_2^{a2} \dots x_n^{an} \tag{5-5}$$

where all terms are defined identically in equations 5-4 and 5-5, with $b_0 = 10^{a_0}$. Regression analyses based on equations 5-4 and 5-5 are routinely used for numerous applications, including extending records and filling in missing data for streamflows and other hydrologic variables.

A random component can be added to the deterministic regression model as follows:

$$y = a_0 + a_1 x_1 + a_2 x_2 + \dots + a_n x_n + t s_y (1 - r^2)^{0.5} \tag{5-6}$$

where t is a random number from an appropriate probability distribution, often assumed to be the standard normal distribution, s_y is the standard deviation of the dependent variable, and r^2 is the correlation coefficient. This regression relationship, applied to the logarithms of the dependent and independent variables, is incorporated in the HEC-4 model noted below.

The computer program HEC-4 Monthly Streamflow Simulation [Hydrologic Engineering Center, 1971] provides capabilities for reconstituting missing data, extending records, and generating synthetic sequences of monthly flows. The MOSS-IV Monthly Streamflow Simulation program is a refined version of HEC-4 [Beard, 1973]. The U.S. Army Corps of Engineers [1993] outlines the computational procedure incorporated in these computer programs. HEC-4 and MOSS-IV fill in missing monthly streamflow data based on streamflows at other nearby gage stations using a multiple linear regression algorithm based on the transformed incremented logarithm of monthly streamflows. A random component is included in order to reproduce the distribution of random departures from the regression model as they are observed in the basic data. The missing dependent value to be estimated is related to values for the same month at all of the stations where such values exist, or values for the preceding month if current month values do not exist. The value for the preceding month at the dependent variable station is always used as one of the independent variables in the regression model.

As discussed above, records are extended based on data for other gages with longer records such that computed and/or gaged flows cover the same period at the different gages. Synthetic streamflow generation, discussed next, involves synthesizing flows sequences which are much longer than any gage record. The generated flows reflect the statistical characteristics of historical flows but do not reproduce the actual flow for any particular time. HEC-4 generates synthetic streamflows for multiple sites using an algorithm similar to its approach for extending records.

5.3 SYNTHETIC STREAMFLOW GENERATION

Reservoir reliability studies have provided a primary impetus for development of techniques of stochastic hydrology. Limited historical data may be extended by synthetic streamflow generation methods to provide the lengthy sequences required by certain types of reservoir system analysis exercises. For example, hundreds or thousands of years of monthly streamflows may be synthesized based on statistical parameters determined from a reasonably long period of record of several decades. Streamflow sequences of any specified length are synthesized based on preserving specified statistical parameters of the historical data, in the sense that the parameter values for an infinitely long sequence of synthetically generated flows would be the same as for the original data. Generated streamflows are called synthetic to distinguish them from historic observations.

Loucks *et al.* [1981], Linsley *et al.* [1982], and Salas [1993] provide concise overviews of synthetic streamflow generation. Bras and Rodriquez-Iturbe [1985] provide an in-depth theoretical treatment of stochastic hydrology. These references present an array of models for synthesizing sequences of streamflows or other hydrologic variables at single or multiple sites, which preserve selected statistical parameters of the historical data. The computer programs HEC-4 [Hydrologic Engineering Center, 1971; U.S. Army Corps of Engineers, 1993] and LAST [Lane and Frevert, 1985] preserve the mean, standard deviation, and lag-1 autocorrelation coefficients.

Synthetic streamflow generation is illustrated by the lag-1 autoregressive Markov model for synthesizing monthly flows for a single station:

$$Q_{i,j} = M_j + r_j(s_j/s_{j-1})(Q_{i,j-1} - M_{j-1}) + t_i s_i (1 - r_j^2)^{0.5} \qquad (5\text{-}7)$$

where subscript i is a counter of the years, and j represents the month, varying from 1 to 12. The subscript i is incremented by one year each time the computations cycle through a generation of 12 months ($j = 1, 2, ..., 12$) of flows. A sequence of flows ($Q_{i,j}$) is synthesized month by month, the flow for each month ($Q_{i,j}$) dependent on the flow just computed for the previous month ($Q_{i,j-1}$). Historical monthly flows are required to estimate 12 sets (for January through December) of means (M_j) and standard deviations (s_j) and the 12 serial autocorrelation coefficients (r_j) between flows in months j and $j - 1$. Random numbers (t_i) are drawn from an appropriate probability distribution. If the distribution of flows is assumed Gaussian, random numbers are drawn from a normal random number generator. If streamflows are log-normally distributed, equation 5-7 is applied to the logarithms of the streamflows, and random numbers (t_i) from the normal distribution are used.

The stochastic model represented by equation 5-7 includes deterministic and random components. The model deterministically regresses flow in a given month ($Q_{i,j}$) with the flow in the previous month ($Q_{i,j-1}$):

$$Q_{i,j} = M_j + r_j(s_j/s_{j-1})(Q_{i,j-1} - M_{j-1}) \qquad (5\text{-}8)$$

This expression represents the expected value of the flow in a month, given the flow in the previous month. A random component:

$$t_i s_i (1 - r_j^2)^{0.5} \tag{5-9}$$

is added to represent the random deviation of flows independent of the correlation with the previous month's flow. Equation 5-7 generates flows for a particular location. The previously cited references outline methods for expanding this model to generate flows at multiple interrelated sites, while preserving the correlation between flows at the different sites.

5.4 WATERSHED (PRECIPITATION-RUNOFF) MODELING

Synthesis of streamflows from rainfall and snow data using a watershed model provides an alternative to using gaged streamflow data. Historical gaged streamflows are typically used in reservoir system modeling studies to the extent feasible. However, the availability of streamflow data is contingent upon gaging stations having been maintained at pertinent locations over a significantly long period of time, and extensive adjustments may be required to remove nonhomogeneities caused by river basin development. Precipitation-runoff modeling for a specified condition of watershed development may be advantageous in planning studies. Also, in support of real-time reservoir operations, watershed models are used to forecast streamflows expected to occur in the near future based on current precipitation measurements. Streamflows during the snowmelt season can be forecast based on snowpack information. Watershed modeling is used most often for simulating flood events for either historical or real-time actual storms or hypothetical design storms.

Watershed models simulate the hydrologic processes by which precipitation is converted to streamflow [Linsley *et al.*, 1982; McCuen, 1989; Singh, 1992; Bedient and Huber, 1992]. The watershed is the system being modeled, with a precipitation hyetograph provided as input and the runoff hydrograph computed. Some models consider only water quantities, while others simulate both water quality and quantity. The source of the water is rainfall or snow melt. Some precipitation is lost through the natural hydrologic processes of interception, depression storage, infiltration, evaporation, and transpiration. The remaining precipitation flows overland and through the soil, collects as flow in swales and small channels, and eventually becomes runoff to streams. Groundwater also contributes to streamflow, largely independent of a particular precipitation-runoff event. Contaminants enter the water and are transported with the runoff. Land use, drainage improvements, storage facilities, and other development activities significantly affect the processes by which precipitation is converted to streamflow. River basins are divided into a number of smaller, more hydrologically homogeneous subwatersheds for modeling purposes. The runoff hydrographs from the individual subwatersheds are routed through stream reaches and combined at appropriate locations.

Watershed models can be categorized as single-event or continuous. Single-event models are designed to simulate individual storm events and have no capabilities for the soil infiltration capacity and other watershed abstraction capacities to be replenished during extended dry periods. Rainfall events may be actual historical or real-time storms, or design storms developed based on intensity-duration-frequency relations or probable maximum precipitation. Continuous models simulate long periods of time which include multiple precipitation events separated by significant dry periods with no precipitation. Continuous watershed models are used to simulate several years of daily (or other time period) streamflows given the corresponding rainfall records and snow information. Runoff water quality parameters often are simulated as well as flows in continuous models.

The HEC-1 Flood Hydrograph Package [Hydrologic Engineering Center, 1990] is an example of an event-type rainfall-runoff model which develops flood hydrographs associated with specified rainfall and/or snow melt events. HEC-1 has been applied in numerous planning and design studies for reservoir projects conducted by the Corps of Engineers and other agencies as well as for various flood plain management and other applications not involving reservoirs. A special version of HEC-1 is also available for use in supporting real-time flood control reservoir system operations. The Streamflow Synthesis and Reservoir Regulation (SSARR) Model is a continuous watershed model that generates sequences of streamflows over a long period of time, which may include dry periods between periods of snow melt and rainfall events [U.S. Army Corps of Engineers, North Pacific Division, 1987]. The SSARR model has been extensively applied by the USACE North Pacific Division and many other entities for many years. HEC-1 and SSARR are discussed further in Chapter 8. Singh [1995] and Wurbs [1995] describe these and other available generalized watershed models.

5.5 TRANSITION PROBABILITY MATRICES

Stochastic reservoir optimization and simulation models have been based on representing the streamflow inflows as probability distributions or stochastic processes, in various formats, which capture the probabilistic characteristics of the adjusted historical streamflows. Summarizing streamflow characteristics as transition probability matrices, reflecting Markov chains, is common in certain types of stochastic reservoir modeling [Loucks *et al.*, 1981; Mays and Tung, 1992]. In a first-order Markov process, the dependence of current flows on past flows is reflected in the immediately preceding time period. A Markov chain is a Markov process in which the flows fall within discrete ranges. In stochastic reservoir optimization and simulation models, inflows are often approximated by Markov chains for simplicity, even though flow is a continuous random variable.

The concept of representing streamflows as a Markov chain described by a transition probability matrix is illustrated using, for simplicity, annual flows as the discrete random variable. Flows are generally more highly correlated for the shorter time intervals typically adopted in modeling studies. The same general modeling approach can be applied to seasonal, monthly, weekly, or daily flows. However, whereas annual flows are described by one transition probability matrix, monthly flows require 12 separate matrices, one for each month of the year. A Markov chain

for each month is defined. Thus, the data and computations are much more voluminous for smaller time intervals.

For each year y, the annual flow Q_y falls within one of n discrete ranges,
called states q_i, where $i = 1, 2, ..., n$. The random variable Q_y has discrete values q_i
with unconditional probabilities p_i, where:

$$\sum_{i=1}^{n} p_i = 1 \qquad (5\text{-}10)$$

The probability of Q_y taking a value or state q_i depends upon $Q_{y\text{-}1}$. This dependence
is expressed by specifying transition probabilities. A transition probability (p_{ij}) is
the conditional probability that the current state is q_j given that the previous state
is q_i.

$$p_{ij} = P(Q_y = q_j | Q_{y-1} = q_i) \qquad (5\text{-}11)$$

The following expression is valid for all i:

$$\sum_{j=1}^{n} p_{ij} = 1 \qquad (5\text{-}12)$$

A transition probability matrix with elements p_{ij} describes the probabilistic characteristics of the annual flows. The probability p_j^y that $Q_y = q_j$ is the sum of the probabilities p_i^{y-1} that $Q_{y-1} = q_i$ times the probability p_{ij} that $Q_y = q_j$ given that $Q_y = q_i$

$$p_j^y = \sum_{i=1}^{n} p_i^{y-1} p_{ij} \qquad (5\text{-}13)$$

The starting Q_y value for the first year affects the flows for several subsequent
years. However, eventually, after several years, the probabilities p_j reach limiting
values called unconditional or steady-state probabilities.

$$p_j = \sum_{i=1}^{n} p_i p_{ij} \qquad (5\text{-}14)$$

The steady-state probabilities can be determined by simultaneously solving equation 5-10 along with equation 5-14 for all of the states j except one.

Example 5–1—Transition Probability Matrix Analysis of Annual Streamflows

Mean annual inflows to a reservoir fall within the following discrete ranges (q_i).

State, i	Annual Flow, q_i (m³/s)
1	10–20
2	20–30
3	30–40
4	40–50

TABLE 5–1   Transition Probability Matrix for Annual Flows of Example 5–1

Flow State i in Year y−1	Flow State j in Year y			
	1	2	3	4
	$P(Q_y = q_j \mid Q_{y-1} = q_i)$			
1	0.5	0.3	0.2	0.0
2	0.3	0.4	0.2	0.1
3	0.1	0.2	0.4	0.3
4	0.1	0.2	0.3	0.4

Historical flows at a gaging station were analyzed, with the flow in each year being compared with the flow in the previous year, to develop the transition probability matrix presented in Table 5–1. The table indicates, for example, that for the years of record with mean annual flows falling between 10 and 20 m^3/s, the previous year's flow was in the same range 50% of the time. If the flow in the prior year $y − 1$ is between 20 and 30 m^3/s ($q_i = q_2$), the frequency in which the flow in the current year y falls between 10 and 20 m^3/s ($q_j = q_1$) is 0.3. As another example of reading the transition probability matrix, given that the flow in a particular year is known to be between 30 and 40 m^3/s ($i = 3$), the probability is 0.3 that the flow during the next year will be 40 to 50 m^3/s ($j = 4$).

Given the streamflow in an initial year, probabilities for subsequent years are estimated based on the transition probability matrix. For example, with a known mean flow of 35 m^3/s during the current year (labeled year 1 in Table 5–2), the probability of flows in future years falling in each of the discrete ranges are determined as follows. For the second year, equation 5-13 is applied for each of the four states.

$$p_1 = (0.00)(0.5) + (0.00)(0.3) + (1.00)(0.1) + (0.00)(0.1) = 0.1$$
$$p_2 = (0.00)(0.3) + (0.00)(0.4) + (1.00)(0.2) + (0.00)(0.2) = 0.2$$
$$p_3 = (0.00)(0.2) + (0.00)(0.2) + (1.00)(0.4) + (0.00)(0.3) = 0.4$$
$$p_4 = (0.00)(0.0) + (0.00)(0.1) + (1.00)(0.3) + (0.00)(0.4) = 0.3$$

TABLE 5–2   Streamflow Probabilities Computed in Example 5–1

Year y	Streamflow State Probabilities P_i			
	P_1	P_2	P_3	P_4
1	0.000	0.000	1.000	0.000
2	0.100	0.200	0.400	0.300
3	0.180	0.250	0.310	0.260
4	0.222	0.268	0.288	0.222
5	0.242	0.276	0.280	0.202
6	0.252	0.279	0.276	0.192
7	0.257	0.281	0.274	0.188
8	0.259	0.282	0.274	0.186
9	0.260	0.282	0.273	0.185
10	0.261	0.282	0.273	0.184
11	0.261	0.282	0.273	0.184

Equation 5-13 is applied for each of the four states for the third year.

$$p_1 = (0.1)(0.5) + (0.2)(0.3) + (0.4)(0.1) + (0.3)(0.1) = 0.18$$
$$p_2 = (0.1)(0.3) + (0.2)(0.4) + (0.4)(0.2) + (0.3)(0.2) = 0.25$$
$$p_3 = (0.1)(0.2) + (0.2)(0.2) + (0.4)(0.4) + (0.3)(0.3) = 0.31$$
$$p_4 = (0.1)(0.0) + (0.2)(0.1) + (0.4)(0.3) + (0.3)(0.4) = 0.26$$

The computations are repeated for each year, and the results summarized in Table 5–2. The steady-state probabilities are reached by about the tenth year. The tenth-year probabilities of 0.261, 0.282, 0.273, and 0.184, respectively, that the annual flow will fall within each of the four ranges, represented by states q_j, are unconditional in that they are the limits eventually reached regardless of the known flow specified for the first year.

Thus, the p_i for each q_i are shown in Table 5–2 for each year, starting with the year with the specified known Q_y and continuing until the steady-state condition is reached. Alternatively, the p_i for just the steady-state condition can be determined by simultaneously solving equation 5-14, for all of the states j except one, and equation 5-10. These equations are developed as follows:

$$p_1 = 0.5p_1 + 0.3p_2 + 0.1p_3 + 0.1p_4$$
$$p_2 = 0.3p_1 + 0.4p_2 + 0.2p_3 + 0.2p_4$$
$$p_3 = 0.2p_1 + 0.2p_2 + 0.4p_3 + 0.3p_4$$
$$p_1 + p_2 + p_3 + p_4 = 1$$

These equations are simplified as follows:

$$-0.5p_1 + 0.3p_2 + 0.1p_3 + 0.1p_4 = 0$$
$$0.3p_1 - 0.6p_2 + 0.2p_3 + 0.2p_4 = 0$$
$$0.2p_1 + 0.2p_2 - 0.6p_3 + 0.3p_4 = 0$$
$$p_1 + p_2 + p_3 + p_4 = 1$$

The set of linear equations is solved to obtain unconditional probabilities p_1, p_2, p_3, and p_4 of 0.261, 0.282, 0.273, and 0.184, respectively.

5.6 RESERVOIR EVAPORATION RATES

Evaporation from the reservoir water surface is a key consideration in modeling conservation storage operations. The impacts of evaporation losses vary greatly among regions and among reservoir projects. For many water supply reservoirs, the annual evaporation volume in a typical year exceeds the volume of water withdrawn for municipal, industrial, agricultural, and other beneficial uses. Evaporation is typically negligible during a flood event, and thus is usually not considered in flood control studies.

Mean annual reservoir evaporation rates in inches are shown in Figure 5–1 for the contiguous United States based on information available from the National Weather Service. Mean annual reservoir evaporation varies from 20 inches (51 cm) in the northeast and northwest to 80 inches (2 m) in the southwest. Reservoir evaporation rates may vary significantly between years, being much higher during drought conditions than during wet years. Evaporation rates vary greatly during the year. In

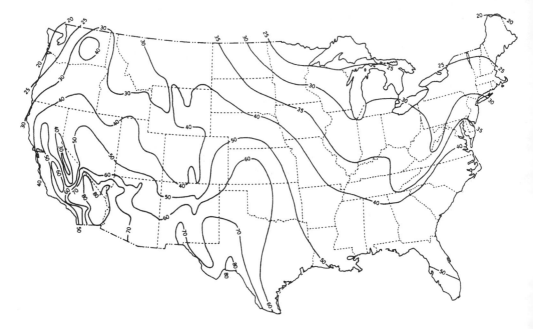

Figure 5–1   Mean annual reservoir evaporation in inches.

regions of extreme temperature differences between hot summers and freezing winters, essentially all of the annual evaporation may occur during the summer months.

Reservoir evaporation volumes are typically computed in a model by multiplying the average water surface area during a computational time interval by an evaporation rate, expressed as a depth. Since evaporation rates during drought years are typically significantly greater than during more normal or wet years, evaporation rates should preferably be based on field measurements for each time interval of the analysis period. However, due to limited data availability, evaporation rates are often provided for each month or season within the year but assumed to remain constant from year to year.

Either gross or net evaporation rates may be included, as appropriate, in the input data for a reservoir model. The gross evaporation rate is the depth of water loss from a unit area of reservoir water surface. Net evaporation rates consist of gross evaporation rates adjusted for precipitation. In a typical modeling application, basin hydrology is represented by naturalized streamflows and net evaporation rates. Without the reservoir, a portion of the precipitation falling over the reservoir site is lost through infiltration and other hydrologic abstractions, and the remainder contributes to streamflow. After the land is inundated by a reservoir, all the precipitation falling on the water surface becomes inflow to the reservoir. Thus, the net evaporation rates often incorporated in reservoir models consist of gross evaporation less effective rainfall, which is the rainfall over the reservoir site less the amount that became runoff to the stream and is reflected in the streamflow data.

An array of techniques for estimating evaporation from water surfaces are described by hydrology textbooks such as [Linsley *et al.*, 1982], [McCuen, 1989], and [Singh, 1992]. The alternative methods involve water budgets, energy budgets, mass transfer equations, aerodynamics equations, and combinations thereof. Data requirements are significant for these evaporation prediction approaches.

Evaporation can be measured with atmometers, also called evaporometers, which are calibrated instruments used to measure the evaporation potential of the air. The different types of atmometers generally utilize a porous bulb that draws water from a container as evaporation occurs from the bulb surface. Use of an atmometer requires establishment of coefficients for correlating evaporation with water loss in the container.

The most widely applied technique for estimating reservoir evaporation involves combining pan evaporation measurements and pan coefficients. Water is maintained in a pan with the water surface exposed to the atmosphere, and the changes in depth due to evaporation and rainfall are periodically measured. A variety of different types of pans have been used. The most common is the standard Class A pan developed by the National Weather Service (formerly the U.S. Weather Bureau) in 1916. The Class A pan is a circular container 47.5 inches (1.21 m) in diameter with vertical walls 10 inches (24.5 cm) deep. The water depth is maintained at eight inches, which is two inches below the rim of the pan. The unpainted galvanized iron or monel sheet metal pan is placed on a wooden platform just above ground level. The water level is measured daily with a hook gage in a stilling well, and water is refilled or removed (in the case of rain) as necessary to maintain a depth of eight inches. Pan evaporation is computed as the difference between observed levels, adjusted for precipitation, if any, measured in a standard rain gage.

Pan evaporation is different from actual lake evaporation due primarily to the energy exchange through the sides and bottom of the pan. Pan coefficients are used to estimate reservoir evaporation rates from measurements of evaporation from pans. *Pan coefficient* is defined as the ratio of lake evaporation (E_R) to pan evaporation (E_P) for the same meteorological conditions:

$$\text{pan coefficient} = E_R/E_P \qquad (5\text{-}15)$$

A mean annual pan coefficient of 0.7 is commonly cited. Annual pan coefficients developed by Farnsworth *et al.* [1982] range from 0.64 to 0.88 for different regions of the United States. Monthly pan coefficient values typically vary with seasons of the year, with winter months being lower than summer. Farnsworth *et al.* [1982] cite southern California, with the pan coefficient varying from 0.64 to 0.68 in the colder winter months to 0.88 in the summer, as an example of extreme seasonal variation for free water surface evaporation representative of shallow reservoirs. Monthly variations in the relationship between pan and reservoir evaporation may be affected significantly more by changes in heat storage in deep reservoirs. In the spring, pans warm up quicker than lakes, resulting in lower pan coefficient values. In the fall, the heat stored in a deep lake from the summer continues to contribute to evaporation, resulting in higher pan coefficients for the fall months.

Pan coefficients have been determined for a few reservoirs by comparing pan evaporation with lake evaporation determined from a detailed water budget analysis. U.S. Geological Survey [1954] studies of Lake Hefner in Oklahoma are an example of this approach. However, pan coefficients are more typically determined by combining pan evaporation measurements and estimates of lake evaporation computed with equations, such as variations of the Penman equation, as a function of wind movement, air temperature, solar radiation, dew point, and other meteorological measurements. Farnsworth *et al.* [1982] and Farnsworth and Thompson [1982] outline such approaches for developing pan coefficient values for different locations.

Farnsworth *et al.* [1982] and Farnsworth and Thompson [1982] provide information for estimating mean evaporation rates for each of the 12 months of the year for any location in the contiguous United States. These publications provide pan evaporation data and estimates of free water surface evaporation, which is defined as evaporation from a thin film of water having no appreciable heat storage. Free water surface evaporation is also referred to as shallow lake evaporation and as potential evapotranspiration. The mean annual free water surface evaporation is also representative of deep lakes, but heat storage effects can be expected to cause variations in individual months.

The evaporation atlas [Farnsworth *et al.*, 1982] includes maps of evaporation observed from Class A pans from May through October, estimates of shallow reservoir evaporation from May through October, estimates of shallow reservoir evaporation for the entire year, and pan coefficients used to convert pan evaporation to shallow reservoir evaporation. Additional information provided by Farnsworth and Thompson [1982] includes tabulations of mean monthly pan evaporation for each of the 12 months of the year for several hundred stations. A procedure is outlined for combining the monthly tabulations and atlas maps to develop monthly reservoir evaporation rates for any ungaged location in the contiguous United States.

Mean monthly pan evaporation rates for selected stations are reproduced in Table 5–3. Approximate values of shallow reservoir evaporation at these locations can be obtained by multiplying the pan evaporation rates by an average pan coefficient of roughly 0.7. More accurate estimates are obtained using pan coefficients provided in the atlas. The National Weather Service publications also outline computational methods for developing more accurate pan coefficients based on measurements of temperature, wind movement, and other meteorological variables.

Daily pan evaporation observations are included in the report *Climatological Data* published monthly by the National Oceanic and Atmospheric Administration on a state-by-state basis. The data are available through the previously discussed National Climatic Data Center in Asheville, North Carolina.

State and local agencies also collect and disseminate evaporation data. For example, the Texas Water Development Board, in conjunction with the Texas Natural Resource Information System, maintains a reservoir evaporation database. The database includes both gross and net evaporation, where net refers to gross evaporation less effective rainfall, which is rainfall not contributing to runoff under without-

TABLE 5–3   Pan Evaporation in Inches [Farnsworth and Thompson, 1982]

Month	Folsom Dam, California	Davis Dam, Arizona	Ft Peck Dam, Montana	Elephant Butte Dam, New Mexico	Whitney Dam, Texas	Wolf Creek Dam, Kentucky	Kerr Dam, Virginia
Jan	0.90	5	—	3.28	2.95	2	—
Feb	1.62	6	—	4.85	3.88	2	—
Mar	3.46	9	—	8.53	6.05	4	—
Apr	5.38	11	—	11.75	7.02	4.68	5.27
May	8.09	14	7.49	14.45	8.46	5.47	6.22
Jun	10.13	16.68	8.68	16.17	10.65	6.35	6.81
Jul	11.46	14.43	10.67	13.64	12.39	6.57	7.20
Aug	10.18	14.62	9.86	11.63	11.38	5.88	6.12
Sep	7.66	11.8	5.88	9.72	8.33	4.58	4.87
Oct	4.96	8.93	3.56	7.70	6.24	3.24	3.37
Nov	2.03	7.45	—	4.75	4.02	2	—
Dec	0.94	5.73	—	3.21	3.12	2	—
Annual, inches (cm)							
	66.81	124.00	46.14	109.68	84.67	49.00	39.86
	(170)	(315)	(117)	(279)	(215)	(124)	(101)

reservoir conditions. Monthly gross and net evaporation rates for each month since January 1940 are provided for each one-degree quadrangle covering the state of Texas. Kane [1967] describes the initial development of the database. The data are updated periodically and are distributed, on request, as computer data files.

5.7   RESERVOIR ELEVATION/STORAGE/AREA RELATIONSHIPS

The relationship between water surface elevation, storage volume, and water surface area is fundamental information characterizing a reservoir, which is required in modeling studies. The storage versus area relationship is used in evaporation computations, since evaporation volumes are a function of evaporation rates and water surface area. The elevation versus storage relationship is required for determining head in hydroelectric power computations. The elevation/storage/area data are developed for specified past, present, or future conditions of reservoir sedimentation. Storage/elevation relations are required to relate storage capacities to elevation-based operating rule curves.

Elevation versus storage and area relationships characterize the topography of a reservoir site and are determined initially from contour maps and later after impoundment from resurveys. The data are typically expressed as a table and also plotted as a graph. An example of elevation versus storage and area curves is presented in Figure 5–2.

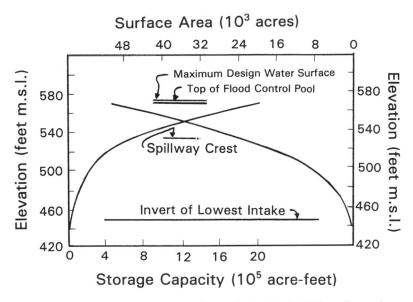

Figure 5–2 Elevation versus storage and area relationships for Whitney Reservoir on the Brazos River, Texas.

The bottom topography of a reservoir changes over time as sediments accumulate. Eventually, all reservoirs will fill with sediment. The time required for complete depletion of the storage capacity may range from a single major flood event to many hundreds of years. The rate at which the storage capacity is reduced by sedimentation depends on the sediment load flowing into the reservoir, the portion of the inflow load that is passed through and the portion trapped in the reservoir, and the density of the deposited sediment. The sediment yield of a watershed depends on land use, soil properties, and numerous other watershed characteristics, flow characteristics of the stream system, and climatic variables. Most of the sediment inflow to reservoirs occurs during high flows, particularly extreme flood events. Since floods occur at random times, sedimentation rates vary greatly over time. The trap efficiency of a reservoir is the percentage of the total inflowing sediment retained in the reservoir. Trap efficiency depends on sediment characteristics and hydraulic characteristics of the reservoir, which include the ratio of storage capacity to inflow rate, reservoir shape, type of outlets, and operating policies.

The location as well as volume of the sediment deposits affect the elevation-storage-area relationship. Sediment deposition results as flow velocities decrease. Thus, deposition of sediment begins with formation of deltas in the upstream reaches of a reservoir with decreases in flow velocities of the inflowing streams. However, deposition also occurs near the dam as well and throughout the reservoir.

Sediment studies are performed during the planning of reservoir projects. A 50- to 100-year sediment reserve capacity is normally included in the sizing of major projects. The Bureau of Reclamation [1987] and USACE [1989] outline an array of methods for predicting sediment yields from the watersheds above reservoirs, estimating reservoir trap efficiencies, predicting the locational distribution of sediment volumes, and adjusting elevation-storage-area curves. Sediment yields can be approximated using mathematical equations, but accurate yield determinations require direct field measurements of sediment loads. These references also describe sediment sampling techniques.

The original elevation versus storage and area relationships for a reservoir project are developed based on a contour map of the site. Resurveys of reservoir bottom topography are made periodically after construction and initial impoundment. The USACE [1987] suggests the scheduling of resurveys at intervals of five to ten years, depending on the magnitude of anticipated sedimentation and probable needs for such information. Resurveys may also be performed after major flood events. However, resurveys are expensive. Numerous federal and nonfederal reservoir projects have been in operation for several decades without ever having a resurvey of bottom topography.

Sedimentation ranges are established during project construction to facilitate later resurveys. A sediment range is simply a fixed line across a reservoir along which elevations are precisely measured. Vertical and horizontal control are established and the ranges marked. A sufficient number of ranges are established to adequately characterize the reservoir bottom topography. With the ranges submerged by the reservoir, echo-sounders are typically used to measure bottom elevations during the resurveys.

6

WATER ACCOUNTING, ROUTING, AND HYDRAULICS

Reservoir/river system analysis models are built around water accounting and routing algorithms. Water budgets are balanced as water is routed through the system. In a reservoir operation model, streamflows provide the inflows to the system. Reservoirs regulate the streamflows through storage and releases. Water flows through river reaches. Water is diverted for beneficial use and is lost through evaporation. Other gains and losses also occur. This chapter outlines fundamental considerations in accounting for the movement, storage, and withdrawal of water in a reservoir/river system. It deals with tracking changes in storage volumes and flow rates. The latter portion of the chapter covers outlet structure and river hydraulics, which involve flow depths as well as flow rates.

6.1 CONSERVATION OF VOLUME

The basic continuity or conservation of volume equation for a reservoir or river reach is

$$S_{t + \Delta t} - S_t = I_{\text{vol}} - O_{\text{vol}} \tag{6-1}$$

where S_t and $S_{t + \Delta t}$ denote the storage volume at the beginning and end, respectively, of an interval of time Δt, and I_{vol} and O_{vol} denote the inflow and outflow volumes during the time period. Alternatively, conservation of volume may be expressed in terms of instantaneous rates as

$$\frac{dS}{dt} = I - O \tag{6-2}$$

where dS/dt denotes the change in storage volume (S) with respect to time (t), and I and O are the volumetric inflow and outflow rates at an instant in time. This equation may be written for a discrete time step (Δt) as

$$\frac{S_{t+\Delta t} - S_t}{\Delta t} = \bar{I} - \bar{O} \tag{6-3}$$

where $\bar{I}$ and $\bar{O}$ are the mean inflow and outflow rates during the time interval Δt. This expression is approximated as

$$\frac{S_{t+\Delta t} - S_t}{\Delta t} = \frac{I_t + I_{t+\Delta t}}{2} - \frac{O_t + O_{t+\Delta t}}{2} \tag{6-4}$$

where I_t, $I_{t+\Delta t}$, O_t, and $O_{t+\Delta t}$ denote the inflow and outflow at the beginning and end of Δt. Equation 6-4 provides the basis for the hydrologic flood routing techniques described later in this chapter.

Modeling of a reservoir operated for conservation purposes is based on a water balance for each time interval expressed as

$$S_{t+\Delta t} = S_t + \text{all inflows} - \text{all outflows} \tag{6-5}$$

or

$$S_{t+\Delta t} - S_t + \text{stream inflows} + \text{other inflows} - \text{withdrawals}$$
$$- \text{releases} - \text{spills} - \text{evaporation} - \text{other losses}$$

Inflows include streams flowing into the reservoir, precipitation falling on the reservoir surface, subsurface flows into the reservoir, and return flows from water use diversions. Lakeside withdrawals and downstream releases are made through controlled outlet facilities for various beneficial purposes. Spills are flows over the spillway of a full reservoir. Evaporation from the water surface is typically a significant loss. Other losses may include seepage through the dam or into the ground. These terms represent inflow and outflow volumes during the time interval Δt. The storage terms S_t and $S_{t+\Delta t}$ are volumes at the beginning and end of the time interval.

Water accounting procedures are based on conservation of mass. Since for most reservoir system analysis applications water is essentially a constant-density fluid, conservation of mass implies conservation of volume as well. The variation of water density as a function of temperature and dissolved constituents is important in water quality modeling, covered in Chapter 12, but is negligible from the perspective of the basic water accounting applications addressed in Chapters 6–11.

6.2 RESERVOIR EVAPORATION AND OTHER GAINS AND LOSSES

Withdrawals/releases for beneficial purposes and streamflows are typically the components of the water budget of primary concern in reservoir system modeling studies. This section addresses other gains and losses. Reservoir evaporation is typically the most important of these other considerations. For some reservoir projects, the amount of evaporation exceeds water supply diversions.

6.2.1 Reservoir Water Surface Losses and Gains

Reservoir evaporation volumes (E_i) for each time period (i) are typically computed in a model by multiplying an evaporation rate (e_i) by the average water surface area (A_i) during the time interval, which is determined as a function of storage.

$$E_i = A_i \, e_i \qquad (6\text{-}6)$$

Typical units are m^3 or acre-feet for E_i; meters or feet for e_i; and m^2 or acres for A_i. The average reservoir water surface area (A_i) during the time interval can be estimated as

$$A_i = (A_t + A_{t+\Delta t})/2 \qquad (6\text{-}7)$$

where the areas A_t and $A_{t+\Delta t}$ at the beginning and end of the time interval Δt are determined as a function of the corresponding storage volumes at the beginning and end of the interval. Alternatively, the beginning-of-period and end-of-period storage volumes may be averaged and then the average storage applied to the storage versus area relationship to estimate the average area. Reservoir evaporation rates and storage/area relationships are discussed in Chapter 5.

Rainfall and evaporation rates are often combined as a net rate. Rainfall and evaporation represent a gain and loss, respectively, which are reflected in the positive or negative sign of the rate values. With natural unregulated streamflows provided as input to a reservoir system analysis model, the corresponding net rainfall minus evaporation rates should reflect precipitation which is not already accounted for in the natural unregulated streamflows. Without the reservoir, a portion of the precipitation falling on the land becomes runoff to the stream, and the remainder is lost as infiltration, evapotranspiration, and other hydrologic abstractions. With the reservoir, all of the precipitation falling on the reservoir water surface is inflow. Net rainfall minus evaporation rates are sometimes adjusted to reflect the difference between rainfall falling on the reservoir water surface and runoff from rain falling on the land area at the site that contributes to streamflow before the reservoir project is constructed. Evaporation from the free surface of the river, prior to reservoir construction, is usually very small relative to reservoir evaporation and thus is typically neglected in the adjustment.

6.2.2 Other Reservoir Losses and Gains

Losses and gains of water to and from the ground under a reservoir are extremely difficult to quantify. Losses to infiltration or seepage and gains from groundwater or bank storage are typically considered negligible and ignored in reservoir system analysis studies. Most reservoirs are constructed at relatively impermeable sites. Permeability of the reservoir bottom tends to decrease over time with sedimentation. The sediment deposits help seal the bottom and prevent seepage. However, seepage, bank storage, groundwater, and other interactions between the reservoir and underlying ground may be significant at a particular reservoir.

Geotechnical considerations are important in site selection studies during the planning of new construction projects. Badly fractured rock, cavernous limestone, or permeable volcanic material at a reservoir site may result in serious leakage. This type of site is avoided if possible. In some cases, grouting has been employed to stop leakage through a few well-defined channels or within a small area of fractured rock.

Projects are typically designed and maintained to minimize leakage through the dam and outlet structures. Thus, leakage is typically not a major loss and is neglected in modeling studies. However, leakage through structures could be significant at some dams.

Unauthorized diversions from reservoirs and rivers are sometimes treated as an unaccounted loss. Farmers, businesses, and other individuals without proper water rights or water supply contracts may pump unrecorded amounts of water that are not reflected in the diversions included in the modeling study.

6.2.3 Channel Losses

Water supply diversions from a river and instream flow needs are often met by releases from reservoirs located great distances upstream. Water released from a reservoir may be partially lost, as it flows through the downstream river channel, as a result of the natural processes of evaporation, transpiration, and bank seepage. Unauthorized pumpage from the river may also occur. These channel losses are difficult to quantify.

The input data for reservoir/river system analysis models typically include unregulated flows developed from gaged streamflow records. Measured streamflows reflect actual channel losses. However, in a model, reservoirs alter streamflows. Channel losses can be expected to vary with changes in streamflows. The significance of channel losses vary with different reservoir/river system modeling applications. In many cases, channel losses are actually negligible. Losses are also often ignored due to the difficulty in quantifying them.

The typical approach for developing channel loss functions has been to analyze stream gage records at upstream and downstream locations. Water budget computations account for all inflows and outflows for the river reach between the gages. Water supply diversions and return flows are recorded. Runoff between the gages is estimated using rainfall-runoff modeling. Historical channel losses or gains are estimated as the remainder of the water budget after accounting for all other inflows and outflows. Losses are then related to streamflow characteristics and river conditions. Losses may be expressed as a variable percentage of streamflow per length of river. The channel loss functions may also be related to factors such as groundwater levels, reservoir release procedures, rate of change of streamflow, and season of the year. Channel loss functions developed for gaged reaches are used for ungaged reaches of the same or similar rivers.

6.3 POWER EQUATION

For projects with hydroelectric power, the reservoir water balance includes releases through turbines for generating energy. Power is related to release rates and other pertinent factors as follows:

$$P = \gamma Q h e \tag{6-8}$$

where:

> P = power ($N \cdot$ m/s or ft $\cdot$ lb/s)
>
> γ = unit weight of water (N/m^3 or lb/ft^3)
>
> Q = flow rate (m^3/s or ft^3/s)
>
> h = head (m or ft)
>
> e = efficiency in converting hydraulic energy to electrical energy

Power is the rate of transferring energy:

$$power = \frac{energy}{time} \tag{6-9}$$

Energy is expressed in metric units of Newton-meter (N $\cdot$ m) or kilowatt-hour (kW $\cdot$ hr) and English units of foot-pound (ft $\cdot$ lb). Power is expressed in units of Newton-meter/second (N $\cdot$ m/s) or foot-pound/second (ft $\cdot$ lb/s). Electrical energy is commonly measured in kilowatts, where a watt is one N $\cdot$ m/s and a kilowatt is 1000 watts. The English unit of horsepower is used for power, where one horsepower equals 550 ft $\cdot$ lb/s. One horsepower is 746 watts.

6.4 WATER BALANCE SIMULATION OF A RESERVOIR

The conservation of volume equation and power equation represent fundamental concepts that are incorporated into the full spectrum of types of reservoir system analysis models, including complex multiple-purpose, multiple-reservoir system simulation and optimization models discussed in later chapters. The basic concepts are illustrated here by a simple application to a single reservoir.

Example 6–1 illustrates the general approach of sequential simulation. A specified set of annual water use requirements are met, to the extent allowed by available storage and streamflows, during a given sequence of unregulated flows representing basin hydrology. The annual water use requirements vary seasonally but are constant from year to year. Basin hydrology is represented by the input reservoir evaporation rates as well as unregulated streamflows. The monthly time step adopted for the simplified example is typical of many actual models used for planning studies involving conservation operations. The hydrologic simulation is limited to 24 months in the example for purposes of brevity. Although short simulation periods of one or two years are appropriate in certain applications, a several-decade hydrologic period-of-record simulation would be more representative of actual practice. The model demonstrates the capabilities of the reservoir project, with the specified operating plan, to meet the specified water use demands during a representative set of inflows.

Example 6–1—Reservoir Simulation

In this illustrative example, a reservoir is simulated by a sequential month-by-month volume balance accounting procedure that combines inflows, water demands, operating rules, and the physical characteristics of the reservoir project. The reservoir is characterized by the elevation versus storage volume and surface area curves previously presented in Figure 5–2. As indicated in Table 6–1, a total cumulative storage capacity of 773 million m^3 (627,000 acre-feet) is provided below the top of conservation pool elevation of 162.5 m (533 ft) above mean sea level. The water surface area at top of conservation pool is 96.3 million m^2 (23,800 acres). The hydropower plant has an installed capacity of 30 megawatts and an efficiency of 0.86. A constant tailwater elevation of 133 m above mean sea level is used in computing head.

The reservoir is operated to meet requirements for water supply diversions through a lakeside pumping plant and pipeline, releases to the river for instream flow needs and water supply diversions from the river at locations downstream of the dam, and hydroelectric power generation. Operating rules are based on the storage allocations delineated in Table 6–1 and Figure 6–1. Downstream water supply releases are also used to generate power. Lakeside water supply diversions are not available to generate power. Operating rules are as follows:

- Spills over the uncontrolled spillway occur any time storage exceeds the total conservation storage capacity of 773 million m^3.
- Hydroelectric energy is generated as long as the storage exceeds 456 million m^3. Energy is generated with spills and releases for downstream water use. Additional releases are made as necessary to meet primary energy targets.
- Releases to meet downstream water diversion and instream flow requirements are made as long as storage exceeds 6.2 million m^3, which represents the inlet of the lowest outlet. Spillway spills and hydroelectric power releases are also used to meet the downstream water use requirements.

TABLE 6–1   Reservoir Pools for Example 6–1

Reservoir Pool Level	Elevation		Storage Volume		Surface Area	
	(m)	(ft)	(10^{6}m^3)	(10^{3}ac-ft)	(10^{6}m^2)	(10^{3}ac)
Top of Conservation Pool	162.5	533	773	627	96.32	23.8
Bottom of Power Pool	158.5	520	456	370	63.54	15.7
Top of Inactive Pool	136.9	449	6.2	5	1.66	0.4

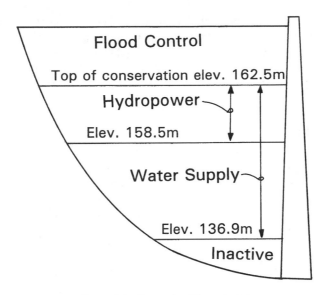

Figure 6–1   Reservoir of Example 6–1.

Diversion, release, and energy targets vary seasonally but are assumed to be constant from year to year. Lakeside diversion and downstream release requirements are tabulated in Table 6–2 as mean monthly discharges in m^3/s. These demands are met as long as the active conservation pool is not depleted. Energy targets, in megawatt-hours per month, also are tabulated in Table 6–2. The monthly energy targets are met as long as water is available from the power pool, which is the portion of the conservation pool above elevation 158.5 m. Re-

TABLE 6–2   Water Use Requirements for Example 6–1

Month	Net Evap (mm)	Lakeside Diversion (m^3/s)	Downstream Release (m^3/s)	Energy (mw-hr)
Jan	9	20	16	2250
Feb	9	20	16	2250
Mar	12	20	16	2250
Apr	24	20	16	2250
May	55	25	22	2250
Jun	107	35	30	3000
Jul	204	38	30	6000
Aug	131	38	30	6000
Sep	107	25	18	3000
Oct	55	25	18	2250
Nov	46	20	16	2250
Dec	27	20	16	2250

leases from this pool are made, if necessary, specifically to meet the indicated primary energy targets. Additional energy is generated with spills and downstream releases for water supply and other instream flow needs. The water and energy requirements represent constant 1995 conditions of population and economic development.

Although a simulation period of 1940–1995 would be realistic for this type of model, for brevity, only the first 24 months of the simulation are presented in Table 6–3. Historical gaged streamflows for the period from January 1940 through December 1941, adjusted to reflect 1995 conditions of river basin development, are tabulated in column 2 of Table 6–3. Mean monthly net evaporation rates for each of the 12 months of the year are reproduced in Table 6 2. Net evaporation is gross evaporation less precipitation. The mean net evaporation rates vary seasonally but are assumed to be constant from year to year in this example. The conservation pool is assumed to be full at the beginning of the simulation.

The simulation was performed with the HEC-5 model described in Chapter 8 [Hydro-

TABLE 6–3   Reservoir Simulation for Example 6–1

Month 1	Stream Inflow (m³/s) 2	Ending Reservoir		Net Evap (m³/s) 5	Downstream Flow (m³/s) 6	Energy (mw-hr) 7	Energy Shortage (mw-hr) 8
		Storage (10³m³) 3	Elevation (m) 4				
1940							
Jan	46	773,000	162.5	0.32	25.7	4751	—
Feb	71	773,000	162.5	0.36	50.6	8464	—
Mar	61	773,000	162.5	0.43	40.6	7507	—
Apr	129	773,000	162.5	0.89	108.1	19,360	—
May	157	773,000	162.5	1.98	130.0	22,320	—
Jun	50	724,059	161.9	3.88	30.0	5322	—
Jul	14	550,991	159.7	6.32	34.3	6000	—
Aug	7	379,158	157.0	3.16	30.0	4827	1173
Sep	42	370,837	156.8	2.21	18.0	2615	385
Oct	33	341,238	156.2	1.05	18.0	2659	—
Nov	54	385,497	157.1	0.92	16.0	2302	—
Dec	50	421,454	157.8	0.58	16.0	2458	—
1941							
Jan	56	474,460	158.7	0.21	16.0	2547	—
Feb	69	553,665	159.8	0.26	16.0	2383	—
Mar	86	686,613	161.5	0.36	16.0	2776	—
Apr	111	773,000	162.5	0.85	56.8	10,007	—
May	135	773,000	162.5	1.98	108.0	19,989	—
Jun	26	662,162	161.2	3.76	30.0	5259	—
Jul	12	482,285	158.8	5.78	35.4	6000	—
Aug	10	319,506	155.8	2.77	30.0	4600	1400
Sep	38	301,668	155.4	1.88	18.0	2475	525
Oct	103	459,390	158.5	1.11	18.0	2714	—
Nov	86	585,741	160.2	1.25	16.0	2564	—
Dec	78	695,985	161.6	0.84	16.0	2803	—

logic Engineering Center, 1982]. A variety of other computer programs are also available for this type of modeling. Simulation results are summarized in columns 3 through 8 of Table 6–3. The water accounting computations are performed for each month in turn. The end-of-month storage becomes the beginning storage for the following month. With reference to equation 6-5 and Table 6–3, the water balance for the first month is as follows:

$$S_{t+1} = S_t + \text{inflow} - \text{diversion} - \text{release} - \text{evaporation}$$

$$773 \times 10^6 \, \text{m}^3 = 773 \times 10^6 \, \text{m}^3 + (46 - 20 - 25.68 - 0.32 \, \text{m}^3 / \text{s})(31 \, \text{days})(86,400 \, \text{sec/day})$$

From equation 6-6, the net evaporation loss during the first month is

$$E_1 = A_1 e_1 = ((96.32 \times 10^6 \, \text{m}^2)(0.009 \, \text{m})) / (31 \, \text{days}) / 86,400 \, \text{sec/day} = 0.32 \, \text{m}^3/s$$

From equations 6-8 and 6-9, the energy generated during the first month is

$$\text{Energy} = \text{power} \times \text{time} = \gamma Q h e (\text{time})$$

$$= (9.81 \text{kN/m}^3)(25.68 \, \text{m}^3/\text{s})(29.5 \, \text{m})(0.86)(31 \, \text{days})(24 \, \text{hrs/day})$$

$$= 4,751,000(\text{kN} \cdot \text{m/s}) - \text{hrs} = 4751 \text{ megawatt-hours}$$

where the head (h) is the reservoir water surface elevation (162.5 m) minus the tailwater elevation (133 m), and the discharge (Q) is the total downstream release including spills.

In the simulation model, evaporation, energy, and end-of-month storage are included in the variables computed for each month. Evaporation and energy during the month are computed as a function of both beginning and end-of-month storage. Thus, an iterative algorithm is required.

The lakeside diversion and downstream release targets are met, without shortage, during each month of the 24-month simulation. The end-of-month storage varies from a full conservation pool of 773,000,000 m³ to a low of 301,668,000 m³ in the month corresponding to September 1941 streamflows. The hydroelectric energy targets are fully met in each month except for August and September of each year. In most months, spills and releases to meet the downstream release targets result in generation of additional energy above the energy targets. Additional releases to meet the primary energy demands occur in July of each of the two years.

6.5 HYDROLOGIC ROUTING

Hydrologic storage routing, also called flood routing, consists of computing the outflow hydrograph for a given inflow hydrograph. Water is temporarily stored as it flows through either reservoirs or river reaches. At an instant in time, the inflow and outflow will be different as water is temporally stored or released from storage. As illustrated in Figure 6–2, the outflow hydrograph has a smaller peak and broader time base than the inflow hydrograph. Storage routing is a prediction of this attenuation effect of temporary storage on the flow hydrograph. Routing computations are applied to both reservoirs and river reaches.

In Example 6–1, with a monthly time interval, the water accounting computations are based on mean monthly inflows and outflows and end-of-month storage,

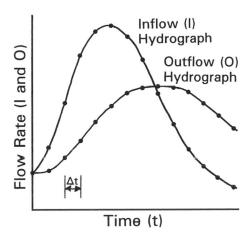

Figure 6-2   Hydrologic routing.

without consideration of instantaneous flow rates. However, in dealing with either instantaneous flows or mean flows over relatively short time intervals, hydrograph attenuation effects are important. Also, in modeling of conservation storage operations, as illustrated by Example 6-1, releases are determined based on water use requirements and reservoir operating rules. Gates are assumed to be operated as necessary to achieve the desired outflows. The following discussion is pertinent to routing through river reaches and reservoirs with either uncontrolled (ungated) outlet structures or gated structures with fixed gate openings.

Routing is most often associated with modeling of flood flows. However, routing may also be important in modeling conservation storage operations using a short computational time step such as a day. For example, water supply diversions often occur at river locations which are great distances below a dam. Several days of travel time may be required for water released from a reservoir to reach the point of diversion from the river. Routing computations are necessary to relate flow versus time at the diversion location to the upstream reservoir releases.

The various hydrologic storage routing methods are based on the conservation of volume or continuity equation in the form of equation 6-4.

$$\frac{S_{t+\Delta t} - S_t}{\Delta t} = \frac{I_t + I_{t+\Delta t}}{2} - \frac{O_t + O_{t+\Delta t}}{2} \tag{6-4}$$

Since there are two unknowns ($O_{t+\Delta t}$ and $S_{t+\Delta t}$) in this equation, it must be combined with some other relationship between storage and discharge. Alternative hydrologic storage routing methods are based on different storage versus discharge relationships.

Two widely used routing approaches, modified Puls and Muskingum, are described here. Modified Puls, with variations thereof, is the conventional standard for routing through uncontrolled (ungated) reservoirs and is also often used for river routing. Muskingum routing is one of a variety of methods that have been widely used for routing flows through river reaches. Modified Puls, Muskingum, and other hydrologic storage routing methods are covered in hydrology textbooks such as [Linsley *et al.*, 1982], [McCuen, 1989], [Bedient and Huber, 1992], and [Singh, 1992].

Muskingum, modified Puls, and other flood routing methods are applied to rivers. Modified Puls routing is normally applied to reservoirs with ungated outlet structures and can also be applied to gated outlet structures for a fixed gate opening. In modified Puls routing, outflow is controlled by a fixed storage versus outflow relationship. For flood control operations of major reservoirs with gated outlet structures, releases are specified by operating rules based on downstream flow conditions. Conservation storage operations are also based on operating rules and water use requirements. Modified Puls or similar routing methods are not used to simulate these types of operation. Rather, the outflows in equation 6-4 are set by operating rules and water management requirements, and water balance computations determine the change in storage.

6.5.1 Modified Puls Flood Routing

The modified Puls approach consists of combining a relationship between storage and outflow with the conservation of volume equation. Equation 6-4 is rearranged with all unknowns on the left of the equal sign as follows:

$$\frac{2S_{t+\Delta t}}{\Delta t} + O_{t+\Delta t} = I_t + I_{t+\Delta t} + \frac{2S_t}{\Delta t} - O_t \qquad (6\text{-}10)$$

This equation is used to compute $2S/\Delta t + O$, which is sometimes called the *storage indication*. A relation between outflow and $2S/\Delta t + O$ is required to determine the outflow. Procedures for developing storage and flow data for determining the outflow versus $2S/\Delta t + O$ relationship are different for reservoir versus river routing. Otherwise, the routing computations are the same.

For a reservoir, storage is determined by site topography, and outflow is determined by outlet structure hydraulics. Reservoir storage volumes are determined based on measuring the areas behind the dam enclosed by contours on a topographic map of the reservoir site. Storage volumes are estimated by accumulating incremental storages between horizontal areas enclosed by contours and the dam.

storage volume = Σ((top area + bottom area)/2)(vertical contour interval)

The reservoir water surface is assumed to be horizontal. Discharge through outlet structures is determined as a function of head, using weir and orifice equations discussed later in this chapter. An elevation versus storage relationship reflecting the topography of the reservoir site is combined with an elevation versus outflow relationship which reflects the hydraulics of the outlet structures. Storage volume (S) and outlet structure discharge (O) are tabulated for discrete values of reservoir water surface elevation. The corresponding values of $2S/\Delta t + O$ are computed to develop a table of O versus $2S/\Delta t + O$.

The approach for developing the table of O versus $2S/\Delta t + O$ is different for river routing. Water surface profiles are computed for a range of assumed flows, using river hydraulics models discussed later in this chapter. Storage volumes for the river reach are determined based on flow depths and cross-sectional geometry of the river.

The O versus $2S/\Delta t + O$ relationship is developed prior to the routing computations. The routing algorithm steps through time, determining the outflow $(O_{t+\Delta t})$ for each time step. Equation 6-10 is solved for $(2S_{t+\Delta t}/\Delta t + O_{t+\Delta t})$. The corresponding outflow $(O_{t+\Delta t})$ is determining by interpolation of the previously developed table of O versus $2S/\Delta t + O$. The computations are illustrated by Example 6-2.

Example 6-2—Modified Puls Routing

Reservoir storage volume (S) and outflow (O) are provided as a function of water surface elevation in Table 6-4. The corresponding values of $2S/\Delta t + O$ are also tabulated for a time interval (Δt) of one hour. The water surface elevation and storage volume are referenced to the crest of an uncontrolled spillway. These are elevations and storages above the spillway crest. The storage volumes are computed from horizontal areas measured from a contour map of the reservoir site, which are also shown in Table 6-4. The outflow through the uncontrolled spillway is determined as a function of head using the weir equation discussed later in this chapter.

A hydrograph of streamflow inflows into the reservoir during a flood event is given in the first two columns of Table 6-5. The flows are tabulated at a Δt of one hour. Modified Puls routing is used to determine the outflow hydrograph shown in the last column of Table 6-5. At time zero, the inflow and outflow are zero, and the water surface is at the spillway crest (elevation zero). The computations step through time. At each time step, the storage indication $2S/\Delta + O$ is determined using the routing equation

$$\left(\frac{2S}{\Delta t} + O\right)_{t+\Delta t} = I_t + I_{t+\Delta t} + \left(\frac{2S}{\Delta t} - O\right)_{\Delta t}$$

TABLE 6-4   Development of Storage Indication Relation for Example 6-2

Elev (m)	Area (ha = 10^4m^2)	Storage (S) (10^4m^3)	Outflow (O) (m^3/s)	$2S/\Delta t + O$ (m^3/s)
0	0	0	0	0
1	16	8	11	55
2	45	38	45	256
3	83	102	102	669
4	128	208	181	1336
5	179	362	283	2294
6	235	569	408	3569
7	296	834	555	5188
8	362	1163	724	7185
9	432	1560	917	9584
10	506	2029	1132	12,404

TABLE 6–5   Modified Puls Routing Computations for Example 6–2

Time (hours)	Inflow (m^3/s)	$I_1 + I_2$ (m^3/s)	$2S/\Delta t - O$ (m^3/s)	$2S/\Delta t + O$ (m^3/s)	Outflow (m^3/s)
0	0	350	0	0	0
1	350	1070	234	350	58
2	720	1660	950	1304	177
3	940	2030	1982	2610	314
4	1090	2150	3116	4012	448
5	1060	1990	4142	5266	562
6	930	1680	4862	6132	635
7	750	1330	5202	6542	670
8	580	1050	5194	6532	669
9	470	850	4956	6244	644
10	380	690	4592	5806	607
11	310	580	4156	5282	563
12	270	490	3708	4736	514
13	220	420	3268	4198	465
14	200	380	2850	3688	419
15	180	330	2480	3230	375
16	150	270	2142	2810	334
17	120	220	1822	2412	295
18	100	180	1564	2042	239
19	80	150	1314	1744	215
20	70			1464	192

The corresponding outflow (O) is determined by interpolation of the O versus ($2S/\Delta t + O$) relation of Table 6.4. The term ($2S/\Delta t - O$), which is used in the routing computations for the next time step, is determined as

$$\left(\frac{2S}{\Delta t} - O\right) = \left(\frac{2S}{\Delta t} + O\right) - 2\,(O)$$

The given reservoir inflow hydrograph and computed outflow hydrograph tabulated in Table 6–5 are plotted in Figure 6–3.

Since outflow is assumed to be a unique function of storage, with modified Puls routing, the peak outflow occurs at the intersection of the outflow hydrograph with the recession side of the inflow hydrograph, as illustrated in Figure 6–3.

Modified Puls works well for routing a flood hydrograph through a reservoir because the flow is controlled by the outlet structures. The outflow through the spillway and outlet works is a well-defined function of storage. The method is routinely applied to uncontrolled (ungated) flood control reservoirs. For a gated outlet structure, a particular O versus $2S/\Delta t + O$ relationship is for a specified gate setting.

A storage versus outflow relationship for a river can be approximated using a water surface profile model such as HEC-2, discussed later in this chapter. Flow depths and corresponding storage volumes in the channel reach are determined for a

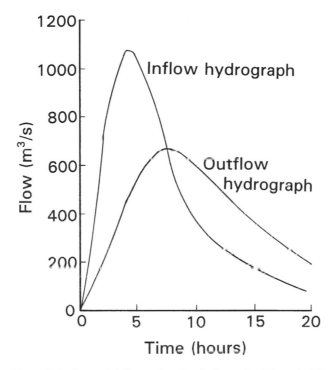

Figure 6–3   Reservoir inflow and outflow hydrographs of Example 6–2.

range of flow rates. Thus, modified Puls is used for river routing as well as reservoir routing. However, as discussed next, the storage versus outflow relationship for a river reach is necessarily approximate.

6.5.2 Muskingum Routing

River routing is complicated by the lack of control structures and the corresponding lack of a unique relationship between storage and outflow. The relationship between storage volume in a reach of river and the flow rate at the downstream end of the reach may change as a flood wave passes through the reach. The Muskingum method addresses this difficulty by relating storage to both inflow and outflow. Thus, the volume of water temporarily stored in the river reach at an instant in time is related to the flow rates at both the upstream and downstream ends of the reach.

Muskingum routing is based on combining equation 6-4 with the following linear relationship between storage (S) and weighted inflow (I) and outflow (O):

$$S = K(xI + (1-x)O) \qquad (6\text{-}11)$$

where x is a dimensionless weighting factor, and K has units of time. The zero to 1.0 weighting factor x has been found to range from 0.1 to 0.3 for most streams, with an average value of about 0.2. The parameter K is a representation of the travel time through the reach.

Combining equations 6-4 and 6-11 results in the following Muskingum routing equation:

$$O_{t+\Delta t} = C_0 I_{t+\Delta t} + C_1 I_t + C_2 O_t \tag{6-12}$$

where

$$C_0 = -\frac{Kx - 0.5\Delta t}{k - kx + 0.5\Delta t}$$

$$C_1 = -\frac{Kx + 0.5\Delta t}{K - Kx + 0.5\Delta t}$$

$$C_2 = -\frac{K - Kx - 0.5\Delta t}{K - Kx + 0.5\Delta t}$$

$$C_0 + C_1 + C_2 = 1 \tag{6-13}$$

Application of the Muskingum routing equation is illustrated in Example 6–3.

Example 6–3—Muskingum Routing

A river reach is characterized by values of 0.2 and 9 hours for the routing parameters x and K. A time interval Δt of four hours is used. A steady flow of 11 m³/s is flowing in the river at time zero at both the upstream and downstream locations. Upstream reservoir releases beginning about this time cause the flows to increase. An inflow hydrograph at the upstream end of the reach is given in Table 6–6. The outflow hydrograph at the downstream end of the reach is computed using the Muskingum routing equation. The coefficients in equation 6-12 are computed as follows:

$$C_0 = -\frac{9(0.2) - 0.5(4)}{9 - 9(0.2) + 0.5(4)} = 0.0217$$

$$C_1 = \frac{9(0.2) + 0.5(4)}{9 - 9(0.2) + 0.5(4)} = 0.4131$$

$$C_2 = \frac{9 - 9(0.2) - 0.5(4)}{9 - 9(0.2) + 0.5(4)} = 0.5652$$

to obtain the routing equation

$$O_{t+\Delta t} = 0.0217\, I_{t+\Delta t} + 0.4131\, I_t + 0.5652\, O_t$$

This equation is applied at each time step to obtain the outflow hydrograph tabulated in Table 6–6. The attenuation affect of temporary storage in the river reach is evident from a comparison of the inflow and outflow hydrographs. The peak flow rate upstream of 93 m³/s occurs at 16 hours. The maximum flow downstream is 77 m³/s at 20 and 24 hours.

The parameters x and K for a river reach are determined from gaged upstream and downstream hydrographs. A parameter calibration option contained in the HEC-1 Flood Hydrograph Package is described in the last section of Chapter 10; it can be used to determine x and K from a set of gaged hydrographs.

TABLE 6–6   **Muskingum Routing for Example 6–3**

Time (hours)	Known Inflow (I) (m^3/s)	Computed Outflow (O) (m^3/s)
0	11	11
4	38	12
8	73	24
12	92	46
16	93	66
20	78	77
24	60	77
28	53	70
32	31	62
36	21	48
40	13	36
44	12	26
48	11	20

Most hydrology textbooks present a procedure based on equation 6-11. The gaged hydrographs are used to develop plots of storage (S) versus ($xI + (1 - x)O$) for different values of x. The plots tend to exhibit a loop. The optimal value of x is selected as the one that minimizes the loop effect. K is estimated as the slope of a straight fit to the plotted values of S versus ($xI + (1 - x)O$).

In the absence of gaged hydrographs, x and K have been roughly approximated as follows. A value for x of 0.2 is considered typical for many rivers and is used in the absence of better information. K represents the travel time through a reach. Travel time (T) is estimated as

$$K - T - \text{reach length/mean velocity}$$

with the mean velocity estimated using river hydraulics techniques discussed later in this chapter.

6.5.3 Comparison of Routing Methods

Example 6–4 presents a simplified problem that can be solved by any hydrologic storage routing method. A hypothetical reservoir or channel reach is assumed to be represented by a simple linear relationship between storage and outflow. With this simplifying assumption, routing can be performed with either the modified Puls method or Muskingum equation, or by substituting the storage-outflow relation directly into equation 6-4. All alternative approaches result in precisely identical solutions for the simple linear storage-outflow relation. However, the Muskingum method allows storage to be linearly related to inflow as well as outflow. In a realistic river-routing situation, the Muskingum parameter x would be nonzero, indicating that storage is dependent on both inflow and outflow. Modified Puls requires a unique relation between storage and outflow, but that relationship can be highly nonlinear. In actual application of modified Puls routing, the reservoir or river reach is characterized by a table of outflow versus storage volume and the corresponding O versus $2S/\Delta t + O$ relationship.

Example 6–4—Hydrologic Storage Routing

An outflow hydrograph is computed for the inflow hydrograph provided in Table 6–7. The inflow hydrograph is routed through a reservoir, river reach, or other container characterized by the following linear relation between storage (S), in m^3, and outflow (O), in m^3/s.

$$S = KO \quad \text{where } K = 32{,}400 \text{ s (9 hours)}$$

The routing is repeated using the following alternative approaches: substitution of the given storage-outflow relation directly into equation 6-4, Muskingum, and modified Puls.

Alternative Solution 1   Since the hydrograph ordinates are spaced at two-hour intervals, Δt in equation 6-4 has a value of two hours. Using the linear storage-outflow equation for the reservoir, terms S_t and $S_{t+\Delta t}$ in equation 6-4 are replaced with $S = KO = (9 \text{ hours}) O_t$ and $(9 \text{ hours}) O_{t+\Delta t}$.

$$\frac{9\,O_{t+\Delta t} - 9\,O_t}{1} = \frac{I_t + I_{t+\Delta t}}{2} - \frac{O_t + O_{t+\Delta t}}{2}$$

This expression is algebraically rearranged to obtain the routing equation.

$$O_{t+\Delta t} = 0.1 I_t + 0.1 I_{t+\Delta t} + 0.8\,O_t$$

Given the inflow hydrograph and outflow for time zero provided in Table 6–7, the outflow hydrograph is computed using this routing equation. For example, the instantaneous outflow rates at two and six hours after time zero are computed as follows:

TABLE 6–7   Alternative Solutions 1 and 2 for Example 6–4

Known Time (hrs)	Computed Inflow (I) (m^3/s)	Outflow (O) (m^3/s)
0	500	500
2	1170	567
4	2400	811
6	4210	1309
8	4780	1947
10	4310	2466
12	3350	2739
14	2230	2749
16	1580	2580
18	1130	2335
20	890	2070
22	720	1817
24	590	1585
26	520	1379

$$O_{2hrs} = 0.1(500 \text{ m}^3/\text{s}) + 0.1(1{,}170 \text{ m}^3/\text{s}) + 0.8(500 \text{ m}^3/\text{s}) = 567 \text{ m}^3/\text{s}$$

$$O_{4hrs} = 0.1(1{,}170 \text{ m}^3/\text{s}) + 0.1(2{,}400 \text{ m}^3/\text{s}) + 0.8(567 \text{ m}^3/\text{s}) = 811 \text{ m}^3/\text{s}$$

Alternative Solution 2 The Muskingum routing equation is based on the following relationship:

$$S = K(xI + (1-x)O)$$

as compared to the simple linear storage-outflow relation of the example problem.

$$S = KO$$

Thus, the Muskingum routing parameters for the example are:

$$x = \text{zero} \quad \text{and} \quad K = 9 \text{ hours}$$

These values are substituted into equation 6 12 as follows:

$$C_o = -\frac{9(0) - 0.5(2)}{9 - 9(0) + 0.5(2)} = 0.1$$

$$C_1 = \frac{9(0) + 0.5(2)}{9 - 9(0) + 0.5(2)} = 0.1$$

$$C_2 = \frac{9 - 9(0) \ 0.5(2)}{9 - 9(0) + 0.5(2)} = 0.8$$

$$O_{t+\Delta t} = 0.1 I_{t+\Delta t} + 0.1 I_t + 0.8 \ O_t$$

This routing equation is applied to compute the outflow hydrograph of Table 6–7.

Alternative Solution 3 Modified Puls routing requires an O versus $2S/\Delta t + O$ relation, which can be developed from the relationship

$$S = KO \quad \text{where } K = 9 \text{ hours}$$

as follows:

$$\frac{2S}{\Delta t} + O - \frac{2 \ (9 \cdot O)}{2} + O = 10 \ (O)$$

$$O = 0.1 \ (\frac{2S}{\Delta t} + O)$$

The modified Puls routing computation are presented in Table 6–8. The following equations are applied at each time step:

$$(\frac{2S}{\Delta t} + O)_{t+\Delta t} = I_t + I_{t+\Delta t} + (\frac{2S}{\Delta t} - O)_t$$

$$O = 0.1 \ (\frac{2S}{\Delta t} + O)$$

$$(\frac{2S}{\Delta t} - O) = (\frac{2S}{\Delta t} + O) - 2 \ (O)$$

TABLE 6–8   Modified Puls Solution for Example 6–4

Time (hrs)	Inflow (m³/s)	$2S/\Delta t - O$ (m³/s)	$2S/\Delta t + O$ (m³/s)	Outflow (m³/s)
0	500	4000	5000	500
2	1170	4536	5670	567
4	2400	6485	8106	811
6	4210	10,476	13,095	1309
8	4780	15,573	19,466	1947
10	4310	19,730	24,663	2466
12	3350	21,912	27,390	2739
14	2230	21,994	27,492	2749
16	1580	20,643	25,804	2580
18	1130	18,682	23,353	2335
20	890	16,562	20,702	2070
22	720	14,538	18,172	1817
24	590	12,678	15,848	1585
26	520		13,788	1379

6.6 RESERVOIR/RIVER SYSTEMS

A reservoir/river system typically includes multiple reservoirs, stream reaches, and water use requirements. River basins may be quite complex, with many reservoirs located on a number of tributaries. The water accounting procedures in reservoir system analysis models trace the movement and storage of water throughout the various components of the system. Computations typically are repeated sequentially by time interval. The storage at the end of a time step becomes the storage at the beginning of the next time step as the accounting procedures step through time. Some optimization models account for storage and flow simultaneously for all time intervals, rather than sequentially.

6.6.1 Flood Control versus Conservation Storage Operations

Simulation procedures vary between flood control and conservation storage operations. The computational time step is a key consideration. Simulation of flood control operations involves short time steps since flows are rapidly changing. A major flood event with a duration of several days or weeks is typically simulated with a one-hour, several-hour, or one-day time interval. Flood routing is required to capture the attenuation effects as the flood water moves through the river/reservoir system. Evaporation is negligible during a flood event.

Water supply, hydroelectric power, and other water use requirements are the driving concern in conservation storage operations. Evaporation losses are important. Longer time steps are adopted in analyzing conservation storage operations.

Although daily or weekly time intervals are common, a monthly interval is more typical for planning studies. Flow attenuation in a river reach, with a travel time of several days or less, is appropriately neglected in a monthly time step model. Mean inflows and outflows during the month are assumed equal. River-routing techniques are not used. In some cases, for long river reaches, storage routing computations may be appropriate in a daily time step model. Modified Puls reservoir routing, with outflows controlled by the hydraulics of ungated outlet structures, is not pertinent for conservation storage operations. Rather, water accounting procedures for a reservoir are based on equation 6-5.

6.6.2 Reservoir/River System Representation

The configuration of a reservoir/river system is represented in a model as a selected set of key locations called *stations, nodes,* or *control points.* Figure 6-4 illustrates a system represented, for modeling purposes, as a set of seven stations or control point locations. Reservoirs are located at three of the stations. Diversion and/or instream flow requirements are specified at six of the seven locations. Conservation pool releases contribute to water supply diversion and instream flow requirements at downstream locations as well as at the reservoir. Flood control pools in the reservoirs are operated based on maximum allowable flows at downstream locations as well as at the dam.

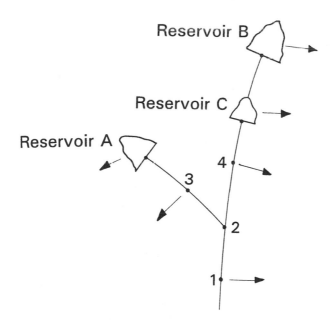

Figure 6–4   Reservoir/river system.

For most reservoir/river system analysis models, the system is operated for sequences of streamflow inflows. Input includes unregulated flows for each time interval at each pertinent location. In the model, water use requirements are met, and reservoirs are operated, while accounting for water storage and flow sequentially during each time interval of the overall period of analysis.

Sequences of unregulated streamflows are provided as model input for each of the seven locations shown in Figure 6–4. The unregulated streamflows represent natural flows unaffected by the reservoirs and water use diversions. At the two most upstream locations, the unregulated flows provide the inflows to reservoirs A and B.

Modeling procedures require a differentiation between total flows and incremental local flows. Incremental local flows represent flows entering the river between node locations. Unregulated flow data are typically developed by adjusting gaged flows to remove the effects of reservoirs and other nonhomogeneities or, in some cases, by precipitation-runoff modeling. The resulting unregulated streamflows are total flows. Incremental local flows are determined as the difference between total flows at adjacent stations. The incremental local flows for station 1 in Figure 6–4 are the total flows at station 1 less the total flows at station 2. The incremental flows at station 2 are determined by subtracting the sum of the total flows at stations 3 and 4 from the flows at station 2. Rapidly varying flood flows tabulated at short time intervals are complicated by the need to incorporate flood routing computations, such as the Muskingum method, in the development of incremental flows. In this case, incremental flows at station 1 are determined as total unregulated flows at station 1 less flows resulting from routing flows from station 2 to station 1 using Muskingum or a similar technique.

Negative values of incremental local flows are often a complicating issue in developing the unregulated streamflow input. The negative increment may be an indication that either actual channel losses exceed gains, or inaccuracies exist in the data or computations. Flows normally increase in a downstream direction, indicating that incremental local flows are positive. In some time periods, unregulated flows at a downstream location may be lower than upstream, indicating a negative incremental flow. This could be due to a number of causes, including channel losses reflected in the stream gage data, temporary storage and travel time effects of flow through the reach, or inaccuracies in gage data or computational adjustments to gage data. Negative incremental flows are more prevalent for a daily interval than a monthly interval. Negative incremental flows are handled in different ways depending on the particular situation.

6.6.3 Reservoir/River System Operations

The complexity of the water accounting computations depends on whether the reservoirs are operated to meet flow or diversion requirements at downstream locations. The following discussion begins with the simpler case of each reservoir being operated to meet requirements associated only with its own location. The complexities of multiple reservoirs operating for multiple locations are then addressed.

Assume that releases from the flood control pools in the three reservoirs of Figure 6–4 are controlled by ungated spillways. Outflow depends solely on the reservoir water surface elevation or head on the outlet structures. The system is simulated for a particular flood event represented by unregulated total flow hydrographs at nodes A and B and incremental local flow hydrographs at nodes 1, 2, 3, 4, and C. A several-hour computational time interval is used in simulating the several-day-long flood event. The simulation consists of the following tasks:

- An inflow hydrograph is routed through reservoir A, using modified Puls routing, to determine the regulated outflow hydrograph at station A.
- The regulated hydrograph at station A is routed from A to 3 using Muskingum, modified Puls, or a similar routing technique. The unregulated incremental local inflow hydrograph at station 3 is added to determine the total regulated flow hydrograph at 3.
- The resulting hydrograph at station 3 is likewise routed to station 2.
- An inflow hydrograph is routed through reservoir B, using modified Puls routing, to determine the regulated outflow hydrograph at station B.
- The regulated hydrograph at station B is routed from B to C using Muskingum, modified Puls, or a similar routing technique. The unregulated incremental local inflow hydrograph at station C is added to the routed hydrograph to determine the total regulated flow hydrograph at C. This is the inflow hydrograph into reservoir C.
- The inflow hydrograph is routed through reservoir C, using modified Puls routing, to determine the outflow hydrograph at station C.
- The hydrograph at station C is routed to station 4, and local flows are added to determine the total flows at station 4.
- The resulting hydrograph at station 4 is likewise routed to station 2.
- The two hydrographs previously routed from stations 3 and 4 are combined at station 2, along with the increment local inflow hydrograph, to obtain the regulated flow hydrograph at station 2.
- The hydrograph at 2 is routed to 1 and combined with local inflows to obtain the hydrograph at station 2.

Thus, the flows for that particular flood are computed for a particular reservoir system configuration and design. This type of simulation can play various roles in a study. For example, the procedure for computation of expected annual economic damages outlined in the last section of Chapter 7 incorporates this system routing procedure. The HEC-1 Flood Hydrograph Package, described in Chapter 8, is an example of a generalized simulation model that performs the computations just described.

This simulation is simplified by the fact that reservoir releases are not dependent on flows at other stations. Computations are performed for one station at a time, working from upstream to downstream. However, another common situation involves gated outlet structures operated by reservoir management personnel in accordance with prescribed regulation plans. As discussed in Chapter 3, flood control operations in major reservoir systems are based on emptying flood control pools as quickly as possible without contributing to flows exceeding specified maximum limits at downstream control points. HEC-5 Simulation of Flood Control and Conservation Systems, described in Chapter 8, simulates this type of operation.

Assume that the flood control pools of the three reservoirs in Figure 6–4 have gated outlet structures which are operated following conventional rules based on the maximum allowable flow rates specified for each downstream station. Releases from reservoir A are constrained by flows at stations 3, 2, and 1. The flows at stations 1 and 2 depend on releases from multiple reservoirs. Flows at downstream locations depend on releases from the reservoirs and incremental local inflows. The flows are affected by the attenuation effects reflected in the routing computations for each reach. HEC-5 and similar models use an iterative procedure in which release decisions are tentative subject to adjustment as the simulation considers each control point of the system in turn.

Likewise, the complexity of models for evaluating conservation storage operations depend on interactions between multiple reservoirs operating for instream flow and diversion requirements at multiple locations. Models of conservation storage operations typically do not include attenuation effects of river routing. Outflows are assumed to equal inflows to the reach for a given time step. Storage changes are limited to reservoirs.

Assume that the diversions at stations 1, 2, 3, and 4 in Figure 6–4 are supplied by incremental local inflows supplemented by releases from the reservoirs as necessary. Various approaches are adopted in different models to handle the interactions between multiple reservoirs and diversion and instream flow requirements at different locations, in the process of accounting for flows and storage throughout the system. The network flow programming models and other optimization models discussed in Chapter 10 have an advantage in this regard because flows and storages at all locations are considered simultaneously. Most simulation models described in Chapter 8, such as HEC-5, consider each location in turn working from upstream to downstream with iterative loops in the procedure to adjust operating decisions. The Water Rights Analysis Package (WRAP) simulation model described in Chapter 9 performs computations for each water demand in turn based on water rights priorities, regardless of location. As each water demand is met, streamflow still available at all locations is adjusted accordingly.

6.7 OUTLET STRUCTURE HYDRAULICS

Until now, this chapter has focused on accounting for water in a reservoir/river system from the perspective of flow rates and storage volumes. Water surface elevations or flow depths are also important in reservoir/river system analysis applications. Velocities may also be of concern. Hydraulic analyses deal with relationships between flow rates, depths, and velocities. The term *rating curve* refers to the relationship between discharge and flow depth or water surface elevation. Rating curves are developed for reservoir outlet structures and for locations on streams.

Rating curves for outlet structures are developed as an integral part of the design of a reservoir project and are available for operational purposes after completion of construction. Basic hydraulic equations are used to develop the stage-discharge relationships. Rating curves for existing structures can also be developed from actual measurements of stage and discharge. For gated structures, a family of rating curves is required to express the water surface elevation versus discharge relationship as a function of gate opening. Rating curves provide fundamental information for reservoir operation. Since stage is much easier to measure than discharge, the discharge from a reservoir is determined by applying the measured water surface elevation to the rating curve. For a measured reservoir level, rating curves are used to select a gate opening or the number of sluices to open to achieve a desired release rate. Rating curves may be used in reservoir operation models to reflect outlet capacity constraints on release decisions.

Spillways, outlet works, and other structures for controlling releases from reservoirs have been constructed in a variety of configurations. Rating curves for most outlet structures can be developed based on various forms of weir and orifice equations. A weir is a notch of regular form through which water flows. An orifice is an opening with closed perimeter through which water flows. If the opening flows only partially full, the orifice becomes a weir. The crest of a weir is the surface over which the water flows.

The discharge versus head relationship for flow over a rectangular spillway is determined using the weir equation,

$$Q = C L H^{1.5} \qquad\qquad (6\text{-}14)$$

where Q is discharge, C is an empirically determined coefficient, L is the length of the weir or spillway crest, and the head (H) is the vertical distance from the crest to the water surface. The weir coefficient (C) is a function of head, spillway shape, and downstream submergence conditions. The effects of abutments and piers on discharge may be taken into account by reducing the net crest length (L) to an effective length. Approach velocity may be considered by including the velocity head in the head (H). Weir coefficients are estimated on the basis of empirical data derived from prototype or laboratory tests.

Flow through a gate opening is represented by the orifice equation,

$$Q = C A \, (2gH)^{0.5} \qquad\qquad (6\text{-}15)$$

where Q is discharge, C is an empirical orifice coefficient, A is the area of the opening, g is the gravitational acceleration constant, and H is the head at the center of the orifice. For a sharp-crested circular orifice, C has a value of about 0.60 for a wide range of heads. In general, the discharge coefficient (C) will depend on head, tailwater conditions, and the configuration of the structure.

An outlet works consists of one or more conduits or sluices through the dam and associated intake and exit structures. Gates and valves are incorporated into an outlet works to control the flow rate. The rating curve depends on flow conditions, such as whether the conduit is flowing full or only partially full. The orifice equation may be valid for flow controlled by a gate in the intake structure with the conduit flowing partially full. In some cases, flow in a partially full conduit is governed by principles of open channel hydraulics. For conduits flowing full, equation 6-15 is applicable if the coefficient (C) is redefined to reflect all head losses in the conduit, intake structure, gates, and other elements of the outlet works.

The basic hydraulic equations, empirical data for coefficients, and other information needed for developing rating curves are provided by the Bureau of Reclamation [1987], U.S. Army Corps of Engineers [1988], and Davis and Sorensen [1984]. Wurbs and Purvis [1991] review available information and document a computer software package for developing rating curves for reservoir outlet structures.

6.8 RIVER HYDRAULICS

Open-channel hydraulics models simulate flow conditions in natural and improved streams and rivers, associated floodplains, and constructed channels. Water surface profiles are needed for many reservoir operation and related water management applications. Water supply diversion intake structures may be inoperative if river stages drop below certain levels. Navigation operations are based on maintaining specified flow depths. River hydraulics is an essential aspect of evaluating reservoir flood control operations. Flood plain management programs and the design of levees and channel improvements are based on water surface profile models. Erosion and sedimentation may also be a significant consideration in design and operation of river control structures. Flow rates and velocities computed with hydraulic models provide basic input required by water quality models.

The fundamentals of open channel hydraulics are covered in textbooks by Chow [1959], French [1985], and Chaudhry [1993]. The U.S. Army Corps of Engineers [1993] provides a comprehensive treatment of river hydraulics. Fread [1993] provides a concise overview of unsteady flow modeling in rivers.

6.8.1 Flow Categorization

Models are categorized based on capabilities for simulating various types of flow conditions. Although the real world is three-dimensional, flow in rivers can be realistically assumed to be either one- or two-dimensional. Flow can also be characterized as steady versus unsteady and as uniform versus nonuniform or varied. Models can also be categorized as fixed- versus movable-boundary.

In one-dimensional flow models, accelerations in any direction other than the longitudinal direction of flow are assumed to be negligible. The water surface is assumed level perpendicular to the direction of flow. River flows that are significantly influenced by abrupt contractions and expansions in channel or floodplain topography, bridge embankments, control structures, and/or tributary confluences may be more realistically modeled with equations that describe the water motion in two horizontal directions.

Flow is also classified based on whether or not flow characteristics (discharge, depth, velocity, etc.) change with location and/or time. Steady versus unsteady refers to whether flow characteristics change with time. Unsteady flows vary with time at a given location. Steady flows are constant over time. Uniform versus nonuniform (varied) refers to whether flow characteristics change with location along the length of the channel. Uniform flow means that discharge, depth, and velocity are the same at each cross-section of the channel reach. Since flow is essentially never simultaneously unsteady and uniform, unsteady flow implies nonuniform flow as well. However, modeling applications are common for steady uniform and both steady and unsteady nonuniform (varied) flow.

In movable-boundary or mobile bed models, the channel geometry and roughness characteristics change during the simulation due to erosion and sedimentation processes. Flow hydraulics and sediment transport are interrelated.

6.8.2 Basic Concepts

Steady uniform flow computations are performed with a uniform flow formula, typically the Manning equation,

$$Q = (\frac{k}{n}) A R^{2/3} S^{1/2} \tag{6-16}$$

where k is 1.0 for metric units and 1.486 for English units, Q is discharge (m^3/s or ft^3/s), R is hydraulic radius (m or ft), S is slope, and n is an empirical roughness coefficient. Steady, gradually varied (nonuniform), one-dimensional water surface profile computations are based on the energy equation,

$$z_1 + y_1 + \frac{v_1^2}{2g} = z_2 + y_2 + \frac{v_2^2}{2g} + h_L \tag{6-17}$$

where the subscripts 1 and 2 refer to cross-sections defining the upstream and downstream ends of a river reach. The channel bottom elevation (z), flow depth (y), and velocity (v) are for specific locations denoted by the subscripts. The headloss (h_L) in the reach between the two sections is estimated from the relation

$$h_L = S L \tag{6-18}$$

where L is the reach length, and S is the slope of the energy line, which is approximated using the Manning equation (equation 6-16). The iterative standard step method solution of the energy equation is commonly applied to streams and rivers, and can also be applied to prismatic man-made channels.

Two alternative general approaches to one-dimensional unsteady flow modeling are commonly applied. The two approaches are often referred to as *hydrologic routing* versus *hydraulic routing*.

Hydrologic storage routing methods, such as the previously discussed Muskingum and modified Puls techniques, predict the hydrograph (flow rate versus time) at a downstream location, given the hydrograph at an upstream location. Hydrologic routing is used to develop hydrographs at pertinent locations in the stream system. Since hydrologic routing involves only flow rates, if the corresponding flow depths and velocities are needed, the routing results must be combined with a hydraulic model. Typically, the depth and velocity is computed only for the hydrograph peaks (maximum discharges). Use of the Manning equation, developed for steady uniform flow, is the simplest and most approximate approach for associating a flow depth and velocity with the peak discharge from the routing model. A somewhat more accurate approach, which incorporates backwater effects, is to use a steady, gradually varied flow model (such as the standard step method solution of energy equation) to compute the longitudinal water surface profile for the given peak discharges at the various cross-sections.

Hydraulic routing is based on the St. Venant equations or simplifications thereto. The St. Venant equations are two partial differential equations, typically solved numerically, which represent the principles of conservation of mass and momentum.

$$\frac{\partial y}{\partial t} + y\frac{\partial v}{\partial x} + v\frac{\partial y}{\partial x} = 0 \tag{6-19}$$

$$\frac{\partial v}{\partial t} + v\frac{\partial v}{\partial x} + g\frac{\partial y}{\partial x} - g(S_o - S_f) = 0 \tag{6-20}$$

The depth (y) and velocity (v) are computed as a function of location (x) along the river and time (t). The energy slope (S_f) is estimated using the Manning equation. S_o denotes the longitudinal channel slope. Dynamic routing refers to solution of the complete St. Venant equations. Other hydraulic routing methods, such as diffusion and kinematic routing, are based on simplifications that involve neglecting certain terms in the momentum equation.

Unlike hydrologic storage routing, hydraulic routing models simultaneously solve for flow rates (or velocities) and depths (or water surface elevations) in the same computational algorithms. Dynamic routing is more accurate than hydrologic routing or simplified hydraulic routing methods because it reflects backwater effects of downstream flow constrictions or flat channel slopes, and acceleration effects of rapid changes in flow over time.

In two-dimensional river modeling, acceleration in the vertical direction is typically assumed to be negligible, with the equations being formulated for the two horizontal dimensions. Two-dimensional modeling may be applied to either steady or unsteady flows. Two-dimensional modeling is much more complex and requires much more input data to describe the channel geometry and flow resistance characteristics than does one-dimensional modeling.

The addition of erosion and sedimentation considerations makes a model significantly more complex. In mobile bed models, the bed geometry and roughness coefficients are interrelated with the flow hydraulics. Movable-boundary models incorporate sediment transport, bed roughness, bed armor, bed surface thickness, bed material sorting, bed porosity, and bed compaction equations as well as the sediment continuity equation, which defines the sediment exchange rate between the water column and bed surface.

6.8.3 Generalized Software Packages

Hydraulic or hydrodynamic computations are an integral component of some of the reservoir/river system analysis models noted in later chapters, particularly the water-quality models of Chapter 12. Most reservoir operation models do not incorporate hydraulics, but rather are used in combination with separate hydraulics models. Generalized river hydraulics software packages are also available that may be applied in combination with reservoir/river system analysis models. The remainder of this chapter describes several such models. The generalized models cited are well-documented public-domain software packages available from either the USACE Hydrologic Engineering Center in Davis, California, or the National Weather Service Hydrologic Research Laboratory in Silver Spring, Maryland. These models have been used extensively by many entities in reservoir system analysis studies as well as many other types of applications. Executable versions are available for MS-DOS-based microcomputers. The FORTRAN programs have been compiled and executed on other computer systems as well. HEC-2 and RAS simulate gradually varied flow based on a standard step method solution of the energy equation (equation 6-17). DWOPER, DAMBRK, FLDWAV, and UNET are dynamic routing models, based on the St. Venant equations (6-19 and 6-20) for analyzing unsteady flow in complex river systems. HEC-6 is a movable bed hydraulic model that simulates erosion and sediment transport. These are all one-dimensional models. FESWMS is a two-dimensional river hydraulics model, with options for either steady or unsteady flow analysis.

6.8.3.1 HEC-2   The HEC-2 Water Surface Profiles program is an accepted standard for developing water surface profiles. The model was originally developed in the 1960s and has evolved through numerous modifications and expansions. A user's manual [Hydrologic Engineering Center, 1991], training documents covering specific model capabilities, and papers and reports on specific applications are available from the Hydrologic Engineering Center. Guidelines for applying HEC-2 are also outlined in several books, including [Hoggan, 1989] and [Bedient and Huber, 1992]. Several universities, as well as the Hydrologic Engineering Center, regularly offer continuing education short courses on applying HEC-2.

Although HEC-2 is a standalone model which has been applied in many different modeling situations, it is often used in combination with HEC-1. The HEC-1 Flood Hydrograph Package is discussed in Chapter 8. A typical HEC-1/HEC-2 application involves predicting the water surface profiles which would result from actual or hypothetical flood events. Precipitation associated with an actual storm, a design storm of specified exceedance frequency, or a design storm such as the probable maximum storm is provided as input to the HEC-1 watershed model. HEC-1 performs the precipitation-runoff and routing computations required to develop hydrographs at pertinent locations in the stream system. Peak discharges from the HEC-1 hydrographs are provided as input to HEC-2, which computes the corresponding water surface elevations at specified locations. HEC-2 is also sometimes used to develop discharge versus storage volume relationships for stream reaches which are used in HEC-1 for the modified Puls routing option. Both HEC-1 and HEC-2 are combined with the HEC-5 reservoir system model, described in Chapter 8, to evaluate flood control operations.

HEC-2 is intended for computing water surface profiles for steady, gradually varied flow in natural or man-made channels. The computational procedure is based on the solution of the one-dimensional energy equation with energy losses due to friction estimated with the Manning equation. This computational procedure is generally known as the *standard step method.* The computations proceed by reach, with known values at one cross-section being used to compute the water surface elevation, mean velocity, and other flow characteristics at the next cross-section. Either subcritical and supercritical flow regimes can be modeled. The effects of obstructions to flow, such as bridges, culverts, weirs, and buildings located in the floodplain, may be reflected in the model.

HEC-2 provides a number of optional capabilities related to bridges and culverts, channel improvements, encroachments, tributary streams, split streams, ice-covered streams, effective flow areas, interpolated cross-sections, friction loss equations, critical depth, multiple profiles, developing storage-outflow data, and calibrating Manning roughness coefficients.

6.8.3.2 HEC-RAS

The USACE Hydrologic Engineering Center is in the process of developing a next generation (NexGen) of hydrologic engineering software. The River Analysis System (HEC-RAS) package, being developed in conjunction with the NexGen project, will provide capabilities for simulating one-dimensional steady or unsteady flow, and also sediment transport and movable boundary open channel flow. The HEC-RAS system will ultimately contain components for steady flow water surface profile computations, unsteady flow computations, and movable boundary hydraulic computations. All three components will use a common geometric data representation and common geometric and hydraulic computation routines. HEC-RAS will be an integrated system, designed for use in a multitasking, multi-user environment. The system will comprise a graphical user interface, separate computational engines, data storage/management components, graphics, and reporting capabilities. An initial version of the HEC-RAS, limited to HEC-2-type steady flow water surface profile computations, was released in 1995 [Hydrologic Engineering Center, 1995].

6.8.3.3 NWS Models   The Operational Dynamic Wave Model (DWOPER), Dam-Break Flood Forecasting Model (DAMBRK), and Flood Wave (FLDWAV) are dynamic routing models developed by the Hydrologic Research Laboratory of the National Weather Service. DAMBRK is a specific-purpose dam-breach model that grew out of the general-purpose DWOPER. The more recent FLDWAV combines the capabilities of DWOPER and DAMBRK. FLDWAV was developed with the intention of replacing DWOPER and DAMBRK, but all three models continue to be widely used. The one-dimensional unsteady flow models are based on an implicit finite difference solution of the complete St. Venant equations. Discharges, velocities, depths, and water surface elevations are computed as a function of time and distance along the river [Fread, 1987, 1988; Fread and Lewis, 1988].

DWOPER is used routinely by the National Weather Service River Forecast Centers and has also been widely applied outside of the National Weather Service. DWOPER has wide applicability to rivers of varying physical features, such as branching tributaries, irregular geometry, variable roughness parameters, lateral inflows, flow diversions, off-channel storage, local head losses such as bridge contractions and expansions, lock and dam operations, and wind effects. An automatic calibration feature is provided for determining values for roughness coefficients. Data management features facilitate using the model in a day-to-day forecasting environment. The model is equally applicable for simulating unsteady flows in planning and design studies.

DAMBRK has been extensively applied by various agencies and consulting firms in conducting dam safety studies. DAMBRK simulates the failure of a dam, computes the resultant outflow hydrograph, and simulates the movement of the flood wave through the downstream river valley. An inflow hydrograph is routed through a reservoir using either hydrologic storage routing or dynamic routing. Two types of breaching may be simulated. An overtopping failure is simulated as a rectangular, triangular, or trapezoidal opening that grows progressively downward from the dam crest with time. A piping failure is simulated as a rectangular orifice that grows with time and is centered at any specified elevation within the dam. The pool elevation at which breaching begins, time required for breach formation, and geometric parameters of the breach must be specified by the user. The DWOPER dynamic routing algorithm is used to route the outflow hydrograph through the downstream valley. DAMBRK can simulate flows through multiple dams located in series on the same stream.

DWOPER does not include the dam breach modeling capabilities of DAMBRK. DAMBRK is limited to a single river without tributaries, and thus does not provide the flexibility of DWOPER in simulating branching tributary configurations.

FLDWAV combines DWOPER and DAMBRK into a single model, and provides additional hydraulic simulation methods within a more user-friendly model structure [Fread and Lewis, 1988]. FLDWAV, like DWOPER and DAMBRK, is based on an expanded form of the St. Venant equations that includes the following hydraulic effects: lateral inflows and outflows, off-channel storage, expansion and contraction losses, mixed subcritical and supercritical flow, nonuniform velocity distribution across the flow section, flow path differences between the flood plain and a sinuous main channel, and surface wind shear. The model can simulate dam breaches in one or several dams located sequentially on the same stream. Other conditions that can be simulated include levee overtopping, interactions between channel and floodplain flow, and combined free-surface and pressure flow. FLDWAV also has a calibration option for determining Manning roughness coefficient values.

6.8.3.4 UNET   UNET simulates one-dimensional unsteady flow through a full network of open channels [Barkau, 1995]. In addition to solving the network system, UNET provides the user the ability to apply several external and internal boundary conditions, including flow and stage hydrographs, rating curves, gated and uncontrolled spillways, pump stations, bridges, culverts, and levees. The dynamic routing model is based on a four-point linear implicit finite-difference solution of the St. Venant equations. UNET is conceptually similar to the National Weather Service DWOPER and FLDWAV models described above. UNET interconnects with the Hydrologic Engineering Center Data Storage System (HEC-DSS), described in Chapter 5. The UNET system also allows use of HEC-2-style cross-section input data files.

6.8.3.5 FESWMS-2DH   The Finite Element Surface-Water Modeling System: Two-Dimensional Flow in a Horizontal Plane (FESWMS-2DH) was developed by the U.S. Geological Survey, Water Resources Division for the Federal Highway Administration [Froehlich, 1989]. The primary motivation for developing FESWMS-2DH was to improve capabilities for analyzing flow at highway bridge crossings where complicated hydraulic conditions exist. However, the generalized model is applicable to a broad range of different steady and unsteady two-dimensional river-modeling situations.

FESWMS-2DH is a modular set of computer programs, which includes DINMOD, the data input module; FLOMOD, the depth-averaged flow solution module; and ANOMOD, the analysis of output module. DINMOD prepares a finite-element network for use by other FESWMS-2DH programs. ANOMOD generates plots and reports of computed values that simplify interpretation of simulation results. FLOMOD solves the vertically integrated conservation of momentum equations and the conservation of mass (continuity) equation, using the Galerkin finite-element method, to obtain depth-averaged velocities and water depth at points in a finite-element network. Energy losses are computed using the Chezy or Manning equations and Boussinesq eddy viscosity concept. The effects of wind stress and the Coriolis force may be included in the simulation. The model is capable of simulating flow through single or multiple bridge openings as normal flow, pressure flow, weir flow, or culvert flow.

6.8.3.6 HEC-6   The HEC-6 Scour and Deposition in Rivers and Reservoirs program [Hydrologic Engineering Center, 1993] is designed to simulate long-term trends of scour and/or deposition in a stream channel resulting from modifying flow frequency and duration or channel geometry. HEC-6 can be used to predict reservoir sedimentation, design channel contractions required to maintain navigation depths or decrease the volume of maintenance dredging, analyze the impacts of dredging on deposition rates, evaluate sedimentation in fixed channels, and estimate maximum scour during floods. HEC-6 features include capabilities for simulating stream networks, channel dredging, and levee and encroachment alternatives.

HEC-6 is a one-dimensional movable-boundary numerical model designed to simulate and predict changes in river profiles resulting from scour and/or deposition over moderate time periods. Although single flood events can be analyzed, the model is oriented toward evaluating trends over a number of years. A continuous flow record is partitioned into a series of steady flows of variable duration. Water surface profile computations are performed for each flow. Potential sediment transport rates are then computed at each section. These rates, combined with the duration of the flow, permit a volumetric accounting of sediment within each reach. The amount of scour or deposition at each section is then computed and the cross-section adjusted accordingly. The computations then proceed to the next flow in the sequence and the cycle is repeated, beginning with the updated channel geometry. The sediment calculations are performed by grain size fraction, thereby allowing the simulation of hydraulic sorting and armoring.

The HEC-6 water surface profile computations are based on a standard step method solution of the one-dimensional steady-state energy equation, as in HEC-2. HEC-6 simulates the capability of a stream to transport sediment, given the yield from upstream sources. Sediment transport includes both bed and suspended load. The computational formulation utilizes Einstein's basic concepts of sediment transport. A number of user-selected alternative transport functions for bed material load are included in the model.

7

RELIABILITY AND RISK ANALYSES

Reliability is a measure of the level of dependability at which water supply, hydro-electric power generation, environmental, flood protection, and other needs can be met. Conversely, risk is a measure of the likelihood of failures in providing these services. System capabilities for satisfying water management requirements must be evaluated from a reliability or risk perspective because streamflow and other variables are characterized by randomness, uncertainty, and great variability. Methods for evaluating yield and reliability for water supply and other conservation purposes are covered in the first part of the chapter; flood risk is addressed later in the chapter.

Definitions of the terms *yield* and *reliability* can be formulated in a variety of ways to provide meaningful information for the particular application of concern. In general, yield is a measure of the amount of water that can be supplied by an unregulated stream, a reservoir, or a river system regulated by multiple reservoirs. Reliability is a measure of the level of dependability at which various yield levels can be supplied. Reliability can be expressed in terms of firm yield, percentage of time target demands can be met, likelihood or probability of meeting target demands during a specified time interval, risk of failure in meeting specified demands, or the likelihood or percentage of time that reservoir storage falls below certain levels. The water use requirements may involve water supply diversions, instream flow needs, generation of hydroelectric power, or maintenance of reservoir storage.

Flood frequency analysis methods provide estimates of the probability of specified flow and volume magnitudes being exceeded during a specified time interval. Expected annual damage analysis methods combine hydrologic, hydraulic, and economic considerations in the evaluation of flooding problems and flood damage reduction measures.

7.1 METHODS FOR ANALYZING UNREGULATED STREAMS

One of the simplest and most informative means of quantifying the yield versus reliability relationship for an unregulated stream is the conventional flow-duration curve, which shows the percentage of time that flows fall within various ranges. Flow-duration relationships are developed by counting the number of periods (typically days or months) for which the mean flow rate equalled or exceeded specified levels during the period of record at a stream gage. Alternatively, the flow-duration relationship may be expressed in terms of flows being less than specified levels. The duration or frequency associated with a given discharge is computed by dividing the number of periods it is equalled or exceeded by the total number of periods in the record. The primary limitation of the flow-duration curve as a method for quantifying yield is that the sequencing of flows is not reflected. The relationship does not indicate whether the lowest flows occurred in consecutive periods or were scattered throughout the record.

A flow-duration curve for the Brazos River at a streamflow gage near the city of Richmond, Texas, is presented in Figure 7–1. This location is about 100 km upstream of the river's mouth and has a drainage area of 117,000 km^2. The flow-duration curve is based on mean monthly gaged flows from October 1922 through December 1984, adjusted to represent unregulated conditions. The flow-duration curve indicates that 50% of the time, the flow is greater than 125 m^3/s. For 30% of the time, the flow exceeds 280 m^3/s.

A flow-duration relationship provides a measure of the amount of water supplied at various levels of reliability for various water uses. Example 7–1 illustrates the application of a flow-duration curve in analyzing the amount of energy which can be produced by a run-of-river power plant.

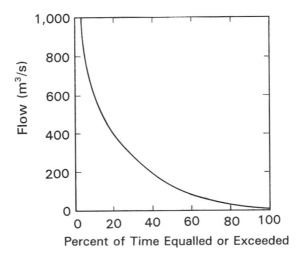

Figure 7–1   Flow-duration curve for the Brazos River near Richmond, Texas.

Example 7–1—Firm and Secondary Power

The flow-duration curve of Figure 7–2 was developed for the site of a proposed run-of-river hydroelectric power project. The plant will have a constant head of 15 m, installed capacity of 100,000 kW, and efficiency of 0.78. The flow-duration curve of Figure 7–2 is used to determine the reliability at which energy can be generated.

The relationship between power and discharge is given by equation 6-8 as follows

$$P = \gamma QHe = (9.81 \text{ kN/m}^3) \, Q \,(15\text{m}) \,(0.78) = (114.8 \text{ kN/m}^2)Q$$

or

$$Q = (0.0087125 \text{ m}^2/\text{kN})P$$

The maximum flow rate that can be used in a plant with an installed capacity of 100,000 kW is

$$Q = (0.0087125 \text{ m}^2/\text{kN}) \,(100,000 \text{ kN} \cdot \text{m/s}) = 871 \text{ m}^3/\text{s}$$

From Figure 7–2, a flow of 871 m^3/s is equalled or exceeded about 23% of the time. Thus, 100,000 kW of power can be generated about 23% of the time.

Firm power is the maximum power that can be provided 100% of the time for the given streamflow record. From Figure 7–2, a flow of 200 m^3/s is available 100% of the time. Thus, firm power is

$$P = (114.8 \text{ kN/m}^2) \,(200\text{m}^3/\text{s}) = 22,960 \text{ kN} \cdot \text{m/s} = 22,960 \text{ kW}$$

Since power is the time rate of producing energy, the firm energy is

$$\text{firm energy} = (22,960 \text{ kW}) \,(8,760 \text{ hours/year}) = 2.01 \times 10^8 \text{ kW} \cdot \text{hrs per year}$$

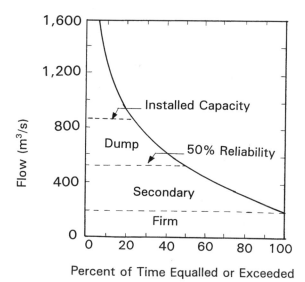

Figure 7–2   Flow-duration curve for Example 7–1.

From Figure 7–2, a flow of 520 m^3/s or greater is available 50% of the time. The area under the flow-duration curve between the 100% reliability flow of 200 m^3/s and the 50% reliability flow of 520 m^3/s is approximately

$$(520 - 200 \text{ m}^3/\text{s})(0.72) = 230 \text{ m}^3/\text{s}$$

The secondary energy that can be provided at least 50% of the time, in addition to the firm energy, is

$$\text{energy} = (114.8 \text{ kN/m}^2)(230 \text{ m}^3/\text{s})(8,760 \text{ hrs/yr}) = 2.31 \times 10^8 \text{ kW} \cdot \text{hrs per year}$$

Low-flow frequency relationships provide another means of analyzing streamflows. This type of analysis is used to quantify the risk of drought conditions. Low-flow frequency curves are developed by determining the minimum flow during periods of various lengths that occurred during each year of the gage record. For example, for each year, the minimum one-day, seven-day, 30-day, and/or 90-day flows are tabulated from the record. Flow versus frequency curves are plotted for each length of time. The percentage of time that the minimum annual 30-day flow is less than a specified magnitude, for example, is determined by counting the number of years in which this was the case.

These and other methods for analyzing the yield provided by unregulated streams are outlined by McMahon and Mein [1986].

7.2 SIMULATION OF RESERVOIR/RIVER SYSTEMS BASED ON HISTORICAL STREAMFLOW SEQUENCES

Reservoir yield-reliability studies are usually based on sequential period-by-period hydrologic period-of-record simulations. The reservoir/river system simulation model combines a specified set of water use requirements with historical streamflows in sequential chronological order. Various reliability indices can be formulated to measure the capabilities of the simulated reservoir/river system to satisfy the specified water use requirements during a postulated repetition of historical hydrology.

As discussed in Chapter 5, sequences of streamflows representing a specified condition of river basin development are typically developed by adjusting gaged flows to remove nonhomogeneities. The adjusted historical flows may represent unregulated or natural conditions or, alternatively, some other condition of basin development. Chapter 6, including Example 6–1, presents the concept of a simulation model consisting of sequential period-by-period water balance accounting as specified water use requirements are combined with historical hydrology. Several such reservoir system analysis models are cited in Chapters 8 and 11. The case study of Chapter 9 includes reliability studies for a multiple-reservoir system. As noted in Chapter 4, various reliability indices can be formulated to concisely summarize the results of a simulation model.

7.2.1 Volume and Period Reliability

Variations of the concepts of volume reliability (R_v), period reliability (R_p), and risk of failure (F_p) are often adopted in modeling studies. These indices are computed from the results of a historical hydrologic period-of-record simulation as

$$R_v = (v/V)\ 100\% \tag{7-1}$$

$$R_p = (n/N)\ 100\% \tag{7-2}$$

$$F_p = (f/N)\ 100\% \tag{7-3}$$

$$F_p = 100\% - R_p \tag{7-4}$$

Volume reliability (R_v) is the ratio of the volume of water supplied (v) to the volume demanded (V) or, equivalently, the ratio of the mean actual diversion rate during the simulation to the mean target diversion rate. Unlike period reliability, volume reliability reflects the shortage magnitude as well as frequency. The shortage volume is the demand target (V) less the volume supplied (v) within the constraints of water availability.

Period reliability (R_p) is the ratio of the number of time periods (n) for which demands could be met to the total number of time periods (N) in the simulation. Period risk of failure (F_p) is the ratio of the number of time periods (f) for which demands could not be met to the total number of time periods (N) in the simulation. The risk of failure (F_p) is the complement of the period reliability (R_p). The period reliability (R_p) represents the percentage of time specified water use requirements are met, or the probability of the requirements being met in any randomly selected period. The risk of failure (F_p) is the percentage of time or the probability that water use requirements will not be fully met.

Reliability estimates are indices reflecting modeling premises and imperfect limited data. A reliability of 100% certainly does not mean that the corresponding yield can be provided with certainty. The reliability estimate is based on historical hydrology, as well as other modeling assumptions. A drought more severe than the drought of record will occur at some unknown future time.

Performance measures of reservoir/river system capabilities to meet demands are devised to fit the particular modeling and analysis application. Various formulations of water use requirements and failures to meet these requirements can be incorporated in simulation studies. Water use requirements may involve water supply diversions, instream flow needs, hydroelectric energy, or maintenance of reservoir storage levels. Period reliability can be formulated in terms of meeting all or at least a specified portion of the demand target. Alternatively, the water use requirements can be defined in terms of meeting demands without reservoir storage falling below specified levels. Reliability can be defined as the percentage of periods during the simulation in which a specified storage level is equalled or exceeded. Water supply failure may be related to necessitating the implementation of emergency demand management measures.

Various definitions of reliability can be formulated for alternative time periods. Monthly periods are typical in simulation modeling studies, but other time intervals are common as well. Period reliability may be defined in terms of meeting an annual diversion target on a yearly basis during a simulation performed using a monthly or weekly computational time interval.

7.2.2 Firm Yield

Firm yield (also called *safe* or *dependable* yield) is a commonly used measure of water supply and hydroelectric power generation capabilities. Firm yield is the estimated maximum release or diversion rate, or hydroelectric energy production rate, that can be maintained continuously during a hypothetical repetition of historical period-of-record hydrology. Firm yield is the draft that will lower the storage in a reservoir or multiple reservoir system to a defined failure level during a hydrologic period-of-record simulation. Period and volume reliabilities, as defined in equations 7-1 and 7-2, are 100% for the firm yield and smaller yields. Yields greater than firm yield have reliabilities of less than 100%. Firm yield estimates reflect all of the premises and assumptions incorporated into the model, including hydrologic data, reservoir characteristics and operating rules, water use scenarios, and impacts of other water use and management activities in the basin. Firm yield is typically expressed in terms of a mean annual rate with monthly or other time interval distribution factors being incorporated in the model to reflect the within-year seasonal variation in water use.

In citing firm yields, it is always important to emphasize that there is no guarantee of being able to supply this amount of water in the future. Firm yield estimates reflect all the premises incorporated in the model, such as the hypothetically assumed repetition of past streamflows. A future drought may be more severe than the worst drought during the period of streamflow gaging records.

Firm yields are commonly determined by repeatedly executing a simulation model with alternative yields specified in an iterative search for the maximum yield that can be maintained, without failure, continuously during the simulation period. For example, assume that for a particular study the yield is defined as a constant diversion from a single reservoir, and a failure occurs if the diversion cannot be fully met due to drawdowns resulting in an empty reservoir. If the yield (diversion) specified in a run of the simulation model is met continuously without reservoir storage being depleted, the diversion is increased for the next run. Eventually a diversion rate is found that just empties the reservoir, resulting in an imminent diversion shortage; this is the firm yield. Several of the generalized simulation models noted in Chapter 8 contain options which automate the iterative search for the firm yield. Variations of the same iterative simulation approach can be applied to complex multiple-reservoir systems with water use requirements and corresponding failures formulated to fit the scope of the particular study.

7.2.3 Firm Yield versus Storage Capacity Relationships

The relationship between firm yield and storage capacity for a reservoir site is fundamental information routinely developed during feasibility studies. As just discussed, simulation models are executed iteratively to determine the firm yield provided by a particular storage capacity or, vice versa, the storage capacity required to provide a specified firm yield. Complex multiple-reservoir, multiple-purpose systems may be evaluated. Optimization models provide another computational approach for determining firm yields for specified storage capacities and vice versa. The use of linear programming in firm yield analysis is discussed in Chapter 10. The simple single-reservoir water supply analysis methods outlined in the following paragraphs illustrate the general concept of a relationship between storage capacity and firm yield.

Rippl presented his graphical mass curve technique for analyzing storage requirements over a century ago [Rippl, 1883]. The Rippl diagram presented in Figure 7–4 is developed in conjunction with Example 7–2. A mass curve is a plot of cumulative inflow to a reservoir. The inflow rate at a given point in time is reflected by the slope of the mass curve. A plot of cumulative demand, for a constant demand rate, is a straight line with a slope equal to the demand rate. If the slope of the cumulative demand curve is greater than the slope of the mass inflow curve at a point in time, reservoir storage is being drawn down. If the mass curve is steeper than the demand curve, the reservoir is filling or, if already full, spilling. Demand curves are drawn tangent to the high points of the mass curve, assuming the reservoir is full at these points. The vertical distance between the demand and mass curves represents storage drawdowns. The required storage capacity is determined as the maximum departure between the cumulative inflow and demand curves.

The following procedure is conceptually similar to the Rippl diagram approach but more convenient computationally. The approach is based on calculating the cumulative total reservoir drawdown at the end of each time interval for a given sequence of reservoir inflows and specified water use demands. The drawdown is the volume of storage that has been depleted or emptied. The required storage capacity is the maximum drawdown to occur during the simulation.

At each time step, the cumulative storage drawdown (SD_t) is determined as follows:

$$SD_t = SD_{t-1} + R_t - I_t \qquad \text{if positive}$$
$$SD_t = 0 \qquad\qquad\qquad\quad \text{otherwise} \tag{7-5}$$

where SD_t and SD_{t-1} are the volume of the reservoir storage drawdown at the end of the current (t) and previous ($t-1$) time intervals, and R_t and I_t are the release and inflow volumes during the interval. A negative value of SD_t indicates spills from a full reservoir, which is handled computationally by setting the negative SD_t to zero.

TABLE 7–1   Data and Computations for Example 7–2

1	2	3	4	5	6	7
		Mean Inflow	Monthly Inflow	Cumulative Inflow	50 m³/s Draft	Draw-down
Month	Days	(m³/s)	(10^6m³)	(10^6m³)	(10^6m³)	(10^6m³)
1	31	46	123	123	134	11
2	28	71	172	295	121	0
3	31	61	163	458	134	0
4	30	129	334	793	130	0
5	31	157	421	1213	134	0
6	30	50	130	1343	130	0
7	31	14	37	1380	134	96
8	31	7	19	1399	134	212
9	30	42	109	1508	130	232
10	31	33	88	1596	134	278
11	30	54	140	1736	130	267
12	31	50	134	1870	134	267
13	31	56	150	2020	134	251
14	28	69	167	2187	121	205
15	31	86	230	2417	134	109
16	30	111	288	2705	130	0
17	31	135	362	3067	134	0
18	30	26	67	3134	130	62
19	31	12	32	3166	134	164
20	31	10	27	3193	134	271
21	30	38	98	3292	130	302
22	31	103	276	3567	134	160
23	30	86	223	3790	130	67
24	31	78	209	3999	134	0
25	31	46	123	4122	134	11
26	28	71	172	4294	121	0
27	31	61	163	4458	134	0
28	30	129	334	4792	130	0
29	31	157	421	5213	134	0
30	30	50	130	5342	130	0
31	31	14	37	5380	134	96
32	31	7	19	5398	134	212
33	30	42	109	5507	130	232
34	31	33	88	5596	134	278
35	30	54	140	5736	130	267
36	31	50	134	5869	134	267

Example 7–2—Firm Yield versus Storage Capacity Analysis

A firm yield versus storage capacity analysis is performed for the sequence of streamflows of
Example 6–1, which are reproduced in column 3 of Table 7–1. The mean monthly reservoir
inflow rates, in m³/s, in column 3 are converted to monthly volumes, in million m³, which
are tabulated in column 4, using the number of days in each month shown in column 2 and
the conversion factor of 86,400 seconds per day. The monthly inflow volumes are converted

to accumulative end-of-month volumes in column 5. This mass curve is plotted in the Rippl diagram of Figure 7–4.

Columns 4, 6, and 7 of Table 7–1 are the terms I_t, R_t, and SD_t, respectively, of equation 7-5. A given constant yield of 50 m³/s is tabulated as monthly volumes in column 6, which reflect the number of days in each month and 86,400 seconds per day. Column 7 is computed using equation 7-5. The required storage capacity is the greatest value in column 7.

A several-decade period of recorded monthly flows is typical for yield studies. However, this simple example is based on only 24 months of gaged streamflows. An infinitely long sequence of inflows is synthesized by assuming that the two-year sequence of flows is repeated every two years. The inflows for months 25 through 36 in column 3 of Table 7–1 is simply a repetition of months 1 through 12. There is no need to repeat the computations any further because a repetitive cycle of drawdowns and refilling is established by the third year and does not change thereafter. The reservoir is assumed to be full at the beginning of the first month. As indicated in column 7 of Table 7–1, the reservoir is full at the end of months 6 and 29, which defines a cycle that repeats beginning with the 30th month.

Table 7–1 indicates that a storage capacity of 302 million m³ is required at this site to provide a firm yield of 50 m³/s. This is the minimum storage that will allow a withdrawal or release of 50 m³/s to be maintained continuously assuming that the two-year sequence of monthly inflows are repeated forever. Note from column 7 that the reservoir is lowered by 11 million m³ during the first month; spills from a full reservoir occur during the second through sixth months; 278 million m³ of the storage capacity is empty by the end of the tenth month; and the reservoir is completely refilled by the 16th month. The maximum drawdown of 302 million m³ occurs at the end of the 21st month, which is September of the second year. This maximum storage depletion represents the capacity required for 50 m³/s firm yield.

Firm Yield versus Storage Capacity Relationship

The firm yield versus stor\ capacity relationship shown in Table 7–2 and Figure 7–3 was developed by repeating the computations of Table 7–1, with a given yield specified in column 6 and the corresponding storage capacity computed as the greatest value in column 7. The mean inflow during the two-year record of 63.5 m³/s represents an upper limit on firm yield. A storage capacity of 590 million m³ is required to provide this maximum possible firm yield

**TABLE 7–2   Example 7–2 Firm
Yield Versus Storage Capacity**

Firm Yield (m³/s)	Storage (10⁶m³)
7	0
20	51
30	112
40	197
50	302
60	463
62	530
63	572
63.5	590

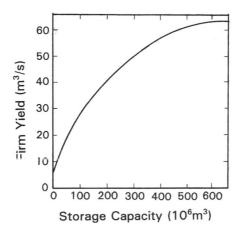

Figure 7–3   Firm yield versus storage capacity relationship for Example 7–2.

of 63.5 m³/s. Increasing the storage capacity above 590 million m³ does not further increase the computed firm yield. The smallest inflow in column 3 of Table 7–1 is 7 m³/s. Thus, a firm yield of 7 m³/s can be provided with zero reservoir storage capacity.

Rippl Diagram

The Rippl diagram provides an alternative approach for performing the analysis. Alternative firm yields of 50 m³/s and 63.5 m³/s are plotted on the Rippl diagram of Figure 7–4, along with the mass curve of cumulative inflows from column 5 of Table 7–1. The firm yields are represented by the slopes of the cumulative demand curves. Required storage is the vertical distance between the demand and inflow curves. The storage capacity is the maximum value of this vertical distance. The reservoir is full at the points in time at which the demand and inflow lines intersect. To the left of these points, inflow rates exceed yield (demand or outflow) rates, as represented by the slopes, and thus spills occur. To the right of the intersection

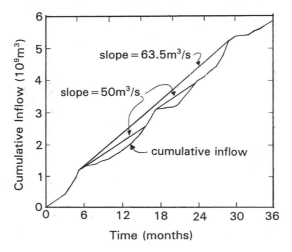

Figure 7–4   Rippl diagram for Example 7–2.

points, yield rates exceed inflows; thus, demands are partially met by withdrawals from storage. For a firm yield (slope of demand line) of 63.5 m^3/s, the maximum vertical difference between the cumulative inflows and demands in Figure 7–4 is 590 million m^3. For a demand line slope of 50 m^3/s, the maximum vertical difference between the cumulative inflow and demand lines is 302 million m^3.

Evaporation

Evaporation and other losses are neglected in the computations of this example. However, net evaporation estimates could be incorporated in the analysis as follows. The computations neglecting net evaporation provide an approximate storage capacity which can be combined with the storage drawdowns for each month (column 7 of Table 7–1) to determine end-of-period storages. Monthly net evaporation volumes are estimated by combining net rates with water surface areas determined as a function of storage. The computations of Table 7–1 are then repeated with the estimates of net evaporation included along with the inflows and water demands.

7.2.4 Combinations of Firm and Secondary Yield

Many different water users may be supplied by the same reservoir or multiple-reservoir system. Water allocation schemes may involve supplying water at different levels of reliability to the various users. For example, a city may be guaranteed a high level of reliability, while agricultural users are supplied from the same reservoir system at a lower reliability level. As discussed in Chapter 3, one or more buffer pools may be designated as a triggering mechanism for curtailing the lower-priority diversions. Full demands are met as long as the reservoir water surface is above the designated buffer level, with certain demands being curtailed whenever the water in storage falls below this level. Likewise, this type of triggering mechanism can be adopted for coordinating demand management measures with water supply operations. Emergency demand management plans are implemented whenever reservoir storage falls below specified levels.

Buffer zone operations allow allocation of withdrawals between firm and secondary yields. Variations of the concept of combining firm and secondary yields are pertinent to a variety of water management applications. The general concept is illustrated by the approach outlined in the following paragraphs.

Reliability is determined by applying equations 7-1 and/or 7-2 to the results of a hydrologic period-of-record simulation. Firm yield is the maximum yield that can be supplied with a volume and period reliability of 100%. Additional secondary yield can be supplied with a reliability of less than 100%. Reservoir operating rules include setting a top-of-buffer-pool level so as to provide a specified firm yield. The secondary yield represents diversions that are curtailed any time storage is below the top of buffer pool.

Wurbs and Carriere [1988] document an analysis of Waco Reservoir on the Bosque River in Texas summarized in Table 7–3. The monthly time-step simulation model used in the analysis is described in Chapter 9. Period reliabilities (equation 7-2) are cited in Table 7–3, but volume reliabilities (equation 7-1) can be computed as well. A reallocation of a portion of the flood control pool to conservation storage

TABLE 7–3   Secondary Yield versus Reliability for Waco Reservoir

Secondary Yield (m³/s)	Percentage of Storage Capacity Below Top of Buffer				
	0%	42%	58%	64%	72%
	Period Reliability (%) for Secondary Yield				
0.37	100.0%	91.8	87.9	82.%	77.8
0.62		90.2	86.6	81.5	76.1
0.93			84.2	79.1	73.1
1.50				78.0	65.6
2.43					59.5

is being considered. The proposed raise in the top of conservation pool increases the conservation storage capacity from 165 million m³ to 230 million m³. This will increase the firm yield by 0.37 m³/s, from 3.28 to 3.65 m³/s. Alternatively, the original 3.28 m³/s firm yield can be combined with a secondary yield greater than 0.37 m³/s but with a reliability of less than 100%.

Secondary yield versus reliability relationships for alternative buffer pool designations are tabulated in Table 7–3. The secondary yields shown in Table 7–3 are in addition to a firm yield of 3.28 m³/s. A release of 3.28 m³/s is maintained continuously during a hydrologic period-of-record simulation. An additional release, represented by the secondary yields of Table 7–3, is maintained as long as the current storage is above the designated buffer pool. The total conservation storage is fixed at 230 million m³. The alternative buffer pools evaluated in Table 7–3 are cited in terms of the percentage of the conservation storage capacity which is below the top of buffer pool. For each secondary yield, there is a buffer pool designation which maximizes the secondary yield reliability. The greatest secondary yield that can be supplied with each buffer pool designation, while still maintaining the 3.28 m³/s firm yield, is shown in Table 7–3. The same secondary yield can be provided by setting the buffer higher, but the reliability is lower.

With the top of buffer pool set at 42% of the total 230 million m³ conservation storage capacity, a secondary yield of 0.62 m³/s is supplied 90.2% of the time. A secondary yield of 0.93 m³/s can be provided with a reliability of 84.2% by allocating 58% of the conservation storage to the buffer pool. A secondary yield of 1.50 m³/s can be provided 78.0% of the time by setting the buffer pool at 64% of the storage capacity. In all cases, the firm (100% reliability) yield of 3.28 m³/s is maintained in addition to the secondary yield.

7.2.5 Other Measures of Reliability

Period and volume reliability, as defined by equations 7-1 and 7-2, are concise indices of the likelihood or percentage of time that water use requirements can be met without failure. Water use requirements and failures to meet these requirements are complex. Simple indices reflect only limited features of water demands and failures to meet water demands. Other indices can also be devised to measure various aspects of capabilities, or lack thereof, for supplying water needs.

The HEC-3 Reservoir System Simulation [Hydrologic Engineering Center, 1981] model includes the following summary measure of failures to meet water supply diversion requirements during a simulation:

$$\text{shortage index} = (\frac{100}{\text{years}}) \sum (\frac{\text{annual shortage}}{\text{annual requirement}})^2 \qquad (7\text{-}6)$$

The reservoir system is simulated using a monthly time step. For each year of the simulation, the diversion shortage is expressed as the ratio of the total volume of shortages during the year to the total volume of target demands. This annual ratio squared is summed over all the years of the simulation and divided by the number of years. The HEC-3 model computes volume and period reliabilities, as defined by equations 7-1 and 7-2 as well as the annual shortage index of equation 7-6, from the results of the monthly time step simulation. Whereas volume reliability measures the total volume of shortages without regard to timing, the shortage index reflects timing as well as magnitude. Squaring of the shortage ratio in equation 7-6 is an expression of the concept that an extremely severe water supply shortage concentrated in one year is much more damaging than the same shortage volume divided over several years.

Performance measures may be formulated that reflect the concepts of vulnerability and resiliency. The precise definition of these criteria and the computational mechanism for their application can vary to fit the scope of the particular study. In general, vulnerability is a measure of the severity of the worst failure. It might be defined as the greatest water supply shortage to occur during any year of a simulation, or perhaps the greatest deficit in meeting a hydroelectric energy production target. Resiliency is a measure of the capability for recovery from failure. Resiliency could be defined as the maximum number of consecutive periods of failure to meet water use demands during a simulation. Alternatively, resiliency may be viewed in terms of the probability of being in a period of no failure given that there was a failure in the previous period [Hashimoto et al., 1982; Moy et al., 1986].

Reliability analyses can be expanded to include economic consequences. For example, economic benefits may be assigned to providing a specified water supply or hydroelectric energy requirement at some high level of reliability along with associating costs or losses with occasional failures to meet the demand requirements.

7.3 RELIABILITY ANALYSES BASED ON SYNTHETICALLY GENERATED STREAMFLOW SEQUENCES

Synthetic streamflow generation is briefly discussed in Chapter 5. Stochastic hydrology techniques [Loucks et al., 1981; Bras and Rodriquez-Iturbe, 1985; Salas, 1993] are available to synthesize streamflow sequences based on preserving selected statistical parameters of the historical data, in the sense that the parameter values for an infinitely long sequence of generated flows would be the same as for the original data. For example, as noted in Chapter 5, the monthly Markov model synthesizes flows characterized by specified values of the mean, standard deviation, and lag-1 autocorrelation coefficient for each of the 12 months of the year. Streamflow is treated as a stochastic process with a random chance component as well some degree of correlation between successive flows.

The historical record represents one out of an infinite number of possible sequences of streamflows that may occur in the future. Synthetic streamflow generation allows any number of equally likely streamflow sequences to be developed. Thus, a broader base is provided for reliability analyses. These methods do not provide additional information regarding the statistical characteristics of streamflows that is not already contained in the historical data. Rather, these are tools for considering the fact that, even if future flows retain the same basic characteristics of past flows, infinite possibilities exist for sequencing of flows in the future.

Depending on the needs of the study and the ingenuity and preferences of the analyst, synthetic streamflow generation can be used in various ways in reservoir/river system reliability studies. For example, for a particular study, reliability could be defined as the likelihood of meeting specified water use requirements, without a failure, during the next 25 years. Two hundred (or some other large number of) 25-year sequences of monthly flows are generated that preserve certain statistical characteristics of the historical streamflow at this site. The reservoir is simulated 200 times using equation 7-5, with each simulation combining the specified water use requirements with one of the 200 equally likely inflow sequences. Each of the 200 simulations results in a computed reservoir storage capac ity required to meet the water demands for that sequence of inflows. The reliability (*R*) associated with a storage capacity of 300 million m^3 is estimated as

$$R_p = (n/N)\ 100\% \tag{7-2}$$

where *N* is 200 and *n* is the number of the 200 simulations that resulted in a required storage capacity of 300 million m^3 or less. In this example, R_p is an estimate of the probability of supplying the specified water demands, without shortage, during a 25-year period.

7.4 STOCHASTIC STORAGE THEORY MODELS

Stochastic storage theory and related models have been addressed extensively in the research literature but applied very little by the agencies that actually construct and operate reservoir systems. This large group of analysis methods is based primarily on the theory presented by Moran [1959] and expanded by Gould [1961]. Klemes [1981] and McMahon and Mein [1986] provide in-depth overviews and cite many significant references.

In terms of practical usefulness, the most important storage probability theory models are described as probability matrix methods [McMahon and Mein, 1986]. Zsuffa and Galai [1987] address probability matrix methods from a practical applications perspective and provide computer programs for implementing the methods. Other methods are of theoretical interest. The mathematics of stochastic storage analysis is complex, necessitating significant assumptions and simplifications. Many of the more sophisticated techniques are severely limited from a practical applications perspective. Klemes [1982] has observed: "This theory has evolved into a highly esoteric branch of pure mathematics which, apart from some elements of the jargon, has very little relevance to the original physical problem. It often solves the wrong problems simply because they are mathematically tractable . . . and that, from the physical point of view, are trivial or irrelevant."

The objective of stochastic storage theory models is to determine the probability distribution of reservoir storage. For a specified water supply release policy and present storage level, the probabilities of the reservoir being at various storage levels at future times are computed. Storage probabilities may be computed at steady state or as a time-dependent function of the starting conditions. Thus, for a given release policy and initial storage content, the probabilities of the reservoir being at various storage levels at future times during the next several months or several years may be estimated. As the analysis period becomes longer, a steady-state condition is reached in which the storage probabilities at a future time are no longer dependent on the starting storage contents.

A nonsteady-state analysis can be useful in developing and implementing reservoir operating plans in which allocations of water to alternative users are made at the beginning of each water year, each irrigation season, or other time period of interest, based on the likelihood of water being available to meet the allocations during the time period. The likelihood of meeting the allocations would be based on the reservoir storage levels existing at the time the allocations are made. Under this type of operating plan, during drought conditions, as significant reservoir drawdowns occur, the allotment of water to the various users for the upcoming irrigation season or other specified time period is reduced accordingly. Storage probability theory models provide useful information regarding the probabilities of the reservoir being emptied by the end of the time period given the known present storage level and assuming different alternative withdrawal rates.

Steady-state probabilities are not dependent on initial storage levels. In this case, storage probability theory models represent an alternative to regular simulation models, using period-of-record or synthetically generated streamflow sequences, for developing yield versus reliability relationships.

The stochastic storage theory models assess system performance based on describing inflows by a probability distribution or stochastic process. The methods are typically applied to a single reservoir, but multiple-reservoir analysis procedures have also been developed. Modeling is performed in two stages. First, a probability distribution function (if the inflows are assumed independent) or stochastic process, such as a Markov chain, is fitted to the historical streamflow record. Then simulation or probability techniques are used to develop the storage versus yield function and corresponding reliability estimators. Discrete probabilities are typically used to approximate the continuous distributions of the inflow process. The assumption of first-order Markovian processes for representing the inflow process of a reservoir has generally been considered in the literature as adequate for most purposes. The development of models incorporating other approaches results in extremely complex transition probability matrices.

Moran [1959] presents various procedures for determining storage probabilities. Numerous other authors have presented solutions or extensions to the basic models formulated by Moran. McMahon and Mein [1986] outline the basic computational procedures and cite many of the key references. A group of Moran procedures are based on considering either time or both time and volume as continuous variables; solutions are complex. Another group of procedures treat time and volume as discrete variables, and application is more practical. A reservoir is subdivided into a number of zones, and a system of equations is developed that approximate the possible states of the reservoir storage. Two main assumptions can be made regarding the inflows and outflows, which occur at discrete time intervals. In a mutually exclusive model, there is a wet period, with all inflows and no outflows, followed by a dry season, with all releases but no inflows. In the more general simultaneous model, inflows and outflows can occur simultaneously. The simultaneous approach is the most practical of the Moran models, but has a number of limitations. Inflows are assumed to be independent, which is not valid for a monthly time period. A constant release rate is typically assumed. A varying release rate can be accommodated if it is storage-, not time-dependent. Thus, seasonality of inflows and releases is not considered. Estimates of the probability of the state of the reservoir can be computed either at steady state or as a time-dependent function of starting conditions.

Gould [1961] modified the simultaneous Moran-type model to account for both seasonality and auto-correlation of monthly inflows by using a transition matrix with a yearly time period, but accounting for within-year flows by using a monthly behavior analysis. Thus, monthly auto-correlation and seasonal release variations can be included. The Gould method, like other probability matrix methods, computes the probability of reservoir storage levels for a given storage capacity and release rate. Storage probabilities can be computed either at steady state or as a time-dependent function of the starting conditions.

Much of the work published in the literature represents modifications or extensions to the basic Moran and Gould models cited above. Zsuffa and Galai [1987] provide in-depth coverage of the probability matrix methods that are probably most pertinent for practical applications.

7.5 FLOOD FREQUENCY ANALYSIS

Estimating probabilities associated with flood flows is routinely performed in many types of planning and design situations, including, but not limited, to reservoir system management applications. Methods for analyzing flood frequencies are covered in most hydrology textbooks, including those by Linsley, Kohler, and Paulhus

[1982], Linsley *et al.* [1992], McCuen [1989], and Bedient and Huber [1992]. Bulletin 17B of the Interagency Advisory Committee on Water Data [1982] outlines flood flow frequency analysis procedures followed by the federal water agencies. The textbooks summarize and discuss Bulletin 17B methodology. The Hydrologic Engineer Center [1992] Flood Frequency Analysis (HEC-FFA) computer program performs the computations outlined in Bulletin 17B.

Frequency analysis methods can be applied to a variety of hydrologic and meteorologic variables. This discussion focuses on annual peak flood discharges. Similar analyses of flood volumes are also noted.

7.5.1 Basic Definitions

The exceedance probability, $P(x)$ of a random variable X is defined as

$$P(x) = \text{Probability } (X \geq x) \tag{7-7}$$

The exceedance probability is the complement of the cumulative probability

$$F(x) = \text{Probability } (X \leq x) = 1 - P(x) \tag{7-8}$$

In flood frequency analysis, the annual exceedance frequency or probability (P) is the probability that a specified flow magnitude will be equalled or exceeded in a given year. The recurrence interval or return period (T) is the average interval, in years, between successive occurrences of events equalling or exceeding a specified magnitude. The recurrence interval T and exceedance probability P are reciprocals of each other.

$$T = \frac{1}{P} \qquad P = \frac{1}{T} \tag{7-9}$$

For example, the 100-year flow has an annual exceedance probability of 0.01, or 1%.

The risk (R) is the probability that a specified magnitude will be exceeded or exceeded in a series of N years.

$$R = 1 - (1 - P)^N \tag{7-10}$$

Example 7–3—Flood Risk

A hydraulic structure is sized for a 50-year recurrence interval design flow rate.
a. What is the probability that the capacity will be exceeded during the 20-year life of the structure?

$$P = 1/50 \text{ years} = 0.02$$

$$R = 1 - (1 - P)^N = 1 - (1 - 0.02)^{20} = 0.332$$

b. What is the probability that the 50-year peak flow rate will be exceeded in the next 50 years?

$$R = 1 - (1 - P)^N = 1 - (1 - 0.02)^{50} = 0.636$$

7.5.2 Probability Distribution Functions

Probabilities are assigned to a random variable by a probability distribution function, with parameters estimated from observed data. Most probability distribution functions for continuous random variables can be expressed as

$$X = \mu + K \sigma \tag{7-11}$$

where μ and σ are the mean and standard deviation of the random variable X, and K is a frequency factor defined by a specific distribution as a function of the probability level $F(x)$ or $P(x)$ of X. For example, the Gaussian or normal distribution has two parameters, μ and σ. The frequency factor K is the standard normal variant z which is related to cumulative probability in a table reproduced in most statistics and hydrology books.

Other probability distributions are based on a logarithmic transformation, with equation 7-11 expressed as

$$\log X = \mu_{\log X} + K \sigma_{\log X} \tag{7-12}$$

For example, with a log-normal distribution, the observed data x_i are transformed to the logarithms of x_i. The sample mean and standard deviation of the logarithms of x_i provide estimates for the parameters $m_{\log X}$ and $s_{\log X}$. The frequency factor K is again provided as the standard normal variant z of the normal probability table. Thus, the log-normal distribution of X is equivalent to applying the normal distribution to the transformed random variable $\log X$.

7.5.3 Flood Flow Frequency Analysis Based on the Log-Pearson Type III Distribution

The log-Pearson type III distribution is used by the federal water agencies for performing flood flow frequency analyses. This probability distribution is recommended, and guidelines for its application are provided in Bulletin 17B of the Interagency Advisory Committee on Water Data [1982]. The distribution has a third parameter, skewness, in addition to the mean and standard deviation. The log-Pearson type III distribution can be expressed in the form of equation 7-12, with values for the frequency factor K tabulated in Table 7–4 as a function of skew coefficient (G). Bulletin 17B provides a more complete K table. If the skew coefficient (G) is zero, the log-Pearson type III distribution is identical to the log-normal distribution.

The sample mean, standard deviation, and skew coefficient are computed for the logarithms of the observed annual peak flows. The skew coefficient is very sensitive to sample size, and estimates from small samples may be inaccurate. Therefore, Bulletin 17B provides generalized skew coefficients as a function of geographic location. Depending on the number of years of gage record, these regionalized skew coefficient values are used either in lieu of or in combination with values computed from observed flows at the particular stream gage.

TABLE 7–4   *K* Values for Log-Pearson Type III Distribution (Source: Interagency
Advisory Committee on Water Data, 1982)

Skew coefficient *G*	Recurrence interval, years							
	1.0101	1.2500	2	5	10	25	50	100
	Percent chance							
	99	80	50	20	10	4	2	1
3.0	−0.667	−0.636	−0.396	0.420	1.180	2.278	3.152	4.051
2.8	−0.714	−0.666	−0.384	0.460	1.210	2.275	3.114	3.973
2.6	−0.769	−0.696	−0.368	0.499	1.238	2.267	3.071	3.889
2.4	−0.832	−0.725	−0.351	0.537	1.262	2.256	3.023	3.800
2.2	−0.905	−0.752	−0.330	0.574	1.284	2.240	2.970	3.705
2.0	−0.990	−0.777	−0.307	0.609	1.302	2.219	2.912	3.605
1.8	−1.087	−0.799	−0.282	0.643	1.318	2.193	2.848	3.499
1.6	−1.197	−0.817	−0.254	0.675	1.329	2.163	2.780	3.388
1.4	−1.318	−0.832	−0.225	0.705	1.337	2.128	2.706	3.271
1.2	−1.449	−0.844	−0.195	0.732	1.340	2.087	2.626	3.149
1.0	−1.588	−0.852	−0.164	0.758	1.340	2.043	2.542	3.022
0.8	−1.733	−0.856	−0.132	0.780	1.336	1.993	2.453	2.891
0.6	−1.880	−0.857	−0.099	0.800	1.328	1.939	2.359	2.755
0.4	−2.029	−0.855	−0.066	0.816	1.317	1.880	2.261	2.615
0.2	−2.178	−0.850	−0.033	0.830	1.301	1.818	2.159	2.472
0	−2.326	−0.842	0.	0.842	1.282	1.751	2.054	2.326
−0.2	−2.472	−0.830	0.033	0.850	1.258	1.680	1.945	2.178
−0.4	−2.615	−0.816	0.066	0.855	1.231	1.606	1.834	2.029
−0.6	−2.755	−0.800	0.099	0.857	1.200	1.528	1.720	1.880
−0.8	−2.891	−0.780	0.132	0.856	1.166	1.448	1.606	1.733
−1.0	−3.022	−0.758	0.164	0.852	1.128	1.366	1.492	1.588
−1.2	−3.149	−0.732	0.195	0.844	1.086	1.282	1.379	1.449
−1.4	−3.271	−0.705	0.225	0.832	1.041	1.198	1.270	1.318
−1.6	−3.388	−0.675	0.254	0.817	0.994	1.116	1.166	1.197
−1.8	−3.499	−0.643	0.282	0.799	0.945	1.035	1.069	1.087
−2.0	−3.605	−0.609	0.307	0.777	0.895	0.959	0.980	0.990
−2.2	−3.705	−0.574	0.330	0.752	0.844	0.888	0.900	0.905
−2.4	−3.800	−0.537	0.351	0.725	0.795	0.823	0.830	0.832
−2.6	−3.889	−0.499	0.368	0.696	0.747	0.764	0.768	0.769
−2.8	−3.973	−0.460	0.384	0.666	0.702	0.712	0.714	0.714
−3.0	−4.051	−0.420	0.396	0.636	0.660	0.666	0.666	0.667

Flood frequency analysis procedures outlined in Bulletin 17B [Interagency Committee on Water Data, 1982] and EM 1110-2-1415 [U.S. Army Corps of Engineers, 1993] are coded in the computer program Flood Frequency Analysis (HEC-FFA) developed by the Hydrologic Engineering Center [1992]. Figure 7–5 is reproduced from the HEC-FFA output for an analysis of annual peak flows from a 65-year record for a stream gage on Chester Creek in Pennsylvania [U.S. Army Corps of Engineers, 1994]. For this stream gage, the observed data provide a random, unregulated, homogeneous series of annual maximum flows. The mean, standard deviation, and adopted skew coefficient of the logarithms of the 65 annual

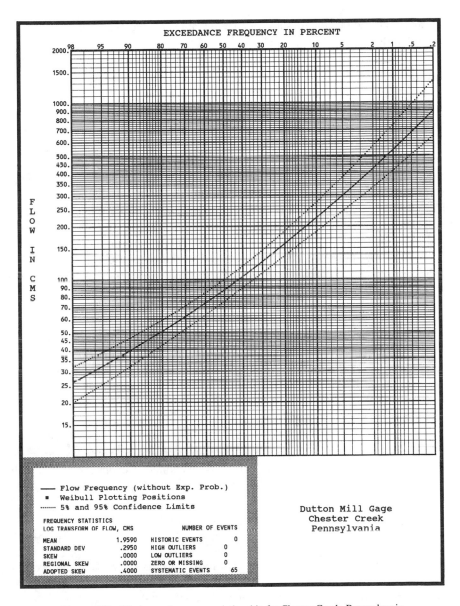

Figure 7–5   Discharge-frequency relationship for Chester Creek, Pennsylvania.

peak flows are 1.959 m³/s, 0.295 m³/s, and 0.40, respectively. The skew coefficient of 0.40 was adopted based on regional values provided in Bulletin 17B.

The 100-year recurrence interval (0.01 exceedance probability) peak flow, for example, can be computed using equation 7-12. From Table 7–4, for a skew coefficient of 0.40, the frequency factor K is 2.615. The estimated mean, standard deviation, and frequency factor are substituted in equation 7-12 to determine the 100-year peak flood flow.

$$\log Q_{100\text{-yr}} = 1.959 + 2.615(0.295) = 2.730$$

$$Q_{100\text{-yr}} = 538 \text{ m}^3/\text{s}$$

The entire frequency-discharge curve is plotted in Figure 7–5.

These computational procedures, based on a limited sample size (65 years of annual peak flows), result in a median discharge frequency relationship. There is an equal 50% chance of the true flow for a given recurrence interval flood being either below or above the estimated value. Bulletin 17B outlines a procedure for adjusting the frequency-flow relationship to reflect expected probability. Bulletin 17B also provides procedures for estimating confidence limits based on describing errors with the non-central t probability distribution. This procedure provides the 90% confidence band shown in Figure 7–5. For example, as previously discussed, the 100-year peak flow is estimated to be 538 m^3/s based on a 65-year record of observed flows. The confidence limits in Figure 7–5 indicate a 5% probability that the true 100-year peak flow is less than 414 m^3/s. There is also a 5% probability that the 100-year flow is greater than 753 m^3/s. Based on 65 years of observed data, the probability is estimated to be 90% that the true 100-year peak flow falls between 414 m^3/s and 753 m^3/s. The confidence limits for the full range of recurrence intervals (exceedance probabilities) are plotted in Figure 7–5. The reader is referred to Bulletin 17B [Interagency Committee on Water Data, 1982] and the HEC-FFA documentation [Hydrologic Engineering Center, 1992] for a description of the computational procedures.

7.5.4 Plotting Position Formulas

A plotting position formula is required to plot observed peak flows versus exceedance probability. Various formulas have been applied. The one most commonly used is the Weibull formula [Linsley *et al.*, 1982; McCuen, 1989; Bedient and Huber, 1992; Singh, 1992],

$$P = \frac{m}{N+1} \qquad T = \frac{N+1}{m} \qquad \qquad (7\text{-}13)$$

where N is the number of years of observation, and m is the rank of the event in order of magnitude, with the largest annual peak flow having $m = 1$. With 65 years of data, for example, the third largest annual peak flow during the 65 years is assigned a recurrence interval of

$$T = \frac{N+1}{m} = \frac{65+1}{3} = 22 \text{ years}$$

and an exceedance probability of

$$P = \frac{m}{N+1} = \frac{3}{65+1} = 0.04545$$

Another alternative is the median plotting position formula [U.S. Army Corps of Engineers, 1993],

$$P = \frac{100\ (m - 0.3)}{N + 0.4} \qquad (7\text{-}14)$$

where m and N are the rank and number of years of record.

Plotting position formulas provide a visual display of the closeness of fit of the analytical probability distribution to the observed data. The plots of observed data should not be extrapolated to estimate infrequent events. The estimates of exceedance probability assigned by a plotting position formula to the largest floods in the observed data may also be highly inaccurate. An analytical probability distribution function, such as the log-Pearson type III, is used to predict extreme events.

7.5.5 Flood Volumes

Streamflow volume-duration-frequency relationships are developed using similar procedures. Flood volume-duration-frequency analyses are often performed for reservoir design and operation studies. The log-Pearson type III distribution is recommended for flood volumes as well as flows [U.S. Army Corps of Engineers, 1993]. Flood volume-duration data is normally obtained from the U.S. Geological Survey WATSTORE database discussed in Chapter 5. The annual maximum volumes are for periods of 1, 3, 7, 15, 30, 60, 120, and 183 days. The previously discussed frequency analysis approach, based on the log-Pearson type III distribution, is applied to an annual series of maximum streamflow volumes for a specified duration. For example, for each year of the stream gage record, the maximum volume to occur during any seven-day period is selected to develop the annual maximum seven-day volume series. The mean, standard deviation, and skew coefficient of the logarithms of the seven-day volumes are used with the log-Pearson type III distribution to develop the exceedance frequency versus volume relationship for the seven-day duration. The procedure is then repeated for other durations.

Balanced hydrographs developed from volume-duration-frequency relationships are used as reservoir inflows in planning studies to size storage capacities and evaluate operating plans. A balanced hydrograph has volumes for specified durations consistent with established volume-duration-frequency relations. For example, a balanced hydrograph for a exceedance probability of 2% is developed so that the peak 1-hour, 6-hour, 12-hour, 24-hour, and 72-hour volumes each corresponds to a 2% exceedance probability.

7.5.6 Regional Flood Frequency Analysis

The previously outlined flood flow frequency analysis procedures require a several-decade series of unregulated homogenous observed annual flows at the location of concern. Gaged flows are often adjusted to remove nonhomogeneities caused by reservoir regulation and other river basin development activities. As noted in Chapter 5, the availability of streamflow data is contingent upon gaging stations having been maintained at pertinent locations over a significantly long period of time, and extensive adjustments may be required to remove nonhomogeneities caused by river basin development. Consequently, alternative flood-flow frequency approaches are often necessary.

A variety of regional flood frequency analysis approaches have been applied in developing a flow frequency curve for an ungaged location based on gaged flows at other locations. For example, data from several gaged watersheds may be used in a regression analysis study to relate the mean of the annual peak flows to watershed characteristics such as area, slope, and land cover [Linsley et al., 1982]. The regression equation is then used to estimate the mean annual peak flow for the ungaged watershed. Using flow data for the gaged watersheds, a generalized regional frequency curve is developed for annual peak flows expressed as the ratio of the peak flow for a year to the mean of the peak flows. The frequency curve for the ungaged location is then developed by combining the estimate of its mean annual peak flow with the generalized frequency curve.

7.5.7 Watershed (Precipitation-Runoff) Models

Watershed models are used, in a variety of roles, in essentially all reservoir system analysis studies involving flood control. As noted in Chapter 5, watershed models simulate the hydrologic processes by which precipitation is transformed to streamflow [Linsley et al., 1982; McCuen, 1989; Bedient and Huber, 1992; Singh, 1992]. The HEC-1 Flood Hydrograph Package [Hydrologic Engineering Center, 1990], described in Chapter 8, is an example of a generalized watershed simulation model that is widely used for flood frequency analyses. Watershed models have the advantage of providing complete runoff hydrographs at all locations of concern as well as peak flows and volumes.

Various precipitation-runoff modeling approaches are adopted for different types of applications. Rainfall intensity-duration-frequency relationships are commonly used to synthesize design storms associated with a specified exceedance probability, which are input to a watershed model to determine the associated flow hydrograph for that exceedance probability. Alternatively, a flood-flow frequency analysis can be performed based on gaged rainfall as follows: The most severe rain storm in each year is selected from the historic record, and the corresponding runoff hydrographs are computed using a watershed model. The log-Pearson type III distribution is then fitted to the resulting annual peak flow series. Instead of working with individual flood events, a continuous watershed model such as the Streamflow Synthesis and Reservoir Operation (SSARR) model, described in Chapter 8, can be used to develop continuous sequences of daily flows covering many years, from rainfall and snow observations.

7.6 EXPECTED ANNUAL FLOOD DAMAGE

Flood-flow frequency, discussed in the previous section, is an integral part of estimating expected annual damages, which in turn is a key focus of economic evaluation of flood control projects. Flood damage reduction benefits are classified as location, intensification, and inundation reduction [Water Resources Council, 1983]. Location benefits refer to facilitating a new economic use of floodplain land, such as shifting from agricultural to industrial use. Intensification benefits result from intensifying the use of floodplain land, such as shifting from lower-value to higher-value crops in response to implementation of a flood control project. Inundation reduction benefits are derived from reducing flood damages to floodplain occupants. Inundation reduction benefits are estimated as the reduction in expected annual damages to result from a particular plan. For example, the benefits to be achieved by a proposed reservoir project are the expected annual damages without the proposed reservoir less the residual expected annual damages that remain if the project is implemented.

Expected annual damage analyses are performed routinely in the evaluation of federal flood control projects and are also often performed for nonfederal projects. The computations are incorporated in several of the reservoir system simulation models described in Chapter 8.

The general approach is outlined in the following section. First, the basic relationships required to determine expected annual damages are described. Evaluation of the impacts of reservoirs, channel improvements, levees, nonstructural measures, watershed development, and other modifications are then addressed. Finally, methods for dealing with uncertainties in the basic data are discussed.

7.6.1 Basic Relationships in Determination of Expected Annual Damages

Economic benefits are evaluated as the reduction in expected annual damages which would result from the implementation of a particular course of action [Hansen, 1987; Davis *et al.*, 1988]. Expected or average annual damage is a probability-weighted average of the full range of possible flood magnitudes and can be viewed as what might be expected to occur, on the average, in any future year. The terms *expected* and *average* annual damage are used interchangeably.

As discussed in statistics textbooks, the expected value of the random value X is determined as

$$E[X] = \int_{-\infty}^{\infty} x f_x(x)\,dx \tag{7-15}$$

where $f_x(x)$ is the probability density function.

The exceedance probability can be expressed as

$$P(X) = \int_x^\infty f_x(x)\,dx \qquad (7\text{-}16)$$

The exceedance probability and probability density are related as

$$\frac{dP(x)}{dx} = f_x(x) \qquad (7\text{-}17)$$

Equations 7-15 and 7-17 are combined to obtain

$$E[X] = \int_{-\infty}^\infty x \frac{dP(x)}{dx}\,dx \qquad (7\text{-}18)$$

Thus, the expected annual damage is determined by integrating the exceedance frequency versus damage function. The integration is performed numerically as illustrated later in Example 7–4.

The basic functional relationships used in estimating expected annual damages are illustrated in Figure 7–6. The frequency-discharge, discharge-stage, and stage-damage relationships are developed from studies involving field surveys and computer modeling. The frequency-damage function is derived from these other three functions. Expected annual damage is the integral of the frequency-damage function.

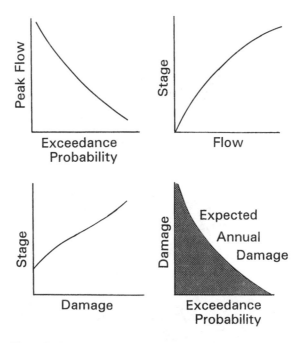

Figure 7–6   Relationships for expected annual damage analysis.

Functional relationships between exceedance frequency, discharge, stage, and damage vary along a river. Consequently, a river system is divided into reaches for analysis purposes. Expected annual damages are computed for each reach and summed. Each reach is represented by an index location. The functional relationships are developed for each index location and represent variables for the entire reach. For example, the floodplain in Figure 7–7 is subdivided into four reaches, each represented by an damage index location.

Flood-flow frequency analysis is discussed in the previous section. The peak discharge versus exceedance frequency relationship describes the probabilistic nature of flood flows, and is developed using standard hydrologic engineering techniques.

The hydraulic relationship between water surface elevation (stage) and discharge is developed based on water surface profile computations, which are discussed in Chapter 6. A stage at an index location corresponds to a water surface profile along the river reach.

The stage-versus-damage relationship represents the damages that would occur along a river reach if flood waters reach various levels. Three alternative approaches for developing stage versus damage relationships involve using historical flood damage data for the study area, engineering cost estimates of damages for assumed flooding levels, and generalized inundation depth versus percent damage functions [James and Lee, 1971]. The most common approach for developing a stage-versus-damage relationship for a river reach involves the use of generalized flooding depth versus percent damage functions which have been previously developed from flood data for a specific river basin, region, or nation. Damage, expressed as a percentage of the market value of property, is a function of depth of inundation. The generalized depth versus percent damage curves are applied to an inventory of the properties located in the particular flood plain to develop a stage-

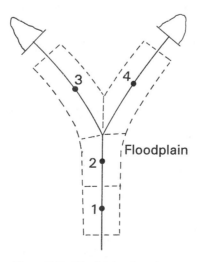

Figure 7–7   Floodplain schematic.

versus-damage relationship. Davis *et al.* [1988], Hansen [1987], and the Institute for Water Resources [1992] review the availability of generic flood damage information. A primary data source in the United States is the Federal Emergency Management Agency, which is the agency responsible for administering the National Flood Insurance Program and disaster response programs.

The frequency-discharge, discharge-stage, and stage-damage relationships are each individually informative in evaluating flooding problems. The frequency-damage relationship derived from these other basic relationships also provides meaningful information. The probability of various levels of damage occurring is defined by this relationship. However, concise indices are typically particularly useful in comparing alternatives in support of decision processes. Expected annual damage provides such an index.

Example 7–4—Expected Annual Damage Computations

Expected annual damages are estimated for the floodplain along a reach of Chester Creek in Pennsylvania [U.S. Army Corps of Engineers, 1994]. The frequency-discharge relationship is plotted in Figure 7–5 and tabulated in Table 7–5. As previously discussed, the frequency-discharge relationship is based on the log-Pearson Type III probability distribution with parameters estimated from gaged streamflow data. The stage-discharge relationship shown in Table 7–5 was developed from a hydraulic analysis of the river reach using the computer program HEC-2 Water Surface Profiles (Chapter 6). Water surface profiles are computed for a range of assumed flow rates and related to the stage at the index location. The stage-damage relationship was developed based on an inventory of property in the floodplain combined with inundation-depth-versus-damage estimates. The field surveys and computer modeling studies required to develop these basic relationships involve a significant amount of work.

The three relationships tabulated in Table 7–5 are combined to develop the exceedance probability-versus-damage relationship tabulated in Table 7–6. The discharge associated with each exceedance probability is obtained from the first relationship in Table 7–5.

TABLE 7–5 Relationships Used in Example 7–4 to Compute Expected Annual Damages

Frequency-Discharge		Stage-Discharge		Stage-Damage	
Exceedance Probability	Discharge (m^3/s)	Stage (m)	Discharge (m^3/s)	Stage (m)	Damage ($1000)
0.002	899	1.97	84	3.35	0
0.005	676	2.39	100	4.27	19
0.01	539	3.39	168	4.57	26
0.02	423	4.07	228	5.18	339
0.05	299	4.58	278	5.49	525
0.10	223	5.50	384	6.10	1100
0.20	158	7.13	606	6.71	2150
0.50	87	7.47	652	8.23	5133
0.80	51	7.75	722	8.53	5654
0.90	39	8.10	838	9.14	6417
0.95	32	8.79	1031	9.45	6592

TABLE 7–6   Numerical Integration of Frequency-Drainage Function
to Determine Expected Annual Damages in Example 7–4.

Exceedance Probability	Damage ($1000)	Probability Increment	Increment Mean Damage ($1000)	Weighed Damage ($1000)
0.002	5290	0.002	5290	10.58
0.005	3830	0.003	4560	13.68
0.01	2045	0.005	2938	14.69
0.02	808	0.01	1426	14.26
0.05	242	0.03	525	15.75
0.10	13.6	0.05	127.8	6.39
0.20	0	0.10	6.8	0.68
			Expected annual damage =	76.03

The stage damage tabulation is interpolated to obtain the corresponding stage. The stage-damage relationship is then interpolated to obtain the corresponding damage. The resulting exceedance-probability-versus-damage relationship is numerically integrated in Table 7–6 to determine the expected annual damages. The expected annual damages are probability-weighted average damages that can be visualized as the area under the exceedance-probability-versus-damage curve. There is a probability of 0.002 that flood damages in any year will equal or exceed $5,290,000. In a given year, the probability is 0.01 that damages will equal or exceed $2,045,000. Expected or average annual damages are $76,030.

7.6.2 Evaluation of Modified Conditions

Figure 7–8 illustrates the reductions in expected annual damages that result from various types of flood damage reduction measures or watershed modifications. Changes in the basin are modeled in terms of impacts on the basic functional relationships. Reservoir storage, reflected in routing computations, modifies the frequency-discharge relationships at downstream locations. Watershed modifications due to urbanization and other land use changes, reflected in precipitation-runoff computations, also result in corresponding changes in the frequency-discharge relationships at downstream locations. Channel improvements, reflected in water-surface profile computations, modify the discharge-stage relationship. Levees and floodwalls are also reflected in the depth to which properties are inundated by a given river discharge. Nonstructural measures such as floodplain regulation, flood-proofing, and removal of buildings reduce the susceptibility of properties to damage. Nonstructural measures are reflected in the stage-damage functions. Any changes in the frequency-discharge, discharge-stage, and stage-damage relationships result in a corresponding change in the frequency-damage function, and thus result in a change in expected annual damage.

To illustrate the analysis approach, consider the floodplain and two flood control reservoirs shown in Figure 7–7. Expected annual damages are included in the evaluation of alternative reservoir designs and/or operating plans. The HEC-1 Flood

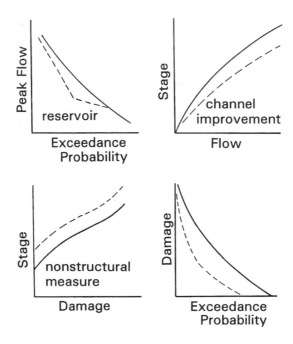

Figure 7–8   Effects of flood damage reduction measures.

Hydrograph Package, HEC-5 Simulation of Flood Control and Conservation Systems, and other models described in Chapter 8 are available to perform the computations involved in developing and integrating the frequency-damage functions.

Damage reaches are delineated, and index locations are selected. The four index locations shown in Figure 7–7 each represent a damage reach. Frequency-discharge, discharge-stage, and stage-damage relationships for each of the index locations are developed as previously discussed for unregulated conditions without the reservoirs.

Other inputs include reservoir characteristics and operating rules, routing parameters for the river reaches, and a set of flood hydrographs representing inflows to the two reservoirs and incremental local flows for each of the four damage index locations. Sets of inflow hydrographs at the six locations are required for a range of flood magnitudes covering the full range from frequent (such as a two- or five-year recurrence interval) to extreme (such as a 250- or 400-year recurrence interval) floods. If available, hydrographs from about ten historical floods could be used. An alternative, more commonly applied, approach included in HEC-1 and HEC-5 consists of providing an inflow hydrograph for each pertinent location, representing one flood, along with a set of about ten multiple-flood ratios. The hydrographs for the single flood could be either based on gaged streamflow or developed with a precipitation-runoff model. Hydrographs for a range of flood events are synthesized by multiplying hydrograph ordinates by multiple-flood ratios. Each multiple-flood ratio represents a different flood event. The ratios are arbitrary as long as a full range of flood magnitudes is adequately covered. The peak flow for each synthesized hydrograph at each location is combined with the frequency-discharge relationship for that location to assign an exceedance probability to each flood. The exceedance probability for a particular flood will vary between locations.

Expected annual damages for a particular reservoir design and operating plan are determined by routing each of the ten or so floods, in turn, through the system. Referring to Figure 7–7, a flood includes inflow hydrographs for the two reservoirs and incremental local inflows at the tributary confluence and damage index locations. Releases from the reservoirs are routed through the downstream river reaches and combined with local inflow hydrographs. The exceedance probability associated with each flood at each location, along with the peaks of the routed hydrographs, are used to develop the required exceedance-frequency-versus-discharge relationships. Thus, the floods are all routed and the frequency-discharge relationships developed for each damage index location. Each flood provides a point on the frequency-discharge curve. For a given frequency flood, the discharge depends on the storage capacity and operating rules of the reservoirs located upstream.

Expected annual damages are then computed for each index location. The frequency-discharge, discharge-stage, and stage-damage relationships are combined to develop a frequency-damage relationship, which is numerically integrated to obtain expected annual damages. A comparative evaluation of alternative plans with different reservoir storage capacities may include expected annual damage estimates for each alternative plan.

7.6.3 Uncertainty in Estimates of the Input Relationships

The expected annual damage evaluation procedures just outlined have been routinely applied for several decades by the Corps of Engineers, Soil Conservation Service, and other federal and nonfederal entities. During the 1990s, the Corps of Engineers established a policy of incorporating additional uncertainty considerations in the analysis procedures [U.S. Army Corps of Engineers, 1994]. Modeling and data uncertainties in developing the basic relationships are explicitly addressed in the analysis.

The frequency-discharge, discharge-stage, and stage-damage relationships illustrated in Figure 7–6 may represent best estimates or most likely values, or may alternatively reflect conservative estimates. Limited data and modeling simplifications necessarily result in uncertainties in developing these relationships. The expanded procedure explicitly considers uncertainty in each of these three relationships. For a given exceedance probability, the peak flow-rate is estimated from the perspective of a probability distribution rather than a single best estimate. Likewise, water surface profile computations involve modeling assumptions and parameter estimates, and thus uncertainty. For a given flow rate, the range of possible stage values is represented by a probability distribution. Likewise, for a given stage, the uncertainty in the damage estimate damage is expressed as a probability distribution. The U.S. Army Corps of Engineers [1994] outlines approaches for quantifying these uncertainties.

Monte Carlo simulation is used to combine the three basic relationships. A Monte Carlo sample consists of: (1) drawing a random number between zero and one from a standard uniform distribution; (2) setting a cumulative or exceedance probability equal to the random number; and (3) determining the corresponding value of the variable based on a specified probability distribution. Many thousand values of annual flood damage are developed based on Monte Carlo sampling and the three basic relationships of Figure 7–6. The expected annual damage is computed by averaging the annual damages. Additional annual damages are computed until the average of the annual damages stabilize at a constant value [U.S. Army Corps of Engineers, 1994].

A single annual damage estimate involves the following steps:

1. An annual exceedance frequency is selected by random sampling from a standard uniform probability distribution. The corresponding discharge is determined from the basic median frequency-discharge relationship.
2. The error in predicting discharge for a given frequency is sampled from a specified probability distribution and included with the discharge.
3. The stage is determined by combining the discharge with error and the basic median discharge-stage relationship.
4. The error in predicting stage for a given discharge is sampled from a specified probability distribution and included in the stage.
5. The damage is determined by combining the stage with the error and the basic median stage-damage relationship.
6. The error in predicting damage for a given stage is sampled from a specified probability distribution and included with the damage.

All of the annual damage values thus computed are averaged, and additional damage estimates are computed following steps 1–6 until the average of all values stabilizes. Steps 1–6 are typically repeated at least several thousand times, with each repetition representing a year. The expected annual damage is estimated by averaging the thousands of values for annual damages.

8

RESERVOIR SYSTEM
SIMULATION MODELS

This chapter reviews several of the numerous reservoir system simulation models that have been reported in the literature. The models cited in this chapter are based on a conventional simulation approach with limited or no use of formal optimization (mathematical programming) algorithms. Optimization models are covered in Chapters 10 and 11. From the perspective of applying models in simulation studies, a number of the models cited in Chapter 11 are applied in the same manner as are the models noted in this chapter, even though an optimization algorithm is used to perform the computations. Many network flow models in particular, and certain other optimization models as well, are appropriately viewed as simulation models and perform essentially the same functions as the more conventional simulation models. Water quantity considerations are addressed in the present chapter, as contrasted with water-quality models, which are covered in Chapter 12.

The sequential simulation models cited in this chapter are based on mass-balance accounting procedures, discussed in Chapter 6, for tracking the movement of water through a reservoir/river system. The computations are repeated sequentially for each time interval of the overall simulation period. A specified water management scenario is combined with a set of streamflow sequences at pertinent locations, representing the hydrologic characteristics of the river basin. For example, a model applied in a water supply planning study might simulate the capabilities of an existing or proposed reservoir system and an associated operating plan to meet projected year 2000 water requirements during a 70-year (840-month) simulation period representing historical 1925–1994 hydrology, using a monthly time interval. A flood control study might simulate system response, for a particular operating plan, to a synthetically generated, historical, or real-time flood event, using a hourly time step.

Various strategies can be adopted for applying simulation models. Series of runs are typically made to compare system performance for alternative system configurations, storage allocations, operating rules, demand levels, and/or hydrologic inflow sequences. System performance may be evaluated by simply observing the computed time sequences of storage levels, releases, streamflows, hydroelectric power generated, water supply diversions, and diversion shortages. Frequency analyses of the storages and flows computed by the simulation model may be performed. The types of reliability and risk analyses described in Chapter 7 are commonly a primary focus of simulation studies. Simulation models provide capabilities to analyze reservoir system operations using hydrologic and economic performance measures such as firm yield, yield versus reliability relationships, reliabilities in meeting specified water management requirements, hydroelectric energy revenues, and flood damages.

8.1 MODEL CATEGORIZATION

For purposes of organizing this chapter, reservoir/river system simulation models are grouped as follows:

- Models developed for specific reservoir/river systems
- Generalized reservoir/river system simulation models
- Generalized reservoir/river system simulation models with precipitation-runoff modeling capabilities
- Generalized reservoir/river system simulation models developed using object-oriented programming
- Models developed using general-purpose commercial software

Early simulation models as well as many of the more recent models are computer programs written in FORTRAN or similar languages for a specific reservoir system. Descriptions of many site-specific simulation models can be found in the published literature. Numerous other models routinely applied in operating reservoir systems throughout the nation have simply never been reported in the literature.

Generalized models are designed for a variety of applications involving any reservoir system in any river basin, rather than developed for one specific system. With a generalized model, the particular reservoir system of concern is described in the input data. Users develop input data files for their particular applications without being concerned with writing computer code. A number of public-domain software packages are readily available for a broad range of reservoir system analysis

applications. Executable microcomputer versions are distributed, and the FOR-
TRAN programs are compiled and executed on larger computer systems as well.

Most of the many reservoir system analysis models currently in use were developed using conventional programming languages, typically FORTRAN. In recent years, models are also being developed using the object-oriented programming
approach, typically with object-oriented versions of the C language. Instructions
and information are coded and stored as objects or modules in this model development environment. Objects can be reused in different programs and subprograms,
and models are easier to modify. Object-oriented programming languages also pro
vide flexible capabilities for developing graphical user interfaces. Reservoir system
models based on object-oriented programming have been implemented primarily on
workstations.

Relatively simple reservoir system analysis models can be constructed using
spreadsheet programs, simulation modeling environments such as STELLA (discussed later in this chapter), or other commercially available software. These
general-purpose desktop computer software packages are widely used in many
areas of education, business, engineering, and science. They can be applied to reservoir system analysis problems along with their myriad of other routine uses.

8.2 MODELS DEVELOPED FOR SPECIFIC RESERVOIR SYSTEMS

Numerous simulation models have been developed for specific reservoir systems,
ranging from relatively simple models of a single reservoir serving a single purpose to
complex models of multiple-purpose, multiple-reservoir systems. Several examples of
models for large-scale river basin systems are noted in the following paragraphs.

8.2.1 Early Simulation Models

Pioneering efforts in computer simulation of reservoir/river systems in the
United States include a study initiated in 1953 by the Corps of Engineers of the operation of the six main-stem reservoirs on the Missouri River [Manzer and Barnett,
1966]. The objective was to maximize power generation subject to constraints im-
posed by specified requirements for navigation, flood control, and irrigation.
Shortly thereafter, both the Corps of Engineers and Bonneville Power Administration conducted simulation studies of hydroelectric power operations on the Columbia
River. The International Boundary and Water Commission simulated a multiple-
purpose two-reservoir system on the Rio Grande in 1954. A simulation study for the
Nile River Basin in Egypt in 1955 considered alternative plans with as many as 17
reservoirs or hydropower sites. The objective was to determine the particular combination of reservoirs and operating procedures which would maximize the volume
of useful irrigation water [Manzer and Barnett, 1966]. Maass *et al.* [1966] and

Hufschmidt and Fiering [1966] discuss the pioneering simulation modeling work of the Harvard Water Program. Simulation models for numerous other river/reservoir systems have been developed since the 1950s.

8.2.2 Missouri River Basin

The Missouri River Division of the Corps of Engineers initiated, in 1989, a several-year study to review the Missouri River Master Water Control Manual, which guides operation of the six main-stem reservoirs described in Chapter 2 [Cieslik and McAllister, 1994]. The study was motivated by a drought in 1987–1992 that affected project uses and environmental resources. The reservoirs are operated for flood control, hydropower, water supply, water quality, irrigation, navigation, recreation, and environmental resources. The system includes a 1200-km-long navigation channel on the Missouri River from its mouth at St. Louis, Missouri, to Sioux City, Iowa.

Reservoir operating criteria involve allocation of storage capacity among uses and release rules during navigation and nonnavigation seasons. Formulation of alternative operating plans focused on the amount of storage reserved for the permanent pool and the resulting size of the carryover multiple use zone available to provide water during droughts, and the quantity and timing of releases for navigation, water supply, irrigation, power production, water quality, flood control, recreation, and environmental resources. Consideration of modifications to release rules involved navigation criteria, including level of service and length of navigation season; water supply needs during the nonnavigation season; seasonal releases from the most downstream dam to improve the river ecosystem; spring and summer releases to protect threatened and endangered birds that nest on river islands; and intrasystem regulation of the upper three reservoirs to provide favorable conditions for fish reproduction.

The Long Range Study (LRS) Model uses a monthly time interval for simulating operation of the system during 96-year (March 1898 to February, 1994] sequences of historical flows at six reservoir nodes and nine other gage locations [Patenode and Wilson, 1994]. An earlier model developed in the 1960s was updated during the 1990s and modified for execution on IBM-compatible desktop computers. The computer programs are coded in FORTRAN. In addition to the system simulation model, a set of supplemental programs is used to process the voluminous simulation results during the process of design, review, and comparison of alternative operating plans.

Two input files are used. One contains historic reach inflow and streamflow depletion data, and the other contains the various constants and variable parameters that define regulation decisions. The historic input file contains annual evaporation rates for the six reservoirs which are distributed monthly by coefficients contained in the parameter input file. Monthly incremental streamflows are combined with depletion factors that adjust the historical flows to current conditions of water use. The parameter input file contains various information to define the sizes and limits of the river and reservoirs, and to establish the guide curves and operating limits of a particular simulation.

The LRS model has been used to simulate and evaluate numerous alternative operating policies. The results of the LRS hydrologic simulation model are used in combination with several other models, including environmental and economic analysis models, in evaluating operations of the mainstem Missouri River system.

8.2.3 Potomac River Basin

The Potomac River Interactive Simulation Model (PRISM) was originally developed by a research team at Johns Hopkins University [Palmer *et al.*, 1982]. A number of water management agencies in the Potomac River Basin participated in drought simulation exercises using PRISM during development and implementation of a regional water supply plan for the Washington Metropolitan Area. The Corps of Engineers modified PRISM for use in certain drought simulation studies [USACE Baltimore District, 1983]. The model has not been actively applied in recent years. PRISM simulates the operation of several reservoirs and allocation of water within the Washington Metropolitan Area. Versions of the model alternatively use a weekly and daily time interval. The model determines the amount of water available to each of the several jurisdictions for given streamflows, demands, water allocations, and reservoir operating rules.

PRISM is designed for use in either a batch mode, where decision strategies are specified by the user prior to model execution, or an interactive mode. When operating in the batch mode, PRISM performs the functions of a regional water supply manager in strict accordance with rules specified by the model user. The interactive model allows participants to engage in a dialogue with the model as it is being executed, thereby changing model parameters and overriding prespecified decision rules. The interactive model represents an attempt to include, in a formal analytical modeling exercise, the process by which water supply management decisions are made.

Measures implemented to meet water supply needs in the Washington, D.C. area have been viewed nationwide as a classic example of optimizing the beneficial use of existing systems as an alternative to construction of additional major reservoir projects. Various systems analysis techniques, including PRISM, were used in developing the plan. The Metropolitan Washington Area Water Supply Study Final Report [USACE Baltimore District, 1983] summarizes the Corps of Engineers study authorized by the Water Resources Development Act of 1974 as well as the studies performed and actions taken by various nonfederal entities. The Corps of Engineers report resulted in a recommendation of no further federal action since the water supply needs could be satisfied by measures being implemented by nonfederal entities.

The study was motivated by the fact that municipal and industrial water needs significantly exceeded supplies. The primary water supply area is the Potomac River Basin and adjacent Patuxent River Basin. Relatively small portions of these watersheds are controlled by five reservoirs, but the single largest source of supply is unregulated flows in the Potomac River. Construction of additional reservoirs was proposed, but was found to be infeasible for various reasons, including lack of public support.

The Interstate Commission on the Potomac River Basin, Corps of Engineers, several water utilities, researchers at Johns Hopkins University, and several committees and task forces all played key roles during several years of studies, resulting in the plan finally implemented. The regional water supply plan included a number of components, involving system operation based on coordination of unregulated flows and withdrawals from the existing five reservoirs, long-term and emergency demand management measures, and construction of a small downstream reregulating reservoir to facilitate improved operation of existing upstream reservoirs. Several contracts and agreements between the various water management agencies were required to implement the plan.

8.2.4 Sacramento and San Joaquin River Basins

The U.S. Bureau of Reclamation [1991] presents descriptions of a number of simulation models used at specific Bureau of Reclamation projects in several western states. Models developed for the California Central Valley and State Water Projects and the Colorado River Basin are noted here. These reservoir/river systems are described in Chapter 2.

The California Central Valley Basin includes the watersheds of the Sacramento and San Joaquin Rivers. As discussed in Chapter 2, the water resources of this basin are managed through the storage and conveyance facilities of the State Water Project and the federal Central Valley Project.

The Projects Simulation Model (PROSIM) was developed by the Bureau of Reclamation to simulate operations of the Central Valley Project and State Water Project [Bureau of Reclamation, 1990; Sandberg and Manza, 1991]. PROSIM uses a traditional water balance approach, with a monthly time step, to simulate a system represented by 50 nodes which include 11 reservoirs. Monthly streamflow data is input at 24 of the nodes for a 57-year (1922–1978] simulation period. A groundwater routine is included in the model to estimate stream accretion from the groundwater basin in addition to the accretion which occurred historically. A routine is included to analyze hydroelectric power production. Reservoirs operate in accordance with storage allocations, rule curves, powerplant discharge capacities, and water demands. Demand for water comes in four basic forms: nonproject demands, project demands, minimum instream flow requirements, and Delta outflow requirements. Rules are provided to allocate water to competing uses and users in accordance with institutionally established requirements.

8.2.5 Colorado River Basin

The Colorado River Basin has a long history of simulation modeling. Studies were performed manually prior to the 1960s. The Colorado River Storage Project model, completed in 1965, was the first computer model of the Colorado River

[Schuster, 1987]. This Bureau of Reclamation model was used to develop annual operating plans for the upper basin reservoirs during the filling of Lake Powell. Over the years, the model was expanded to include lower-basin reservoirs, power-plants, salinity, and operating criteria. A second model, called the River Network Model, was developed in 1973 to evaluate the salinity impacts resulting from water resource development and salinity control projects and to aid in establishing salinity standards for the Colorado River. The Colorado River Simulation System (CRSS), described in the following paragraphs, stemmed from these prior models, motivated by a need to have a flexible, comprehensive model of the Colorado River Basin that would incorporate all areas of interest, including legislative requirements.

The CRSS, originally developed by the Bureau of Reclamation during the 1970s and subsequently revised and updated, simulates operations of the major reservoirs in the Colorado River Basin for water supply, low-flow augmentation, hydroelectric power, and flood control [Schuster, 1987]. The CRSS is a set of computer programs, data files, and databases used in long-range planning. The monthly-time-interval historical hydrologic period-of-record model reflects operation of the system in accordance with a series of river basin compacts, laws, and agreements collectively called the Law of the River. Salt concentrations are also included in the simulation.

The main component of the CRSS is the Colorado River Simulation Model (CRSM). The CRSM is a water and salt accounting program. Historical monthly streamflows are input at pertinent locations, flows are routed through the system, and water supply commitments are met. Salt is introduced through inflows and re turn flows, and is routed through the system with the water. The model includes runoff forecasting, reservoir operations (rule curves, evaporation, bank storage, and sediment accumulation), flood control regulations, operating strategies of the system (shortage and surplus strategies), hydroelectric power generation, and legislative requirements.

The hydrology database contains the flow and salt data for the basin. The demand database contains diversion data. Other components of the CRSS include computer programs that process output from the CRSM and data in the hydrology and demand databases.

8.3 GENERALIZED RESERVOIR/RIVER SYSTEM SIMULATION MODELS

Generalized operational reservoir system simulation models are designed to be readily applied to a variety of different systems. The term *generalized* means that the computer model is designed for application to a range of problems dealing with reservoir systems of various configurations and locations, rather than developed for a specific reservoir system. Users develop the input data required for their particular systems in a specified format, and execute the model without being concerned with developing or modifying the actual computer code.

Operational means that a model is reasonably well documented and tested, and is designed to be used by professional practitioners other than the original model developers. Generalized models should be convenient to obtain, understand, and apply, and should work correctly, completely, and efficiently. Documentation, user support, and user friendliness of the software are key factors in selecting a model. The extent to which a model has been tested and applied in actual studies is also an important consideration.

A number of generalized operational reservoir system simulation models have been reported in the literature. Several are cited below. Executable versions of most of these models are available for IBM-compatible microcomputers operating under MS-DOS. The FORTRAN programs have been executed on other computer systems as well. The generalized models, with documentation, can be obtained by contacting the developer or sponsoring agency. Most of the models were developed at federal agencies or universities and are in the public domain, with a handling fee charged for distribution.

8.3.1 HEC-3 and HEC-5

Feldman [1981] provides an overview of generalized computer programs available from the Hydrologic Engineering Center (HEC), including HEC-3 and HEC-5. Upon request, the HEC provides copies of their computer program and publications catalogs and instructions for obtaining HEC products. Their address and telephone are: Hydrologic Engineering Center, U.S. Army Corps of Engineers, 609 Second Street, Davis, California 95616, 916-756-1104.

The HEC-3 Reservoir System Analysis for Conservation computer program simulates the operation of a reservoir for conservation purposes such as water supply, low-flow augmentation, and hydroelectric power. HEC-3 is documented by a user's manual [Hydrologic Engineering Center, 1981] and other publications available from the HEC. Various Corps of Engineers offices and other entities have modified HEC-3 for different applications. HEC-3 and HEC-5, discussed next, have similar capabilities for simulating conservation operations. However, HEC-3 does not have the comprehensive flood control capabilities of HEC-5. The Hydrologic Engineering Center is no longer distributing HEC-3 because essentially all of its capabilities have been duplicated in HEC-5. HEC-3 and HEC-5 were applied in the case study presented in Chapter 9.

The HEC-5 Simulation of Flood Control and Conservation Systems computer program is probably the most versatile model available in the sense of being applicable to a wide range of both flood control and conservation operation problems. HEC-5 has been used in a relatively large number of studies, including studies of storage reallocations and other operational modifications at existing reservoirs as well as feasibility studies for proposed new projects. The program is also used for real-time operation. An initial version released in 1973 has subsequently been greatly expanded. The FORTRAN programs have been run on various computer systems. Executable MS-DOS-based microcomputer versions of the model are available. The HEC-5 package includes several utility programs to aid in developing input data files and analyzing output. Alternative versions of the model are available which exclude and include water quality analysis capabilities. The water-quality version is discussed in Chapter 12. The HEC-5 user's manual [Hydrologic Engineering Center 1982, 1989] provides instructions for its use. Various publications regarding the use of HEC-5 available from the Hydrologic Engineering Center include training documents covering various features of the model, and reports and papers documenting specific applications of the model in actual reservoir system analysis studies. The model is also included in the inventory of short courses taught by the Hydrologic Engineering Center.

HEC-5 simulates the sequential period-by-period operation of a multiple-purpose reservoir system for input sequences of unregulated streamflows and reservoir evaporation rates. Multiple reservoirs can be located in essentially any stream tributary configuration. The program uses a variable time interval. For example, monthly or weekly data might be used during periods of normal or low flows in combination with daily or hourly data during flood events. The user specifies the operating rules in HEC-5 by inputting reservoir storage zones, diversion and minimum instream flow targets, and allowable flood flows. The model makes release decisions to empty flood control pools and to meet user-specified diversion and instream flow targets based on computed reservoir storage levels and streamflows at downstream locations. Seasonal rule curves and buffer zones can be included in the operating rules. Multiple-reservoir release decisions are based on balancing the percentage of storage depletion in specified zones. Hydrologic flood routing options include modified Puls, Muskingum, working R&D, and average lag. HEC-5 has various optional analysis capabilities, including computation of expected annual flood damages and single-reservoir firm yields for water supply and hydroelectric power.

8.3.2 IRIS and IRAS

The Interactive River System Simulation (IRIS) model was developed with support from the Ford Foundation, United Nations Environment Program, International Institute for Applied Systems Analysis, and Cornell University [Loucks *et al.*, 1989 and 1990]. Users interested in obtaining the model should contact the developers:

D. P. Loucks	M. R. Taylor
Civil and Environmental	Resources Planning Associates, Inc.
Engineering	Langmuir Building, Suite 231
Cornell University	Cornell Business & Technology Park
Ithaca, NY 14853-3501	Ithaca, NY 14850
607-255-4896	607-257-4305

Model development was motivated largely by the objective of providing a useful tool for water managers responsible for negotiating agreements among individuals and organizations in conflict over water use. IRIS operates in a menu-driven microcomputer or workstation environment with extensive use of computer graphics for information transfer between machine and user. The configuration of the system is specified by "drawing in" nodes (reservoirs, inflow sites, junctions, and other key locations) and interconnecting links (river reaches, canals). IRIS simulates a water supply and conveyance system of essentially any normal branching configuration for input streamflow sequences, using a user-specified time step. Hydroelectric power and water-quality features are included. System operating rules include reservoir releases specified as a function of storage and season of the year, allocation functions for multiple links from the same node, and storage distribution targets for reservoirs operating in combination. The model allows the user to interactively change the operating rules during the course of a simulation run. Another unique feature allows several alternative sets of inflow sequences to be considered in a single run of the model. Model output includes time series plots of flows, storages, energy generated, and water-quality parameters at any node or link in the reservoir/river system, and probability distribution displays of magnitude and duration of shortages or failure events.

The Interactive River-Aquifer Systems (IRAS) model was developed as an extension of IRIS [Loucks *et al.*, 1995]. IRAS is a generalized program for analyzing regional surface water and groundwater management systems. IRAS is designed to be an interactive, flexible system for addressing problems involving interactions between ground and surface waters and between water quality and quantity. The model predicts the range and likelihood of various water quantity, quality, and hydropower impacts, over time, associated with alternative design and operating policies, for portions or entire systems, of multiple rivers and ground water aquifers. Simulations are based on mass balances of quantity and quality constituents, taking into account flow routing, seepage, evaporation, consumption, and constituent growth, decay, and transformation, as applicable. A variable computational time step is used. IRAS provides a menu-driven graphics-based user interface.

8.3.3 Water Rights Analysis Package (WRAP)

The Water Rights Analysis Package (WRAP), developed at Texas A&M University, is designed to simulate management and use of the streamflow and reservoir storage resources of a river basin, or multiple basins, under a prior appropriation water rights permit system. The model is designed for simulation studies involving a priority-based allocation of water resources among many different water users. Water use diversions and reservoir storage facilities may be numerous, and the allocation system complex. The WRAP model can be used to evaluate operation of a particular multiple-reservoir, multiple-user system while considering interactions with numerous other water management entities which also hold permits or rights to store and withdraw water from the river system. Multiple-reservoir release decisions are based on balancing the percent depletion in user-specified storage zones. The original model described by Wurbs and Walls [1989] has been significantly expanded [Wurbs *et al.*, 1993]. A version with capabilities for considering salinity was subsequently developed [Wurbs *et al.*, 1994].

A monthly time interval is used, with no limit on the length of the period of analysis. Input requirements include naturalized streamflows, reservoir evaporation rates, water rights data (including permitted diversion and storage amounts, use types, return flow requirements, and priorities), hydroelectric power plant characteristics, and reservoir characteristics and operating policies. Output includes diversions, shortages, hydroelectric energy generated, streamflow depletions, unappropriated streamflows, reservoir storages and releases, and reliability statistics. The WRAP model is discussed further in Chapter 9.

8.3.4 MITSIM

The MITSIM model was originally developed at the Massachusetts Institute of Technology [Strzepek *et al.*, 1979]. Early versions of the model were used in studies of the Rio Colorado in Argentina and the Vardar/Axios project in Yugoslavia and Greece. Subsequent versions have been applied in a number of studies. The model has been updated and adapted to changing computer environments at the Center for Advanced Decision Support for Water and Environmental Systems [Strzepek *et al.*, 1989]. MITSIM provides capabilities to evaluate both the hydrologic and economic performance of alternative river basin development plans involving reservoirs, hydroelectric power plants, irrigation areas, and municipal and industrial water supply diversions. A river/reservoir/use system is conceptualized as a collection of arcs and nodes. A variable computational time interval is used. The model assesses system reliability in meeting demands. Economic benefits and costs can also be evaluated. Benefits are divided into long-term benefits and short-term losses. Optional displays of net economic benefits and benefit-cost ratios for the entire river basin and/or subregions within the basin can be included in the output.

8.3.5 USACE Southwestern Division Model

A generalized reservoir system simulation model, called SUPER, was developed by the USACE Southwestern Division [Hula, 1981]. SUPER performs the same types of hydrologic and economic simulation computations as HEC-5, including comprehensive flood control analyses. The Southwestern Division model uses a one-day computation interval, whereas HEC-5 uses a variable time interval. Details of handling input and output data and various computational capabilities differ somewhat between HEC-5 and SUPER. The division and district offices in the Southwestern Division have applied the model in many studies over a number of years. In particular, the Reservoir Modeling Center in the Tulsa District, has routinely used SUPER to simulate various major USACE reservoir systems located throughout the Southwestern Division.

8.3.6 USACE North Pacific Division
Hydropower Models

The Hydro System Seasonal Regulation (HYSSR), Hourly Load Distribution and Pondage Analysis (HLDPA), and Hydropower System Regulation Analysis (HYSYS) models were developed by the USACE North Pacific Division. The models are described in the Hydropower Engineer Manual [U.S. Army Corps of Engineers, 1985]. The user's manuals are available from the North Pacific Division. HYSSR is a monthly sequential routing model designed to analyze the operation of reservoir systems for hydroelectric power and snowmelt flood control. It has been used to analyze proposed new reservoirs and operations of existing systems in the Columbia River Basin and several other river basins. HLDPA is a hourly-time-interval planning tool designed to address such problems as optimum installed capacity, adequacy of pondage for peaking operation, and impact of hourly operation on nonpower river uses. HYSYS is a generalized model designed to support real-time operations.

Information regarding these models and the SSARR, discussed next, can be obtained by contacting U.S. Army Corps of Engineers, North Pacific Division, Attn: CENPD-EN-WM (HES), Portland, OR 97208-2870, 503-326-3758.

8.4 GENERALIZED RESERVOIR/RIVER SYSTEM SIMULATION MODELS WITH PRECIPITATION-RUNOFF MODELING CAPABILITIES

The previously cited reservoir/river system simulation models start with streamflows provided as input data. As discussed in Chapter 5, streamflow inflow sequences are typically developed from stream gage records. Streamflows may also be synthesized from precipitation data. Watershed models are used to determine streamflows for inputted precipitation data and watershed parameters. Singh [1995] and Wurbs [1995] describe several generalized watershed (precipitation-runoff) simulation models used for a broad range of applications. The SSARR, HEC-1, HMS, and BRASS models, discussed next, are particularly closely associated with reservoir system analysis applications. The SSARR watershed component is a continuous model which can be used for synthesizing long sequences of streamflows resulting from multiple precipitation events and snowmelt. The watershed modeling features of other models are designed for simulating individual flood events.

8.4.1 Streamflow Synthesis and Reservoir
Regulation (SSARR) Model

The Streamflow Synthesis and Reservoir Regulation (SSARR) model was developed by the North Pacific Division for streamflow and flood forecasting and reservoir operation studies. A program description and user's manual [USACE North Pacific Division, 1987] documents the current version of the model. Various versions of the model date back to 1956. The SSARR model was originally developed in support of Corps of Engineers studies involving planning, design, and operation of water control projects in the Pacific Northwest. Further development was motivated by operational river forecasting and management activities of the Cooperative Columbia River Forecasting Unit, sponsored by the National Weather Service, Corps of Engineers, and Bonneville Power Administration. Subsequently, numerous river systems in the United States and abroad have been modeled with the SSARR by various agencies, universities, and other organizations. An executable version of the FORTRAN77 programs is available for MS-DOS-based microcomputers. The model has been compiled and executed on various other computer systems as well.

The SSARR modeling package is composed of two components: a watershed model, and a river system and reservoir operation model. The watershed model simulates rainfall-runoff, snow accumulation, and snowmelt-runoff processes. The reservoir/river system model routes flows through river reaches and through controlled and uncontrolled reservoirs.

SSARR is a continuous watershed model designed for large river basins. The computational time step may be varied from 0.1 hour to 24 hours. Streamflows are generated from rainfall and snowmelt runoff. Rainfall data are provided as input. Snowmelt is computed based on input data regarding snow depth, elevation, air and dewpoint temperatures, albedo, radiation, and wind speed. Snowmelt options include the temperature index method and the energy budget method. Application of the model begins with a subdivision of the river basin into hydrologically homogeneous subwatersheds. For each subwatershed, the model computes base flow, subsurface or interflow, and surface runoff. Each flow component is delayed according to different processes, and all are then combined to produce the total subwatershed outflow hydrograph. The subwatershed outflow hydrographs are routed through stream reaches and reservoirs, and combined with hydrographs from other subwatersheds to obtain streamflow hydrographs at pertinent locations in the river system.

A hydrologic storage routing technique is applied to both reservoirs and stream reaches. The routing method developed for the SSARR conceptually treats a river reach as a cascade of reservoirs. Reservoir release rules may be specified a function of either pool elevation, changes in pool elevation, storage, changes in storage, or outflow. Inflows may also be routed through uncontrolled (ungated) reservoirs. Streamflows may be routed as a function of multivariable relationships involving backwater from tides or reservoirs. Diversions may also be included in a simulation.

8.4.2 HEC-1 Flood Hydrograph Package

HEC-1 provides an extensive package of optional computational methods [Hydrologic Engineering Center, 1990]. Precipitation-runoff modeling and flood routing represent the central focus of the package, but other related capabilities are provided as well. In addition to the basic watershed modeling capabilities, the HEC-1 package includes several other optional features, involving partially automated parameter calibration, multiplan-multiflood analysis, dam safety analysis, economic flood damage analysis, and flood control system optimization. The economic flood damage analysis capabilities include computation of expected annual damages as outlined in Chapter 7. The parameter calibration and flood damage reduction system optimization options are noted in the last section of Chapter 10.

The HEC-1 Flood Hydrograph Package includes capabilities for routing hydrographs through reservoirs controlled by either ungated outlet structures or a regulation schedule that is fixed independent of conditions at downstream locations. It does not contain capabilities like HEC-5, described earlier in this chapter, to make release decisions based on flows at downstream control points. HEC-5 does not include the precipitation-runoff modeling capabilities of HEC-1.

Watershed models are often used in combination with other models. For example, HEC-1 synthesizes streamflow hydrographs, that are from rainfall and/or snowmelt provided through the HEC Data Storage System, described in Chapter 5, as input to the HEC-5 reservoir/river system simulation model. The HEC-2 Water Surface Profiles model, described in Chapter 6, is often used in combination with HEC-1. For specified flood events, peak discharges computed with HEC-1 are input to HEC-2, which computes the corresponding water surface profiles.

User's manuals, training documents, and papers and reports on specific applications are available from the Hydrologic Engineering Center. Several universities, as well as the Hydrologic Engineering Center, conduct annual several-day-long short courses for practicing engineers on the application of HEC-1 and HEC-2. HEC-1 and HEC-2 are also used in regular undergraduate and graduate hydrology and hydraulics courses at several universities, and are discussed in textbooks such as [Bedient and Huber, 1992]. References on applying HEC-1 and HEC-2, such as the book by Hoggan [1989], are also available. The HEC list of software vendors includes a number of firms and universities that offer various forms of HEC-1 and HEC-2 user assistance.

HEC-1 provides flexible options for developing and/or inputting precipitation data, which may reflect snowmelt as well as rainfall. Synthetic, historical, or real-time rainfall events may be simulated. A HEC-1 precipitation-runoff modeling application typically involves dividing a watershed into a number of subwatersheds. Runoff computations are performed for each subwatershed. Precipitation volumes are converted to direct runoff volumes using one of the following optional methods: the Soil Conservation Service curve number method, initial and uniform loss rate, the exponential loss rate function, the Holtan loss rate function, or the Green and Ampt relationship. Runoff hydrographs are computed from the incremental runoff volumes using either the unit hydrograph or kinematic routing options. An unit hydrograph may be input to HEC-1. Alternatively, the model includes options for developing synthetic unit hydrographs using either the Soil Conservation Service, Snyder, or Clark methods. The kinematic wave watershed routing option included in the HEC-1 package is usually associated with small urban watersheds.

Watershed modeling also involves routing hydrographs through stream reaches and reservoirs. HEC-1 uses modified Puls routing for reservoirs. The following river-routing options are provided: Muskingum, Muskingum-Cunge, modified Puls, working R&D, average lag, and kinematic wave. The modified Puls and

Muskingum routing methods are described in Chapter 6. Working R&D routing combines features of modified Puls and Muskingum.

8.4.3 Hydrologic Modeling System (HMS)

The Hydrologic Engineering Center is in the process of developing a next generation (NexGen) of hydrologic engineering software. HEC-1 is being replaced by the Hydrologic Modeling System (HMS), which will continue to provide all the capabilities of HEC-1 as well as certain additional capabilities. The NexGen project encompasses several other modeling areas as well. The models will include a graphical user interface, data management, graphics, and reporting capabilities as well as the computational engines. Object-oriented programming, discussed in Section 8.5, is a key feature of the NexGen model development effort. The HMS is coded in C++ and FORTRAN90 and designed for interactive use in a Windows environment on either microcomputers or workstations.

8.4.4 Basin Runoff and Streamflow Simulation (BRASS) Model

The Basin Runoff and Streamflow Simulation (BRASS) model was originally developed by the USACE Savannah District to provide flood management decision support for operation of a reservoir system in the Savannah River Basin [McMahon et al., 1984; Colon and McMahon, 1987] It is used for flood forecasting and other flood management decision support activities. The model is generalized for application to other river basins.

BRASS is an interactive hydrologic/hydraulic simulation model that combines dynamic streamflow routing with aspects of continuous and event rainfall-runoff modeling. For given precipitation input, runoff hydrographs from various subbasins are developed and routed through the stream/reservoir system. This includes storage routing through gated reservoirs and dynamic streamflow routing. BRASS incorporates the National Weather Service (NWS) Dynamic Wave Operational (DWOPER) program for streamflow routing. As discussed in Chapter 6, the NWS DWOPER model computes discharges and water surface elevations based on a numerical solution of the St. Venant equations [Fread, 1987].

8.5 GENERALIZED RESERVOIR/RIVER SYSTEM ANALYSIS MODELS BASED ON OBJECT-ORIENTED PROGRAMMING

Most of the models cited in this chapter are coded in FORTRAN, which is based on a traditional structured approach to programming. The alternative approach of object-oriented programming has become increasingly popular in recent years. In object-oriented programming, a program is treated as a collection of objects which can be reused in different programs and subprograms. Generalized models can be developed that provide an interactive user-friendly graphical environment for users to build models for their particular applications by selecting from a library of pre-coded objects. If needs surface for additional types of objects, a programmer can conveniently expand the library. Object-oriented versions of the C programming language were used to develop the models described next. These generalized modeling systems are designed to provide flexible frameworks for building models for particular reservoir/river system studies. They represent pioneering efforts in applying object-oriented programming to reservoir system analysis.

8.5.1 River Simulation System (RSS)

The River Simulation System (RSS) was developed at the Center for Advanced Decision Support for Water and Environmental Systems (CADSWES) at the University of Colorado, under the sponsorship of the Bureau of Reclamation [CADSWES, 1992]. The generalized modeling system is designed to be adaptable to any reservoir/river system. The interactive graphics-based software package is designed to run on workstations using the UNIX operating system. The model combines advanced computer graphics and data management technology with river/reservoir system simulation capabilities. Several commercial software products are incorporated within the RSS to manage input data and analyze output.

The actual reservoir/river system simulation component of the RSS package is an object-oriented model written in the C and C++ programming languages. River/reservoir systems are represented within the model by node-link components. The user builds a model of a particular system by selecting and combining objects. Preprogrammed instructions for performing computations and data-handling functions are associated with each object. For example, the user could select a reservoir object, hydropower object, or diversion object to represent a system component, which results in the model performing certain computations associated with these particular objects. The user defines reservoir system operating rules using "English-like" statements following a specified format. In general, the user can develop a model for a particular system using the preprogrammed objects and functions provided by the RSS. However, the object-oriented program structure also facilitates a programmer altering the software to include additional objects or functions as needed for particular applications.

8.5.2 CALIDAD

CALIDAD [Bureau of Reclamation, 1994] is also an object-oriented programming model designed for a workstation environment. Initial applications included simulation of the California Central Valley Project [Boyer, 1994]. CALIDAD was developed in the C programming language and one of its object-oriented extensions, Objective C. The graphical user interface was developed in C using Motif and the X Intrinsics Libraries.

CALIDAD simulates a river basin system and determines the set of diversions and reservoir releases which best meets the management objectives and institutional constraints. A user-specified water use scenario is supplied for sequences of streamflow inflows. In the computational algorithms incorporated in the model, user-defined management and institutional constraints are handled using a heuristic technique, called tabu search, to determine permissible diversion and reservoir releases. If the system is overconstrained, the tabu search selects a release schedule using weighting factors provided by the user. Ongoing research and development includes investigation of more computationally efficient solution schemes.

A model for a particular river basin is created by combining a collection of objects representing system features such as stream inflows, reservoirs, municipal or irrigation demand sites, or hydroelectric power plants. CALIDAD provides a palette of precoded objects which can be used for building models for different river basins. Additional objects may be programmed and added to the library as needed. An object developed for a particular application can then be used for other modeling projects as well. Both computational algorithms and data are associated with each object. The physical data for system features such as reservoir storage capacities and streamflows may be entered as object data.

The interactive graphical model building approach is illustrated by the computer monitor screen reproduced in Figure 8–1. The user has specified the configu-

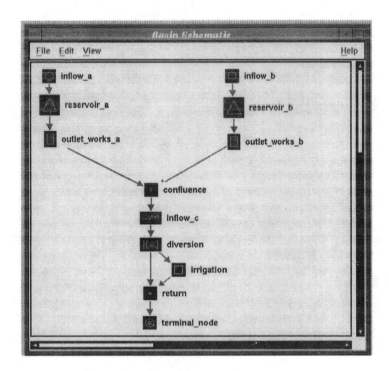

Figure 8–1   CALIDAD basin schematic window.

ration of his particular reservoir/river system by selecting and combining objects from the palette provided by the CALIDAD modeling environment. The collection of objects includes two reservoirs with associated inflows and outlet works. The reservoirs are located on tributaries. Water is diverted for irrigation use at a location below the confluence of the streams. Various input data are required for each object. For example, inflow objects require sequences of streamflows. Data input for a reservoir object include storage capacities and a table defining the relationship between elevation versus storage and area.

Institutional constraints and management objectives, called *rules* in the model, are also treated as data and entered through a separate rules editor. A computer monitor screen with the rules editor is reproduced as Figure 8–2. The user interactively develops operating rules by relating variables associated with objects using arithmetic and logical operations provided by the editor. The user also assigns

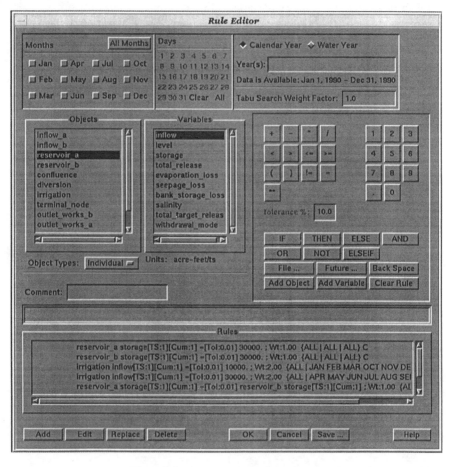

Figure 8–2   CALIDAD rule editor window.

weighting factors representing the relative importance of each rule. During the computations for a particular time step of the simulation, it may not be possible to simultaneously meet all of the rules. The weighting factors are used in the computational algorithm to determine which rules to violate whenever conflicts occur.

8.5.3 Power and Reservoir System Model (PRSYM)

PRSYM is a river, reservoir, and hydropower modeling framework being developed jointly by CADSWES, Bureau of Reclamation, Tennessee Valley Authority, Electric Power Research Institute, and Western Area Power Administration [Shane *et al.*, 1996]. Development of this object-oriented software package was initiated in 1993. It is written in C++.

PRSYM is similar to RSS and CALIDAD in that a library of objects is available for many modeling methods, and additional objects can be developed as needed for particular applications. Each object has computer code that performs appropriate modeling functions. Objects are used to represent reservoirs, powerplants, river reaches, diversions, river confluences, and other system components. Using a screen workspace, users graphically select and link icons associated with the objects needed to represent their particular reservoir/river systems.

Calculations are managed by controller objects. A basic simulation controller performs mass balance and other simulation computations. A rule-based simulation controller combines the basic simulator with policy rules that are specified by the user in an interpreted language. A linear programming optimization controller allows specification of system operating policies in terms of prioritized constraints that can include economic objectives. A water quality simulation controller allows modeling of temperature, salinity, and dissolved oxygen.

The general-purpose PRSYM is designed particularly for analyzing hydroelectric energy production from reservoir systems with non-power objectives and environmental constraints. Initial applications of the object-oriented modeling framework include constructing models for the Tennessee Valley Authority System and the Bureau of Reclamation Colorado River System [Shane *et al.*, 1996].

8.6 GENERAL PURPOSE COMMERCIAL SOFTWARE

A myriad of commercially available software products may be pertinent to reservoir system analysis applications, in regard to providing capabilities for data management, statistical analyses, and graphical displays, as well as constructing the actual simulation model. The discussion here focuses on developing the reservoir/river system simulation model itself rather than pre- and post-processor programs. Spreadsheet programs and object-oriented system simulation environments are particularly applicable for constructing relatively simple models.

8.6.1 Spreadsheets

Numerous spreadsheet/graphics/database programs have been introduced into the market since the early 1980s. The more popular programs include Lotus 1-2-3, Quattro Pro, and Excel, marketed by Lotus Development Corporation, Borland International, and Microsoft Corporation, respectively. The three programs have similar capabilities, and all are available for desktop computers operating under the various popular operating systems. These and many other spreadsheet programs are extensively used in business, engineering, science, education, and other professional fields.

Water management professionals recognized the potential of electronic spreadsheets soon after they were first marketed. These software packages are routinely applied in a variety of water resources planning and management applications. Spreadsheet programs have the advantage of applying the same familiar software to many different applications. A reservoir system analysis problem can be addressed using software which is already being used in the office for other purposes as well. For relatively simple applications, spreadsheets provide capabilities for developing complete reservoir system analysis models. Spreadsheets are also used to manage input and output data for other reservoir operation models.

Spreadsheet programs play a broader role when used in combination with other reservoir operation models. For example, Ford [1990] describes a reservoir operation model called ResQ, which is designed to be used in combination with the user's choice of spreadsheet program. ResQ is an interactive menu-driven model, written in QuickBASIC, which simulates the operation of a single reservoir. The computational routines in ResQ are patterned after the previously discussed HEC-3, but are limited to one reservoir. The simulation results are written to an output file to be read by a spreadsheet program, which can then be used to perform various statistical analyses and create graphical presentations.

8.6.2 Object-Oriented Simulation Modeling Environments

STELLA [High Performance Systems, 1992] is a particularly notable example of a commercially available general-purpose computer program that is well suited for developing reservoir system simulation models. Object-oriented simulation environments like STELLA do not have as vast a market as spreadsheet programs, but are used in a broad range of applications in education, business, science, engineering, and other professional fields. STELLA is marketed by High Performance Systems, Inc., Hanover, New Hampshire, 603-643-9636. STELLA is an acronym for "Systems Thinking, Experiential Learning Laboratory with Animation."

Karpack and Palmer [1992] used STELLA to analyze the water supply systems of the cities of Seattle and Tacoma, Washington. STELLA was applied to several reservoir systems in conjunction with the National Study of Water Management During Drought [Werick 1993; Keyes and Palmer, 1993]. The Hydrologic Engineering Center [1994] used STELLA to simulate the Missouri River system.

STELLA is an object-oriented modeling package designed to simulate dynamic (time-varying or otherwise changing) systems characterized by interrelated components. The user builds a model for a particular application, using the operations and functions provided, and designs the tabular and/or graphical presentation of simulation results. The microcomputer model is graphically oriented and relies on standard Macintosh and Windows operations involving extensive use of a mouse.

A model is developed using STELLA by combining four types of icons or objects: stocks, flows, converters, and connectors. Stocks accumulate flows and are used as state variables to reflect dynamic time-varying characteristics of the system. Numerical integration methods are used to solve the mass or volume balance at each stock. The value or amount associated with a stock can change in each time period in response to flows into and out of the stock. For example, if a reservoir system is being modeled, stocks can represent reservoir storage, which is a time-varying function of STELLA flow objects representing stream inflows, water supply diversions, reservoir releases, and evaporation. Converters are used to store mathematical expressions and data. Connectors provide a mechanism to indicate the linkages between stocks, flows, and converters. A system representation may consist of any number of stocks, flows, converters, and connectors. STELLA provides a number of built-in functions, which are used in developing the logic and mathematics for the particular application

9

A RESERVOIR SYSTEM
SIMULATION CASE STUDY

A series of analyses of a system of 12 multiple-purpose reservoirs in the Brazos River Basin is used as an illustrative case study. Water supply capabilities are evaluated for alternative water use scenarios and system operating strategies. Although the focus is on water supply, hydroelectric power and flood control are also considered.

No single case study can reflect the extreme diversity of water management problems addressed by reservoir operators throughout the nation and world, and of the analysis approaches they are applying to their systems. Each reservoir system and each modeling application are unique. However, the case study serves to illustrate, in some detail, some of the basic concepts outlined in the previous chapters.

9.1 OVERVIEW OF THE STUDY

This chapter is based on a series of studies conducted at Texas A&M University and sponsored by the Texas Water Resources Institute, Brazos River Authority, Texas Water Development Board, Texas Advanced Technology Program, U.S. Geological Survey, and U.S. Army Corps of Engineers. The studies are documented by [Wurbs and Carriere, 1988] and [Wurbs *et al.,* 1988, 1993, 1994].

The primary objective of the Brazos River Basin simulation studies was to quantify the reliability at which various amounts of water can be supplied by an existing system of 12 reservoirs with alternative water use scenarios and management strategies. The investigation included numerous simulation runs and a variety of analyses of simulation results. Alternative reservoir operating policies evaluated in the study involved coordination of releases from multiple reservoirs and unregulated flows, permanent and seasonal reallocation of storage capacity between flood control and conservation purposes, joint use of storage for water supply and hydroelectric power, and minimizing the adverse impacts of salinity on water supply reliability. The improvements in water supply reliabilities that could be achieved by construction of a system of salt control impoundments to reduce salinity loads were also analyzed. Numerous alternative operating plans were formulated and evaluated. Sensitivity analyses and evaluations of various modeling approaches and premises were also performed. Key aspects of these studies are briefly summarized in this chapter.

9.1.1 Reservoir/River System

The Brazos River Basin is described in Chapter 2. As indicated by Figure 2–15, the 118,000 km^2 basin stretches across Texas. Climatic, hydrologic, physiographic, and economic characteristics vary greatly from the upper basin in the flat, arid High Plains to the central basin in the rolling wooded agricultural lands of central Texas to the lower basin in the coastal plain near the cities of Houston and Galveston. Mean annual precipitation varies from 41 mm in the extreme upper basin to 130 mm in the lower basin. Much of the industrial, municipal, and agricultural use of the waters of the Brazos River and its tributaries occur in the lower basin and adjoining coastal area.

The photograph of the Brazos River in Figure 9–1 was taken from Old Washington State Park looking across the Brazos River to the mouth of the Navasota River. This location is downstream of the 12 reservoirs, but upstream of many of the water users. The Navasota River confluence is about 220 km downstream of Whitney Dam, the most downstream of the three reservoirs on the Brazos River. Limestone is the only reservoir on the Navasota River. Five of the 12 reservoirs, including Stillhouse Hollow Reservoir on the Lampasas River, shown in Figure 9–2, are located on tributaries of the Little River, which is one of several major tributaries of the Brazos River. The landscape along the Brazos River and tributary rivers in the Central Texas region consists of pastures, woods, cultivated fields, and small communities.

The case study focuses on evaluating the water supply capabilities of the system of 12 multiple-purpose reservoirs shown in Figure 2–16 and Table 2–18. As discussed in Chapter 2, this system is owned and operated by the Brazos River Authority (BRA) and the Fort Worth District of the U.S. Army Corps of Engineers (USACE). For modeling purposes, the spatial configuration of the river system is represented by a set of control points, shown in the basin map of Figure 9–3 and the system schematic of Figure 9–4. A control point is specified at each of the 12 USACE/BRA reservoirs and Hubbard Creek Reservoir. Five other control points are located at key stream gaging stations. In the models, the location of reservoirs, diversions, return flows, streamflows, and other features are specified by control point. The spatial configuration of the reservoir/river system is represented in the

Figure 9–1 The mouth of the Navasota River can be seen to the far right in this photograph of the Brazos River taken at Old Washington State Park, the site of the 1836 capitol of the Republic of Texas.

Figure 9–2 Stillhouse Hollow Reservoir is on the Lampasas River, which is a tributary of the Little River. (Courtesy Fort Worth District, USACE)

Figure 9-3   Reservoirs and the stream gages adopted as control points for the models.

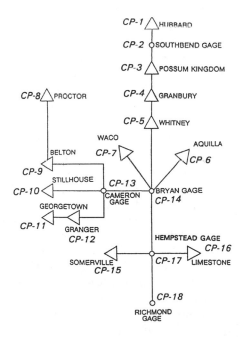

Figure 9-4   System schematic showing 18 control points.

models as shown schematically in Figure 9–4. The majority of the analyses are based on these 18 control points. Some analyses include extra control points and/or do not use all of those shown.

9.1.2 Computer Simulation Models

HEC-3, HEC-5, WRAP, and RESSALT are generalized reservoir/river system simulation models used in the studies. This chapter focuses on HEC-3, HEC-5, and WRAP, which are described in Chapter 8. The economic analysis described in the last section of the chapter is based on monthly and daily time steps, respectively, for simulating water supply and flood control operations. All of the other analyses were performed based on a monthly time step and an 85-year hydrologic simulation period.

The HEC-5 Simulation of Flood Control and Conservation Systems model provides flexible modeling capabilities applicable to a broad range of reservoir system analysis applications. The generalized simulation model has been widely applied in many studies of different river basins performed by various organizations. Only a portion of the optional capabilities were used in the Brazos River Basin study. HEC-5 is described by a user's manual [Hydrologic Engineering Center, 1982, 1989] and other publications available from the Hydrologic Engineering Center. The MS-DOS-based microcomputer package includes several utility programs along with the HEC-5 simulation model, interconnected by a menu program. The Brazos River Basin simulation study was performed using a VAX minicomputer.

HEC-3 was also used in the study. As noted in Chapter 8, essentially all the capabilities of HEC-3 are also contained in HEC-5. The monthly time step simulation of conservation storage operations could be performed with either the simpler HEC-3 or the much larger and more complex HEC-5.

Wurbs, Karama, Saleh, and Ganze [1993] describe a generalized simulation model called RESSALT that was developed in conjunction with the Brazos River Basin study. RESSALT incorporates a salt balance in the monthly simulation. Salt loads are provided as input along with the streamflows. Like the salinity version of WRAP, RESSALT declares diversion shortages if salt concentrations exceed specified limits. RESSALT is patterned after HEC-3, reads an HEC-3 input file, and performs many of the same computations. RESSALT does not include all of the HEC-3 optional capabilities, but, unlike HEC-3, contains salinity features.

The Texas A&M University Water Rights Analysis Package (TAMUWRAP) consists of several versions of the Water Rights Analysis Program (WRAP) simulation model and a program, called TABLES, used to organize and summarize WRAP output [Wurbs, Dunn, and Walls, 1993]. WRAP simulates surface water management under a prior appropriation water rights system.

Unlike HEC-5, WRAP has no flood control capabilities. From the perspective of simulating conservation storage operations, WRAP and HEC-5 have many similar features, including capabilities for using the same Brazos River Basin 85-year sequences of monthly unregulated streamflows and net evaporation rates. The

WRAP input file format is patterned after HEC programs. However, there are significant differences. HEC-5 computations, including those related to meeting water use requirements, proceed from upstream to downstream. WRAP meets water use requirements based on priorities specified in the input. HEC-5 has limits on the number of control points, reservoirs, and other features that can be included in the model. For example, the HEC-5 version used in the study is limited to 20 reservoirs. WRAP is designed for a water rights system involving numerous reservoirs and water use requirements. The Brazos River Basin simulation includes about 600 reservoirs and 1200 diversions. A salinity version of WRAP incorporates simple salt balances. HEC-5Q, discussed in Chapter 12, is a sophisticated water-quality version of HEC-5.

9.1.3 Measures of System Performance

The voluminous HEC-5 and WRAP simulation results include, for each month of a 85-year hydrologic record, reservoir storages, releases, and evaporation, unregulated and regulated streamflows; diversions and shortages; and hydroelectric energy and shortages. The WRAP model also outputs streamflow depletions and unappropriated flows. The salinity version of the model computes salt loads and concentrations. These monthly time series data are summarized in a variety of formats, including summary statistics, time series plots, frequency or duration curves or tables, and reliability indices.

Period and volume reliabilities provide a concise index of capabilities for meeting specified demands. These indices are cited throughout the chapter. Volume reliability (R_v) and period reliability (R_p) were previously defined by equations 7-1 and 7-2.

$$R_v = (v/V)\ 100\% \tag{7-1}$$

$$R_p = (n/N)\ 100\% \tag{7-2}$$

Volume reliability (R_v) is the percentage of the mean rate of some specified target diversion (V) that is reflected in the actual mean diversion rate (v) during the simulation. The difference between v and V represents shortages resulting from water supplies being insufficient to meet demands in some months. In this study, period reliability (R_p) is the percentage of the 1020 months of the simulation for which a target demand is fully met. Thus, N is 1020, and n is the number of months that some specified target demand is satisfied without a shortage. Firm yield is the maximum diversion rate that can be maintained with period and volume reliabilities of 100%.

Evaluation of reliabilities for meeting specified water use requirements is a major focus of the study. Three alternative approaches are adopted for defining water use requirements in the simulation model: (1) hypothetical yields, (2) specified water use scenarios representing actual past (year 1984) or projected future (year 2010) water use, and (3) permitted water rights. Traditionally in Texas, firm yield has been the primary measure of water availability used by federal, state, and local water agencies. Firm yield and other yields associated with yield versus reliability relationships are hypothetical diversion or instream flow requirements representing potential rather than actual historical or projected future water use. The second alternative approach consists of formulating water use scenarios representing actual water use during the past year of 1984 and projected future water use for the year 2010. The third approach for representing water use in the simulation model consists of assuming that all water users use the full amounts to which they are legally entitled by their water rights permits.

9.1.4 Multiple-Purpose Reservoir System Operations

The study focuses on capabilities for supplying water for municipal, industrial, agricultural, and other uses. However, the multiple-purpose reservoir system is actually operated for flood control, hydroelectric power, and recreation, as well as water supply.

As indicated by Table 2–18, the nine Corps of Engineers reservoirs contain flood control pools with storage capacities that are much larger than the conservation pools. Designated top of conservation pool elevations separate the flood control and conservation pools in each reservoir. The Corps of Engineers is responsible for flood control operations. The Brazos River Authority has contracted for most of the conservation storage, the most notable exception being the sizable inactive pool in Whitney Reservoir. The other three reservoirs were constructed by the Brazos River Authority and contain no flood control storage. Flood control operation is based on emptying the flood control pools as quickly as possible without contributing to flows exceeding specified levels at the dams and downstream gages. For example, releases are curtailed whenever flows below Whitney Dam exceed 708 m^3/s or flows at the Bryan or Richmond gages exceed 1700 m^3/s. Allowable flows in the Bosque River below Waco Dam vary from 85 m^3/s to 850 m^3/s, depending on the level of storage in the Waco Reservoir flood control pool. The allowable release rate increases as the flood control pool fills.

A time step of a day or less is required to accurately simulate flood control operations. In most of the monthly time step simulations, the flood control pools were assumed to be empty at the end of each month. In other cases, flood control operations were roughly approximated by including specification of the maximum allowable flows at each control point in the monthly-time-step HEC-5 simulations. The expected annual damage computations of the last study discussed in this chapter were computed by HEC-5 using a daily time step.

Hydroelectric energy is generated at Whitney and Possum Kingdom Reservoirs, which are shown in Figures 9–5 and 9–6. Power generation at Possum Kingdom Reservoir is largely incidental to water supply, with flows through the turbines being limited primarily to releases for downstream water supply diversions. Therefore, the simulation model included no releases from Possum Kingdom Reservoir specifically

Figure 9–5   Whitney is the largest reservoir in the Brazos River Basin. (Courtesy Fort Worth District, USACE)

Figure 9–6   Possum Kingdom is the most upstream of the three reservoirs on the Brazos River. (Courtesy Brazos River Authority)

for hydropower. Whitney Reservoir has a large inactive pool that provides head for hydropower and an active conservation pool which is shared by water supply and hydropower. The Brazos River Authority has contracted with the Corps of Engineers for 22% of the active conservation pool for use for water supply. The Southwestern Power Administration markets the power generated by the Corps of Engineers project to the Brazos Electric Power Cooperative. The simulated Whitney hydroelectric power operations in the model represent an approximation of the contractual commitments that guide the actual operations. In the HEC-5 and WRAP simulations, as long as storage is above the top of inactive pool, the following monthly energy generation targets are met: 6 gigawatt-hours in July and August, 3 gigawatt-hours in June and September, and 2.25 gigawatt-hours in each of the other eight months of the year. These monthly energy requirements total 36 gigawatt-hours per year.

9.1.5 Organization of the Chapter

The case study is presented as a series of four studies:

1. A conventional yield study to estimate firm yields and relationships between yield and reliability for the 12 reservoirs
2. Simulation of system capabilities for meeting specified water use scenarios representing past (year 1984) and projected future (2010) demands
3. A water supply reliability study considering water rights and salinity
4. An economic evaluation of a reallocation of storage capacity in Waco Reservoir from flood control to water supply

The chapter is organized based on devoting a separate section to each of the four studies. Development of the basic hydrologic input data common to these studies is also covered as a separate section.

The first two simulation modeling exercises consist of applying conventional reliability analysis methods to a system of 13 reservoirs. One additional reservoir is included in the model to capture its effects on inflows to other reservoirs in the 12-reservoir system. The 13 reservoirs account for all of the flood control capacity and over 70% of the conservation storage capacity in the basin. However, the effects of numerous other smaller reservoirs in the basin are not considered. HEC-3 and HEC-5 are used in these studies.

The third simulation study addresses water rights and salinity, which are both important considerations in evaluating water supply reliabilities. The WRAP model is used. Water supply reliabilities of the 12-reservoir system are evaluated while considering the fact that over a thousand entities, owning about 600 reservoirs, hold water rights permits to use the waters of the Brazos River and its tributaries.

Natural salt pollution significantly constrains water use in the Brazos River Basin as well as in several other river basins in the southwestern United States.

Along much of the length of the Brazos River, the water is unsuitable for most beneficial uses much of the time due to high salinity. In the first three studies, alternative simulation modeling exercises were performed without and with consideration of the effects of salt concentrations on water supply reliabilities. However, salinity is discussed here primarily in conjunction with the third study.

The fourth study is an economic evaluation of plans for reallocating storage capacity in one of the reservoirs. An approach is presented for analyzing the economic tradeoffs between flood control and water supply. Expected annual flood damages are computed with HEC-5 based on daily time interval streamflow data from historical flood events. Average annual water supply losses are estimated based on assigning dollar values to annual shortages determined based on a monthly time step simulation. Both historical and synthetically generated streamflows are used.

9.2 BASIC HYDROLOGIC DATA

The input files for the HEC-3, HEC-5, WRAP, and RESSALT simulation models include unregulated streamflows for each control point and net evaporation rates for each reservoir location for each month of a 1020-month hydrologic period-of-record simulation period extending from January 1900 to December 1984. Input files for RESSALT and the salinity version of WRAP also include unregulated total dissolved solids, chloride, and sulfate loads for each control point for each of the 1020 months of the simulation. Storage-area relationships are provided for the 13 reservoirs included in the HEC-3, HEC-5, and RESSALT models, and the 592 reservoirs included in the WRAP model.

9.2.1 Unregulated Streamflows

The unregulated flows were developed by adjusting gaged flows to remove the historical effects of reservoirs and water use. Regression analyses were performed to extend gage records with different lengths to cover the 1900–1984 simulation period. Flows at gages located some distance below Aquilla and Limestone Reservoirs were transferred to the dam sites based on ratios of drainage areas, using equation 5-3. Streamflow gages are located near enough to the other dams that adjustments were not required. Gaged monthly flows at the Richmond gage, on the lower Brazos River, are shown in Figure 9–7.

About 140 stream gage stations in the Brazos River Basin are maintained by the U.S. Geological Survey. The records are available through the USGS WATSTORE as well as the Texas Natural Resource Information System (TNRIS) computer databases. Monthly flows for the 18 gages represented by the control points of Figures 9–3 and 9–4, along with flows for a number of other gages, were obtained through the TNRIS. Records of monthly reservoir storage contents for 21

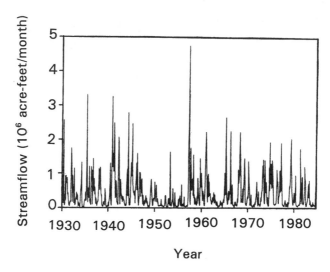

Figure 9–7   Monthly flows at the Richmond gage on the Brazos River.

major reservoirs and diversions for two major canal systems were also obtained from the TNRIS.

Two unregulated streamflow data sets were combined. Unregulated flows covering the period 1940–1976 had previously been developed by the Texas Water Commission (predecessor of the Texas Natural Resource Conservation Commission). As part of the Texas A&M University study, recorded flows were adjusted and extended to develop an alternative set of flows covering the entire 1900–1984 simulation period. The Texas Water Commission data covering the period 1940–1976 were adopted, and the other set was used to cover the remainder of the 1900–84 simulation period.

The Texas A&M University data set was developed by adjusting recorded flows by removing the effects of 21 of the largest reservoirs and major water use diversions through two major canal systems. A FORTRAN computer program was written to perform the computations based on equations 5-1 and 5-2. The Texas Water Commission data set had been similarly developed by adjusting the gaged flows considering numerous additional smaller reservoirs and water use diversions. The two sets of flows for the common 1940–1976 period were found to be very similar, indicating that the differences between the gaged and adjusted unregulated flows are due primarily to the effects of relatively few large reservoirs and diversions rather than the numerous smaller ones.

The streamflow gages all have different lengths of record. Ten gages have records beginning before 1925, and the others begin after 1925. Records extend back to 1899 for two of the gaging stations. Some of the gages have gaps, with records missing for various reasons. The MOSS-IV Monthly Streamflow Simulation model [Beard, 1973], available from the Texas Water Development Board, was used to fill in gaps and extend the records to cover the period 1900–1984. As noted in Chapter 5, the MOSS-IV regression analysis is based on equation 5-6. Several gages in addition to the 18 reflected in Figures 9–3 and 9–4 were used in the data synthesis.

9.2.2 Salt Loads

Input requirements for RESSALT and the salinity version of WRAP include salt loads along with the unregulated streamflows. A water-quality sampling program conducted by the U.S. Geological Survey in support of Corps of Engineers natural salt control studies [McCrory, 1984] provided the data used for developing the salt loads required for the simulation studies. Salt concentrations vary greatly, both temporally and spatially. Mean monthly observed total dissolved solids (TDS) concentrations at the Richmond gage are plotted in Figure 9–8.

Unregulated total dissolved solids, chloride, and sulfate loads were developed for each month of the 1900–1984 simulation period at each control point location. The USGS sampling program was conducted primarily during the period 1964–1986. For stations on the mainstream Brazos River, measured discharges and salt loads were adjusted to remove the storage and evaporation effects of selected major reservoirs. Regression analyses were used to fill in any missing data during the period 1964–1984. Thus, complete sequences of unregulated discharges and loads, covering the period 1964–84, were estimated for the main-stem stations. The 1964–84 data were then used to develop salt load versus discharge regression equations to which unregulated flows were applied to obtain loads for the period 1900–63. For stations on the better-quality tributaries with relatively small salt concentrations, long-term mean measured concentrations were combined with the un-

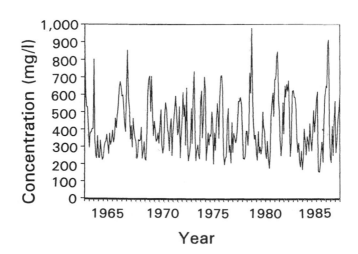

Figure 9–8   Total dissolved solids concentrations at the Richmond gage.

regulated discharges to obtain the monthly loads. The mean concentrations were adjusted to remove the effects of evaporation in selected major reservoirs.

9.2.3 Reservoir Evaporation Rates

Monthly gross and net reservoir evaporation rates for the period January 1940 through December 1986 were obtained from the Texas Natural Resources Information System (TNRIS). Kane [1967] describes development of this TNRIS database. The net reservoir surface loss rate is the gross evaporation loss rate less the effective rainfall rate, which is the rainfall less the estimated amount of runoff without the reservoir. The data are provided on a one-degree quadrangle basis. For reservoirs extending across quadrangle boundaries, the evaporation data for the adjoining quadrangles were averaged for the simulation study. The evaporation database extends back to January 1940. Average 1940–1986 values for each of the 12 months of the year were used in the simulation models for the period prior to January 1940.

9.2.4 Reservoir Elevation versus Storage and Area Relationships

The Brazos River Authority and Corps of Engineers provided elevation versus storage, area, and outlet discharge capacity data for each of their reservoirs. Pool elevations, storage capacities, and other information regarding operating rules were provided as well.

Reservoir storage capacities change over time due to sedimentation. Sediment reserve capacities were specified for the reservoirs during preconstruction planning. For the present study, the agencies provided elevation-storage-area curves for the following conditions: initial at the time of impoundment; resurveyed for reservoirs resurveyed since construction; and ultimate at future times, varying for the different reservoirs from 50 to 100 years after initial impoundment. The estimates of sediment accumulation reflected in the ultimate condition were determined during preconstruction planning based on methods outlined by Borland and Miller [1958] and the Texas Water Development Board [1959]. For purposes of the present study, linear interpolation was applied to the Corps and BRA initial (or resurveyed) and ultimate condition elevation-storage-area relations to develop alternative data for years 1984 and 2010 conditions of sedimentation.

The conservation storage capacities for the reservoirs are presented in Table 9–1 for both 1984 and 2010 conditions of sedimentation. Table 9–1 also includes the dates of initial impoundment and resurveys, if any, of the reservoir bottom topography. Only three of the reservoirs have been resurveyed since initial impoundment. The total storage capacity below the top of conservation pool is shown in Table 9–1, which includes inactive storage at some of the reservoirs.

The Texas Water Development Board [1973] provides information regarding the physical characteristics of 36 major reservoirs in the Brazos River Basin, including elevation-storage-area curves. This information was used in the previously dis-

TABLE 9–1   Storage Capacity Below Top of Conservation Pool

Reservoir	Initial or Resurveyed ($10^6 m^3$)	1984 Sediment ($10^6 m^3$)	2010 Sediment ($10^6 m^3$)	Initial Impoundment	Sediment Resurvey
Hubbard Creek	392	380	371	1962	—
Possum Kingdom	703	671	589	1941	1974
Granbury	189	169	140	1969	—
Whitney	773	739	708	1951	1959
Aquilla	65	64	58	1983	—
Waco	188	165	134	1965	—
Proctor	73	58	39	1963	—
Belton	552	528	459	1954	1975
Stillhouse	291	278	259	1968	—
Georgetown	46	45	43	1980	—
Granger	81	79	70	1980	—
Somerville	197	190	180	1967	—
Limestone	278	269	264	1978	—
Total	3828	3635	3314		
12 Reservoirs	3436	3255	2943		

cussed computations which adjusted the gaged flows to reflect unregulated conditions.

The WRAP model includes 592 reservoirs. Storage capacities specified in the water rights permits are available from a database maintained by the Texas Natural Resource Conservation Commission. Reservoir storage volume versus water surface area data are difficult to obtain for the numerous smaller reservoirs. Storage-area tables were obtained for 36 of the largest reservoirs in the basin, which account for over 90% of the total storage capacity of the 592 reservoirs. A generic storage-area relationship, developed from data available for several relatively small reservoirs, was adopted for the 556 smaller reservoirs.

9.3 ESTIMATES OF FIRM YIELD AND RELATIONSHIPS BETWEEN YIELD AND RELIABILITY

The first simulation exercise involves applying conventional methods to develop estimates of firm yield and yield versus reliability relationships. Both HEC-3 and HEC-5 were used to perform the computations. HEC-3 and HEC-5 provide similar capabilities for simulating conservation storage operations, and either of the models can be used for this type of yield study.

Capabilities are evaluated for meeting hypothetical yields during an assumed occurrence of historical hydrology represented by the 1900–1984 sequences of monthly unregulated flows and net evaporation rates. Yields are expressed as a mean annual diversion rate. However, in the model, diversion targets vary during the 12 months of the year to reflect seasonal variations in water use. A set of 12 monthly water use distribution factors was estimated based on historical water use data. These factors are multiplied by the mean annual yield rate to obtain the diversion target for each of the 12 months of the year.

Analyses are performed for individual reservoirs and systems of multiple reservoirs. An individual reservoir yield is a lakeside diversion from a particular reservoir. Multiple-reservoir system yields are formulated as a diversion from a specified river location that is met by releases from reservoirs located upstream. Alternative simulations are performed for storage capacities corresponding to years 1984 and 2010 conditions of sediment accumulation.

9.3.1 Firm Yields for Individual Reservoirs

Individual reservoir firm yields for 1984 sediment conditions are presented in Table 9–2. The firm yields are expressed as mean annual diversion rates in m^3/s and millon m^3/year. In the model, the diversion requirement is constant from year to year but varies over the 12 months of the year. The firm yields can be compared with the mean reservoir inflows and active conservation storage capacities, which are also tabulated in Table 9–2.

The firm yield is the maximum mean annual diversion rate that can be maintained continuously during the 1900–84 simulation period that represents historical hydrology. This diversion rate just empties the reservoir. End-of-month storages are plotted in Figure 9–9 for a simulation of Somerville Reservoir, shown in Figures 3–3 and 9–10, with a diversion equal to its firm yield of 1.73 m^3/s. The storage hy-

TABLE 9–2   Individual Reservoir Firm Yields

Reservoir	Active Storage ($10^6 m^3$)	Mean Inflow (m^3/s)	Firm Yield (m^3/s)	Firm Yield ($10^6 m^3$/yr)
Hubbard Creek	380	4.5	1.61	51
Possum Kingdom	671	31.8	11.40	360
Granbury	117	28.8	2.35	74
Whitney	293	43.8	5.18	163
Aquilla	64	2.9	0.71	22
Waco	165	12.8	3.28	103
Proctor	58	4.5	0.85	27
Belton	528	17.0	5.01	158
Stillhouse	278	8.6	3.06	97
Georgetown	45	2.5	0.65	21
Granger	79	6.3	0.96	30
Somerville	190	9.2	1.73	55
Limestone	268	8.6	2.83	89
Total	3136	181.3	39.62	1250
12 Reservoirs	2756	176.8	38.01	1199

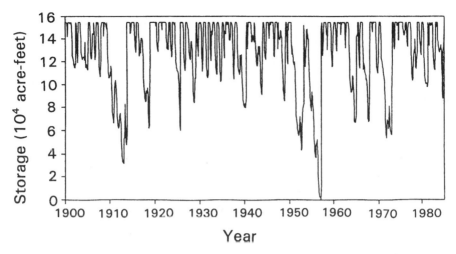

drograph shows that the drought of the 1950s is the critical drawdown period that controls the firm yield. The reservoir is emptied during spring 1957 and completely refilled by the flood of May 1957.

The following scheme is used to reflect the effects on reservoir inflows resulting from reservoirs located upstream. In determining the firm yield for a particular

Figure 9–10   Somerville Reservoir is located on Yequa Creek. (Courtesy Fort Worth District, USACE)

reservoir, any reservoirs located upstream are included in the simulation with diversions equal to their firm yields. For example, referring to Figure 9–4 and Table 9–2, the firm yield of 5.01 m³/s for Belton Reservoir is computed with Proctor Reservoir, included in HEC-5 with a diversion of 0.85 m³/s. Likewise, when computing the firm yield for Granbury Reservoir, its inflows are affected by Hubbard Creek and Possum Kingdom Reservoirs and their diversions of 1.61 and 11.4 m³/s. Hubbard Creek Reservoir is not a component of the 12-reservoir USACE/BRA system. It is included in the simulation because of its impact on inflows to the other reservoirs. Hubbard Creek is the largest reservoir in the basin that is not owned by the BRA or USACE. The numerous smaller reservoirs in the basin are neglected for now, but are considered later in the chapter.

9.3.2 Firm Yields for Multiple-Reservoir Systems

The following operating plan is reflected in the multiple-reservoir system firm yields cited in Table 9–3. Hubbard Creek Reservoir, with a diversion equal to its firm yield, is included in the model and affects inflows to Possum Kingdom Reservoir, but otherwise is not counted in the 12-reservoir system firm yields shown in Table 9–3. Local use diversions of 1.1 m³/s and 3.28 m³/s are specified at Whitney and Waco Reservoirs, to approximate their actual operation as local-use reservoirs. Whitney Reservoir is also operated for hydroelectric power, as outlined earlier. The other ten system reservoirs release to meet the downstream diversion. Multiple-reservoir release decisions are based on maintaining approximately the same percentage of depletion of conservation storage capacity in each of the ten reservoirs. The system firm yield is the sum of the local diversions from Whitney and Waco Reservoirs and the downstream system diversion.

The system diversion is alternatively located at the control point, indicated in the

TABLE 9–3   Comparison of Individual Reservoir and System Firm Yields

Control Point	Individual Reservoir Firm Yield (m³/s)	System Firm Yield		System/Individual	
		Excluding Local Flows (m³/s)	Including Local Flows (m³/s)	Excluding Local Flows (%)	Including Local Flows (%)
1984 Condition of Sedimentation					
Cameron gage	10.5	11.4	13.9	109%	132%
Bryan gage	33.5	41.3	46.7	123%	140%
Richmond gage	38.0	48.0	64.1	126%	169%
2010 Condition of Sedimentation					
Cameron gage	9.7	10.8	13.6	112%	140%
Bryan gage	31.3	39.5	45.1	126%	144%
Richmond gage	35.8	45.8	66.8	128%	173%

first column of Table 9–3. The diversion is met by releases from reservoirs located upstream. System firm yields are shown alternatively excluding and including local flows that enter the river upstream of the diversion location but downstream of the most downstream dams. Incremental flows between adjacent control points are computed by the model as the difference between unregulated flows provided in the input data for the respective control points. The local flows are all incremental flows between the location of the diversion and the upstream reservoirs. One set of system firm yields is based on supplying the diversion requirement strictly with reservoir releases without using local flows. Another set of firm yields is based on meeting the diversion with local flows whenever available, supplemented by reservoir releases as necessary.

Table 9–3 demonstrates the significant contribution to system firm yield of the unregulated flows entering the river below the dams. For example, for 1984 sediment conditions, with the previously discussed operating plan, a diversion at the Richmond gage has firm yields of 48.0 and 64.1 m³/s, without and with use of the local flows. Thus, the local flows increase the firm yield provided by the reservoirs by a third. The flows entering the Brazos River below the dams are little or none some of the time, but significant much of the time. Without reservoir storage these flows provide zero firm yield, but used in combination with reservoir releases, greatly contribute to system firm yield. This is particularly important in the Brazos River Basin, because much of the actual water use is from diversions from the lower reach of the river.

The firm yields of Table 9–3 also illustrate the significance of operating multiple reservoirs in coordination. The multiple-reservoir system firm yields are compared with the sum of the firm yields of the individual reservoirs included in the system. For example, the system above the Cameron gage includes five reservoirs with firm yields from Table 9–2 that total to the 10.5 m³/s cited in Table 9–3. The corresponding multiple-reservoir system firm yield without and with use of local flows, respectively, are 109% and 132% of the sum of the individual reservoir firm yields. The system firm yield associated with placing a diversion at the Richmond gage in the model, excluding and including local flows, is 126% and 169% of the sum of the 12 individual reservoir firm yields. Even neglecting local flows, a diversion at the Richmond gage results in a system firm yield of 126% of the sum of the individual reservoir firm yields.

Multiple-reservoir system operations increase firm yields, or yields associated with a specified reliability, because the critical drawdown periods for the individually operated reservoirs do not perfectly coincide. Operated individually, one reservoir may be completely empty and unable to meet its demands, while significant storage remains in the other reservoirs. At other times the other reservoirs are empty. System operation balances storage depletions and shares risk.

9.3.3 Firm Yield versus Storage Capacity

Firm yield versus storage capacity relationships are useful in investigating reallocations of storage capacity between water supply and flood control pools. Individual reservoir firm yields for alternative storage capacities are presented in Table 9–4 for

TABLE 9–4 Storage Capacity Versus Firm Yield Relationships

	Conservation Storage Capacity as a Percent of Actual Capacity								
Reservoir	: 50	: 80	: 90	: 110	: 120	: 130	: 150	: 175	: 200
	Firm Yield as a Percent of Firm Yield for 100% Actual Capacity								
Aquilla	60	88	96	104	108	116	124	128	132
Waco	69	91	95	103	106	109	112	121	126
Proctor	60	83	93	107	110	113	117	120	127
Belton	76	90	95	105	110	115	120	125	130
Stillhouse	77	94	97	103	105	106	111	116	120
Georgetown	65	87	91	104	104	109	113	117	122
Granger	66	91	95	105	109	114	123	134	139
Somerville	64	85	95	103	107	110	115	123	131

several of the reservoirs for 1984 conditions of sedimentation. Conservation storage capacity is expressed as a percentage of the actual capacity of the reservoir as storage capacity is currently allocated. Likewise, firm yield is expressed as a percentage of the firm yield for the current actual capacity. For example, from Table 9–2, Waco Reservoir has a storage capacity of 165 million m^3 and firm yield of 3.28 m^3/s. Table 9–4 indicates that a 50% increase in the storage capacity will result in a 12% increase in firm yield. A reallocation of 33 million m^3 of the flood control pool of Waco Reservoir to water supply will increase the conservation storage by 20% and the firm yield by 6%.

9.3.4 Yield versus Reliability

The 12-reservoir system firm yield of 64.1 m^3/s cited in Table 9–3 represents, for this operating plan, the maximum diversion that can be met with a computed reliability index of 100%. As indicated in Table 9–5, increasing the diversion target at the Richmond gage results in some shortages, and thus period and volume reliabilities of less than 100%. For example, a diversion target of 80.1 m^3/s is supplied with period and volume reliabilities of 96.76% and 97.16, respectively.

TABLE 9–5 Reliability for a Diversion at the Richmond Gage Control Point

Diversion Target		Shortage Periods (months)	Mean Shortage (m^3/s)	Reliability	
(% Firm Yield)	(m^3/s)			Period (%)	Volume (%)
100%	64.1	0	0.00	100.00	100.00
105%	67.3	10	0.42	99.02	99.38
110%	70.5	15	0.80	98.53	98.86
125%	80.1	33	2.27	96.76	97.16
150%	96.2	68	5.68	93.33	94.10
175%	112.2	110	10.55	89.21	90.60
200%	128.2	157	17.04	84.61	86.71

9.4 SIMULATION OF SPECIFIED WATER USE SCENARIOS

The yields just discussed represent a hypothetical potential, rather than actual, demand. An alternative approach is to incorporate a water-use scenario in the model that represents actual water use. Alternative scenarios based on water-use records for the year 1984 and future projections for the year 2010 were adopted for the study. The Texas Water Development Board collects water-use data and prepares water-use projections in conjunction with its statewide water resources planning responsibilities. The historical water use and future projections are developed by county, by type of use, and by surface versus groundwater sources. The Brazos River Basin encompasses all or portions of 74 counties. Brazos River water is also transported to an adjoining coastal basin for use in four other counties. For the simulation study, water-use data were aggregated by control point. The 18 control points shown in Figure 9–3 plus several others were adopted. The water use varies over the 12 months of the year. Return flows were specified for the different types of water use at each control point. Return flows include water returned to the stream system from withdrawals from both surface and groundwater sources. The 1984 and 2010 water use requirements were combined with 1984 and 2010 conditions of reservoir sedimentation, respectively. Whitney hydroelectric power operations were based on the previously cited energy demands. The effects of flood control operations on water supply were roughly approximated in the monthly-time-step HEC-5 simulation by including specified maximum allowable target flows at pertinent control points.

The 1984 net diversion requirements total 31.9 m³/s, with return flows totalling 5.1 m³/s. About half of the total diversion is assigned to the most downstream control point, with no return flow to the Brazos River. Much of the diversion at the downstream control point represents water transported to the adjoining coastal basin. The 1984 water use is distributed among types of use as follows: municipal 27%, manufacturing 33%, steam electric cooling 14%, irrigation 19%, and livestock 7%. The 2010 diversion requirements total 91 m³/s with return flows of 9.9 m³/s.

The two alternative sets of water-use requirements are combined with the 1900–1984 monthly unregulated streamflow and net evaporation sequences. In the model, the 1984 and 2010 water-use requirements vary over the 12 months of the year, but are constant for each year of the 85-year simulation. Thus, the model simulates capabilities for meeting either 1984 or 2010 water demands during 85-year sequences of monthly flows and net evaporation rates representing historical hydrology.

HEC-5 simulations were performed for numerous alternative storage allocations and management strategies to develop an understanding of system response to the different operating plans. The 1984 and 2010 water-use scenarios were also incorporated in the RESSALT model to evaluate the effects of salinity. The voluminous simulation results were analyzed and displayed in various ways. Two alternative HEC-5 runs are summarized in Table 9–6 and Figures 9–11 and 9–12.

TABLE 9–6   System Water Balance for Alternative Water Use Scenarios

Scenario (year)	System Inflow (m³/s)	Flow to Gulf (m³/s)	Reservoir Evaporation (m³/s)	Net Diversion (m³/s)	Diversion Shortages (m³/s)
1984	223	181	15.9	26.8	0.14
2010	223	139	13.8	75.9	5.24

A water balance for the river basin is presented in Table 9–6 in terms of mean flows during the 1900–1984 simulation period. With the 1984 water-use scenario, the total mean unregulated inflows to the system of 223 m³/s (plus a difference in storage contents between the beginning and end of the 85-year simulation equivalent to less than 1 m³/s averaged over the 85 years) is accounted for as follows. The mean flow to the Gulf of Mexico is 181 m³/s. Reservoir evaporation losses account for 15.9 m³/s. The net diversions less return flows account for the remaining 26.8 m³/s. Diversion shortages averaging 0.14 m³/s occur at an isolated upstream control point, due to lack of access to storage available in the system reservoirs. From a system-wide perspective, 1984 water demands can be met continuously during a repetition of historical 1900–1984 hydrology without greatly depleting storage. Significant shortages occur at some locations with the simulation of 2010 water use requirements.

Storage hydrographs for the total storage in the 13 reservoirs are plotted in Figures 9–11 and 9–12, respectively, for 1984 and 2010 conditions of water use and reservoir sedimentation. Note from Table 9–1 that the total conservation storage capacity in the 13 reservoirs is 3635 and 3314 million m³ (2.95 and 2.69 million ac-ft) respectively, for 1984 and 2010 conditions of sedimentation.

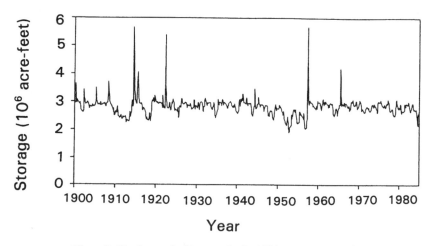

Figure 9–11   Storage in 13 reservoirs for 1984 water-use scenario.

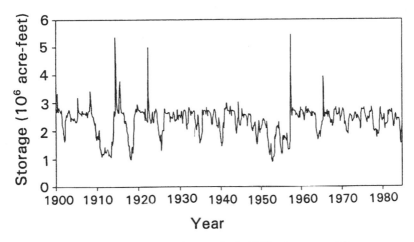

Figure 9–12   Storage in 13 reservoirs for 2010 water use scenario.

9.5 SIMULATION ANALYSES CONSIDERING WATER RIGHTS AND SALINITY

Alternative simulation analyses, without and with incorporation of salinity constraints, were performed using the Water Rights Analysis Program (WRAP) described in Chapter 8 and earlier in the present chapter. As discussed in Chapter 2, during the 1970s and 1980s, the state of Texas completed an adjudication process to consolidate all surface water rights into a prior appropriation permit system, which is administered by the Texas Natural Resource Conservation Commission. Water rights have become a driving concern in water management in Texas.

Natural salt pollution is another significant consideration in evaluating water supply capabilities in the Brazos River Basin and other major river basins in the Southwest. Most of the salt load of the Brazos River originates from groundwater emissions within an area of about 3900 km^2 in the Salt Fork and adjacent watersheds in the upper basin some distance upstream of Possum Kingdom Reservoir. Salt springs and seeps and runoff from salt flats in this area result in high salt loads in the mainstream Brazos River. Possum Kingdom Reservoir, shown in Figure 9–6, has extremely high concentrations. The quality of the river improves greatly in the lower basin due to dilution by tributaries.

9.5.1 Water Rights

The over 1200 rights to store and/or divert water from the Brazos River and its tributaries are included in the model. The water rights include storage in 592 reservoirs. Each diversion and reservoir is assigned a control point location in the model. The totals of the diversion and storage rights aggregated to each control

TABLE 9–7   Pertinent Data by Control Point Location

Control Point	Watershed Area (Km²)	Mean Flow (m³/s)	Mean TDS (mg/l)	Diversion Rights (m³/s)	Storage Capacity (10⁶m³)
1. Hubbard Res	2820	4.5	150	2.41	458
2. Southbend Gage	58,700	28.9	1430	7.70	584
3. Possum Kingdom	61,700	35.0	1200	6.81	770
4. Granbury Res	66,900	45.6	1050	3.40	265
5. Whitney Res	70,600	64.9	700	2.69	1123
6. Aquilla Res	2510	2.9	247	0.27	65
7. Waco Res	4290	12.8	141	2.71	205
8. Proctor Res	3270	4.5	127	1.56	127
9. Belton Res	9170	18.3	127	7.82	562
10. Stillhouse Res	3420	8.6	127	1.87	292
11. Georgetown Res	642	2.5	141	0.54	46
12. Granger Res	1910	7.0	141	0.50	81
13. Cameron Gage	18,300	50.3	141	1.28	4
14. Bryan Gage	102,000	152.	386	4.35	96
15. Somerville Res	2610	9.2	98	1.03	217
16. Limestone Res	1750	8.6	131	2.07	299
17. Hempstead Gage	114,000	205.	311	4.13	120
18. Richmond Gage	117,000	223.	308	38.21	112
				89.33	5426

point are tabulated in Table 9–7. In the prior appropriation system, if all demands cannot be met, water is allocated based on the priority dates specified in the water rights permits. A diversion right may be met by one or several reservoirs or be run-of-river. In some cases, multiple diversion and/or storage rights are associated with a single reservoir.

Diversion rights totaling 89.3 m³/s are distributed among types of use as follows: municipal 55%, industrial 23%, irrigation 19%, and other uses 3%. In the model, monthly water-use distribution factors and maximum allowable salt concentrations are specified as a function of type of use. Hydroelectric power is generated at Whitney Reservoir, with the energy target of 36 gigawatt-hours/year being lower-priority than all diversion rights in the basin. As indicated in Table 9–7, 43% of the total diversion rights are assigned to the Richmond gage control point, which represents the most downstream reach of the river. Much of the water diverted from the lower Brazos River is transported to the adjacent coastal region between the cities of Houston and Galveston. Over a thousand water districts, cities, companies, and individual citizens hold one or more water rights permits. Brazos River Authority permits account for 32% of the total diversion rights. A major portion of the BRA diversions occur in the lower basin, supplied by releases from multiple reservoirs.

Storage capacity in 592 reservoirs totals 5,426 million m³. The 12 BRA/USACE reservoirs contain 63% of the total conservation storage capacity in the basin. Hubbard Creek Reservoir accounts for an additional 8.4% of the basin storage capacity. Thus, about 71% of the storage capacity of the 592 reservoirs is contained in the 13 largest reservoirs.

9.5.2 Water Rights Model

In the simulation study, all water rights holders are assumed to use the full amounts specified in their permits, subject to water availability, during the 1900–1984 hydrologic simulation period. For each month of the simulation, the WRAP model performs the water accounting computations for each right, in turn, on a priority basis. The water rights input file includes the priority date specified in the permit for each right. The computations proceed by month and, within each month, by water right, with the most senior right in the basin being considered first. Diversion shortages are declared whenever insufficient streamflow and/or reservoir storage is available to meet the diversion target. With the salinity version of the model, diversion shortages are also declared whenever the salt concentrations of the streamflow or reservoir storage exceed maximum allowable concentrations. Various rules can be specified for using combinations of unregulated streamflows and reservoir releases to meet diversion and instream flow requirements. Multiple-reservoir system operations are based on balancing the percentage full of specified storage zones in each reservoir included in the system for a particular right. Input for each water right includes the location, annual water use amount (or hydroelectric energy demand), storage capacity, priority date, type of water use, and return flow specifications. A set of 12 monthly water use distribution factors is provided for each type of water use.

9.5.3 Complexities and Issues in Administering and Modeling a Water Rights System

The implementation of a permit system and the adjudication of water rights have resulted in a manageable allocation of surface water resources in Texas. However, many issues remain to be resolved. Allocating a highly variable water resource to numerous water managers and users, who use the water for a broad range of purposes, is necessarily complex. The simulation modeling study included identification and examination of key aspects of river basin management and associated water availability modeling under the water rights system. Considerations noted in the following discussion include defining supply failures, allocating water shortages among numerous water users, priority of reservoir storage relative to diversion rights, return flows, multiple-reservoir system operations, water quality constraints, and assessing reliability.

The simulation model allows each right, in priority order, to meet its diversion target and refill reservoir storage, as long as yet-unappropriated flows are available, during each month of the 1900–1984 simulation period. Shortages are declared for a particular right if streamflow and/or storage is not available. The diversion shortages computed by the WRAP model represent a general index of supply failures. In the real world, supply failures involve emergency demand management measures, negotiation of resource reallocations, or similar actions. In the model, a junior right is not allowed to make any diversion which would adversely affect a more senior right. In an actual real-world situation involving insufficient water supply, the shortages are shared, to some degree, by water users regardless of the relative seniority of their rights. However, detailed drought contingency plans have not yet been developed in Texas. Also, in the simulation modeling study, supply shortages are not declared until the pertinent reservoir storage is empty. In actuality, water management problems will occur, and emergency actions be taken, long before reservoir storage is completely depleted.

Although some recently issued water rights have addressed return flows, most permits do not specify the amount of the diversion to be returned to the stream. Return flows can significantly impact the availability of water to downstream users. The WRAP input file includes a return flow location and factor, reflecting the percentage of the diversion returned to the stream, for each diversion right. Limited historical data compiled by the Texas Water Development Board and Texas Water Commission (TNRCC's predecessor) were used to estimate return flow factors as a function of type of use for different regions of the basin. In most cases, flows were assumed to be returned at the next downstream control point. Since the Richmond gage is the most downstream control point, the very large diversions here contribute no return flows in the model.

Assigning priorities to maintaining reservoir storage levels relative to diversion rights is another important issue. Reservoir operation in Texas is based on providing long-term storage as protection against infrequent but severe droughts. The right to store water is as important as the right to divert water. If junior appropriators located upstream of a reservoir diminish inflows to the reservoir when it is not spilling, reservoir dependable yield is adversely affected. Each drawdown could potentially be the beginning of a several-year critical drawdown which empties the reservoir. Thus, protecting reservoir inflows is critical to providing a dependable water supply. On the other hand, forcing appropriators with rights junior to the rights of the reservoir owner to curtail diversions to maintain inflows to an almost full, or even an almost empty, reservoir is difficult and not necessarily the optimal use of the water resource. If junior diversions are not curtailed, the reservoir will likely refill later anyway, without any shortages occurring.

Handling of the storage aspect of water rights is not yet precisely defined in Texas. Water rights grant the right to both divert and store water. Water rights permits in Texas include a single priority date. Priorities are not specified separately for diversions and storage. Reservoir storage priorities have a very significant effect on simulation results. For example, the diversion rights associated with the 12 BRA/USACE reservoirs have high reliabilities because of the large storage volumes. These water supply reliabilities are highly dependent on protecting storage from the effects of junior appropriators.

The following scheme was somewhat arbitrarily adopted in the simulation study as being a reasonable approach. The major reservoirs are filled to 80% of capacity with priorities associated with the water rights and then to 100% capacity with priorities junior to all diversion rights in the basin.

9.5.4 Salinity Constraints

Normally in Texas, the quality of water supply diversions has not been explicitly considered in the process of evaluating water rights applications and issuing permits. Water quality considerations have typically focused on including restrictions on new water rights permits to maintain instream flows. Although salinity is widely recognized as being an important concern in the Brazos and other river basins, there really are no commonly accepted practices for incorporating salinity considerations in assessment of water supply reliabilities in the state. However, the Brazos River Basin study includes formulation of modeling exercises to contribute to a better understanding of the impacts of salinity.

In the Brazos River Basin simulation study, salinity is treated as a constraint to the use of diverted water for off-stream uses. Alternative WRAP simulations are performed with maximum concentration levels specified for each salt constituent for each type of water use. Diversion shortages are declared by the model any time salt concentrations exceed the allowable limits. Total dissolved solids (TDS), chloride, and sulfate are the constituents used in the model. As previously discussed, WRAP input includes loads of each constituent for each month of the 1900–1984 simulation period at each control point, along with streamflows. Concentrations exhibit great variability, both temporally and spatially. Salt accounting computations in WRAP are simple mass balances based on the assumptions that the salts are conservative, and are instantaneously and uniformly mixed in reservoirs and river reaches. Reservoir evaporation removes water but not salt. Diversions remove both water and salt.

The simulation study did not include establishment of acceptable salt concentration limits for particular water uses; rather, alternative model runs were made to demonstrate the sensitivity of simulation results to a range of assumed maximum allowable concentrations.

Tolerable or acceptable concentration limits for various types of water use are difficult to precisely define. The tolerance to infrequent short periods of high salinity may be significantly different than to more constant long-term high salinity levels. Acceptable salt concentration limits for irrigation vary greatly depending on various factors, such as the type of crop and relative amounts of rainfall versus supplemental irrigation. Salinity impacts and tolerance limits also vary greatly with different types of industrial water use. The U.S. Environmental Protection Agency secondary drinking water standards suggest a TDS limit of 500 mg/l. The state of Texas uses a 1000 mg/l TDS drinking water criterion. Incorporation of salinity limits in water supply reliability studies is further complicated because the water rights system allows significant flexibility for shifting between types of use each year.

9.5.5 Proposed Salt Control Impoundments

The simulation study also includes an analysis of the impacts of a system of three salt control impoundments previously proposed by the Corps of Engineers. Natural salt pollution control studies conducted by the USACE involved formulation and evaluation of a comprehensive array of strategies for dealing with the salinity problem in the Brazos River Basin [McCrory, 1984]. Much of the salt load in the Brazos River originates from relatively small subwatersheds in the upper basin. A number of the alternative plans consist of systems of salt control dams to contain the runoff in the primary salt source areas. A plan consisting of three impoundments was recommended but later found to be economically infeasible for federal implementation. The proposed dams would impound runoff from Croton, Salt Croton, and North Croton Creeks, which are small tributaries of the Salt Fork of the Brazos River located in the primary salt source area of the upper basin about 150 km upstream of the Southbend gage shown in Figure 9–3. The impounded water would be partially lost over time due to evaporation, with the remaining brine being permanently stored. The Corps of Engineers salt impoundment plan is designed to prevent all flows and loads from passing the three proposed dam sites [McCrory, 1984].

In the present simulation study, the salt impoundment plan is represented by removal of all flows and salt loads at the sites of the three proposed salt control dams. The basic data sets of 1900–1984 monthly unregulated streamflows and salt loads are discussed earlier in this chapter. The proposed salt impoundment plan is represented by an alternative set of flows and loads for all control points on the main-stem Brazos River adjusted to remove all flows and salt loads originating above the sites of the proposed salt control dams.

9.5.6 Simulation Results

The simulation study involved numerous executions of the WRAP model for alternative water management strategies, system operating rules, and modeling assumptions [Wurbs et al., 1994]. Table 9–8 illustrates the relative magnitude of various quantities involved in a base run simulation without considering salinity. A water balance for the river basin for a particular month, year, or the entire 1900–84 simulation period can be expressed as follows:

storage change = unregulated flows + return flows
$$- \text{ evaporation} - \text{diversions} - \text{unappropriated streamflows}.$$

Basin total quantities are tabulated in Table 9–8 for 1900–1984 means and for years 1956 and 1957, which represent extreme drought and flood conditions. Remember that, in the model, present water rights diversion targets and storage capacities are combined with 1900–1984 historical natural streamflows and reservoir evaporation rates. The most severe drought during the 1900–84 hydrologic simulation period began in 1950 and ended with one of the largest floods of record in April 1957. The

TABLE 9-8   Basinwide Water Balance

Quantity	Mean (m³/s)	1956 (m³/s)	1957 (m³/s)
Unregulated streamflow	223	36	586
Available return flow	18	16	17
Evaporation	23	24	16
Diversion	84	68	85
Unappropriated flows	134	0	386
Change in storage	0	−40	116

driest year of the simulation, with the smallest unregulated streamflows and greatest model diversion shortages, is 1956. Interestingly, the following year, 1957, has the highest unregulated streamflow of the 85 years. Table 9–8 shows the great variation in mean unregulated streamflows from 36 m³/s in 1956 to 586 m³/s in 1957, compared to a 1900–84 mean of 223 m³/s.

Of the total diversion rights of 89.33 m³/s, the mean 1900–84 diversion is 83.36 m³/s for a volume reliability of 93.32%. Volume reliabilities for diversion rights aggregated by control point are tabulated in Table 9–9 for base run 1, previously described, and several other runs, noted next. The diversion rights or targets are also shown. The corresponding information for the diversion rights associated with the 12 BRA/USACE reservoirs are tabulated in Table 9–10 and also noted in the discussions to follow. The BRA diversion rights at the Richmond gage control point are supplied by unregulated flows supplemented by multiple-reservoir releases.

9.5.7 Reliability versus Allowable Salt Concentration Limits

The TDS concentration versus reliability relationships shown in Tables 9–9 and 9–10 are based on the hypothetical assumption that the indicated TDS concentration limits are applied to all the diversion rights in the basin. Specified maximum allowable TDS concentration limits incorporated in the alternative simulation runs of Tables 9–9 and 9–10 range from constraining all diversions to a very stringent TDS limit of 500 mg/l (run 2), to the other extreme of specifying no limits at all (run 1). The volume reliability estimates for the total of all the diversion rights in the basin range from 66.69% for the 500 mg/l TDS limit to 93.32% if salinity is not considered. For the aggregated total of all the BRA diversion rights, the reliability ranges from 64.90% to 98.22%. For just the BRA diversions assigned to the Richmond gage control point in the model, the aggregated reliability is 86.58%, 98.79%, and 100.00%, respectively, for TDS constraints of 500 mg/l, 1000 mg/l, and 2000 mg/l.

Run 5 in Tables 9–9 and 9–10 represents conditions that would exist with the construction of the previously proposed system of three salt control impoundments

TABLE 9–9   Volume Reliability by Control Point for Alternative Simulation Runs

Control Point	Rights (m³/s)	Reliability (%) for Alternative Model Runs				
		1	2	3	4	5
1. Hubbard	2.41	78.15	87.21	86.76	82.57	86.44
2. South Bend	7.70	76.45	0.24	0.24	28.10	0.35
3. Possum K.	6.81	99.15	0.68	0.99	70.97	4.66
4. Granbury	3.40	95.85	1.31	2.28	87.47	24.16
5. Whitney	2.69	90.76	3.03	51.78	92.74	83.49
6. Aquilla	0.27	99.15	99.13	99.88	99.87	99.88
7. Waco	2.71	95.42	95.65	95.59	95.49	95.56
8. Proctor	1.56	75.19	75.33	75.27	75.21	75.26
9. Belton	7.82	89.93	90.07	90.06	89.99	90.06
10. Stillhouse	1.87	96.69	97.04	96.97	96.84	96.95
11. Georgetown	0.54	99.84	99.88	99.87	99.85	99.87
12. Granger	0.50	96.44	97.15	96.98	96.64	96.90
13. Cameron	1.28	89.09	92.43	92.02	90.56	91.61
14. Bryan	4.35	90.38	63.16	90.50	91.77	91.69
15. Somerville	1.03	99.89	99.93	99.93	99.93	99.93
16. Limestone	2.07	99.46	99.52	99.52	99.52	99.52
17. Hempstead	4.13	93.01	80.35	94.04	94.10	94.65
18. Richmond	38.21	97.53	86.48	97.37	98.06	97.70
Basin Total	89.33	93.32	66.69	74.82	87.25	77.11

Run 1—salinity not considered
Run 2—500 mg/l TDS constraint
Run 3—1000 mg/l TDS constraint
Run 4—2000 mg/l TDS constraint
Run 5—1000 mg/l TDS constraint with salt impoundments

in the primary salt source areas of the upper basin. The model input for run 5 is identical to run 3, except the input unregulated streamflows and salt loads have been adjusted to reflect the proposed salt impoundment plan.

9.5.8 Multiple-Reservoir Systems

Water rights permits in Texas are granted for individual reservoirs rather than multiple-reservoir systems. All of the diversion rights held by the Brazos River Authority were granted in conjunction with each of the 12 individual reservoirs. However, the Brazos River Authority and Corps of Engineers operate their reservoirs as a system. A large portion of the BRA diversion rights involve withdrawals from the lower Brazos River which are supplied by excess streamflows and releases from multiple reservoirs. Excess flows consist of unappropriated unregulated flows entering the river below the upstream dams and uncontrolled reservoir spills. The BRA holds an excess flows permit which allows diversion of unregulated streamflows from the lower Brazos River, in lieu of reservoir releases, as long as no other water users are adversely affected. The BRA permits have been modified to allow multiple-

TABLE 9–10   Volume Reliability For Brazos River Authority Diversion Rights

Control Point	Rights (m³/s)	Reliability (%) for Alternative Model Runs				
		1	2	3	4	5
3. Possum K.	5.99	100.00	0.68	0.99	71.15	4.66
4. Granbury	2.15	97.23	1.33	2.28	88.88	24.02
5. Whitney	0.72	70.45	2.70	47.81	77.16	73.51
6. Aquilla	0.26	99.28	99.24	100.00	100.00	100.00
7. Waco	2.31	100.00	100.00	100.00	100.00	100.00
8. Proctor	0.77	83.71	83.72	83.72	83.71	83.72
9. Belton	3.92	97.40	97.41	97.41	97.41	97.41
10. Stillhouse	1.55	100.00	100.00	100.00	100.00	100.00
11. Georgetown	0.53	100.00	100.00	100.00	100.00	100.00
12. Granger	0.43	100.00	100.00	100.00	100.00	100.00
15. Somerville	1.03	99.96	100.00	100.00	100.00	100.00
16. Limestone	1.83	99.94	100.00	100.00	100.00	100.00
18. Richmond	6.71	100.00	86.58	98.79	100.00	99.26
Total	28.20	98.22	64.90	69.10	91.64	72.30

Run 1—salinity not considered

Run 2—500 mg/l TDS constraint

Run 3—1,000 mg/l TDS constraint

Run 4—2,000 mg/l TDS constraint

Run 5—1,000 mg/l TDS constraint with salt impoundments

reservoir releases as long as the total of the diversions specified in the individual reservoir permits are not exceeded. Flexibility has also been provided for shifting between types of water use as well.

The various aspects of multiple-reservoir system operation addressed by Wurbs et al. [1994] include use of excess flows in combination with reservoir releases; balancing multiple-reservoir releases; effects of tributary versus main-stem reservoir releases on salinity in the lower Brazos River; and balancing local versus system diversion reliabilities.

The simulation study demonstrates that multiple-reservoir system operations, particularly use of excess flows in combination with reservoir releases, are very beneficial in maintaining water supply reliabilities in the Brazos River Basin. For example, the yield versus reliability relationships for scenarios 1 and 2 in Table 9–11 illustrate the effects of using excess flows along with the permitted diversion rights associated with the reservoirs. A diversion right of 31.3 m³/s at the Richmond gage is met with volume reliabilities of 99.45% and 97.68%, respectively, with (scenario 1) and without (scenario 2) use of excess flows. The nine-reservoir system firm (100% reliability) yields are 27.4 m³/s and 20.7 m³/s, respectively, with and without use of excess flows.

The yield versus reliability relationships tabulated in Table 9–11 are for a municipal diversion at the Richmond gage control point supplied by releases from nine BRA/USACE reservoirs. These yields represent hypothetical diversions at the Rich-

TABLE 9–11　Yield Versus Reliability Relationships for a Diversion
at the Richmond Gage Supplied by Nine BRA/USACE Reservoirs

Yield (m³/s)	Reliability (%) for Scenario			
	1	2	3	4
24.4	100.00	99.50	84.78	89.31
27.4	100.00	98.81	84.15	90.40
31.3	99.45	97.68	83.00	90.42
39.1	97.76	95.66	82.01	89.95
46.9	96.23	93.17	80.60	89.79

Scenario 1—salinity is not considered

Scenario 2—salinity not considered and excess flows not used

Scenario 3—500 mg/l TDS constraint

Scenario 4—500 mg/l TDS constraint with salt control impoundments

mond gage control point in the model and replace diversion rights of 24.4 m³/s associated with the nine reservoirs. In scenarios 1 and 3, the Richmond gage diversion is met by excess flows supplemented as necessary by releases from the nine reservoirs. In scenario 2, the Richmond gage diversion is met solely by reservoir releases without using excess flows. Limiting a 27.4 m³/s target diversion from the lower Brazos strictly to releases from the upstream reservoirs reduces the reliability from 100% to 98.81%. Salinity is not considered in scenarios 1 and 2. Diversions are constrained to a maximum allowable TDS concentration of 500 mg/l in scenario 3. A TDS concentration limit of 500 mg/l reduces the reliability for the 27.4 m³/s diversion target from 100% to 84.15%. Specifying a stringent allowable TDS concentration of 500 mg/l will reduce the firm yield to zero.

9.5.9 Yield versus Reliability

Water management decisions necessarily require qualitative judgment in determining acceptable levels of reliability for various situations. Trade-offs occur between the amount of water to commit for beneficial use and the level of reliability that can be achieved. Beneficial use of water is based on ensuring a high level of reliability. However, limited resources may have to be allocated to many competing users. If water commitments are limited as required to ensure an extremely high level of reliability, the amount of streamflow available for beneficial use is constrained, and most of the water flows to the ocean or is lost through reservoir evaporation much of the time.

Reliabilities are not very sensitive to changes in diversion amounts. Conversely, yields change greatly with relatively small changes in reliability. Referring to Table 9–11, for example, without considering salinity, increasing the nine-reservoir system diversion demand 25% from 31.3 m³/s to 39.1 m³/s reduces the reliability from 99.45% to 97.76%. Adding a relatively large 3.9 m³/s to the firm yield of 27.4 m³/s results in a yield of 31.3 m³/s which still has a relatively high reliability of 99.45%. If diversions are constrained by specifying maximum allowable salt concentration limits, reliabilities are even less sensitive to changes in yield. If the Richmond gage yield of Table 9–11 is constrained to a TDS concentration of 500 mg/l, the salinity constraint controls the reliability with almost no variation of reliability for different yield magnitudes.

The amount of water supplied from the Brazos River Basin can be increased significantly by accepting somewhat higher risks of shortages or emergency demand reductions. Yield versus reliability estimates are not highly precise and can vary significantly with incorporation of different but still reasonable assumptions in the model. Depending on the diversion location and allowable salt concentrations for a particular water use, water availability may be controlled more by quality than quantity.

9.6 ECONOMIC EVALUATION OF A STORAGE REALLOCATION

Wurbs and Cabezas [1987] present an economic evaluation procedure which is applied to a proposed reallocation of storage capacity in Waco Reservoir from flood control to municipal and industrial water supply. The economic analysis compares the trade-offs between flood control and water supply in commensurate units of dollars. Average annual flood losses are computed using the conventional damage-frequency method outlined in Chapter 7. Unlike traditional practices, water supply is treated analogously to flood control, with economic consequences of water shortages being quantified. Average annual water supply losses, in dollars, are estimated by developing a water shortage versus economic loss function, which is then applied to water shortages computed by a hydrologic simulation. The water shortage versus loss function reflects emergency demand management and supply augmentation measures. Average annual water supply losses are estimated for a given demand level. Long-term demand management strategies are reflected in the water demand projections.

Construction of Waco Reservoir was completed by USACE in 1965. As indicated in Tables 2–18 and 9–1, the original water supply and flood control storage capacities were 188 and 708 million m^3, respectively. The conservation pool is committed entirely to supplying water for the city of Waco and adjacent smaller cities. Figure 9–13 shows the reservoir with the water surface at the level of the original top of conservation pool.

The analysis of Waco Reservoir reported by Wurbs and Cabezas [1987] and summarized here was a university research effort building upon a prior study performed by the Corps of Engineers. During the early 1980s, at the request of the City of Waco and the BRA, the USACE investigated the feasibility of increasing the water supply storage capacity. Studies resulted in a recommendation to raise the designated top of conservation pool by 2.13 m (7.0 ft), converting 58.6 million m^3 of flood control storage to water supply. A reallocation of this magnitude fell just

Figure 9–13 The Waco Reservoir storage reallocation involves raising the top of conservation pool. (Courtesy Awes S. Karama)

within the discretionary authority of the USACE to implement without requiring Congressional approval. A contract between the BRA and Corps of Engineers has been executed. A water rights permit for the reallocation was granted in the early 1990s. Reallocation of recreation facilities is required prior to actually implementing the pool raise.

The flood control pool of Waco Reservoir will contain a flood with an estimated recurrence interval slightly greater than 100 years without releases contributing to downstream flooding. The reallocation will reduce the protection to containing an approximately 80-year recurrence interval flood. The firm yield will be increased by about 27%. The actual determination of the amount of storage capacity to reallocate was largely based on providing an estimated firm yield in excess of projected water needs. For this particular reservoir, the results of the economic analysis are consistent with the objective of maintaining a firm yield just in excess of demand.

Over the past several decades, detailed economic evaluation procedures have been developed for flood control primarily because the Flood Control Act of 1936 and subsequent policy statements have required a benefit-cost justification for federal projects. Municipal and industrial water supply projects, on the other hand, have been developed based on the concept of meeting needs at least cost. However, the limited experience of this case study indicates that supply losses could likely be estimated as meaningfully as flood losses if a comparable level of effort were to be devoted to the development and application of economic evaluation methods for water supply; realizing, of course, that both water supply and flood control analyses involve significant estimations and judgments, and are necessarily approximate.

9.6.1 Average Annual Flood Damages

The damage-frequency method of estimating expected annual flood damages is outlined in Chapter 7. The expected annual damage option of HEC-5 was used to perform the computations. The scope of the study did not allow conducting the field studies necessary to obtain detailed data required for an accurate flood damage evaluation. Consequently, the HEC-5 input data were developed based on approximations and information available from prior Corps of Engineers basinwide planning studies.

Nine reservoirs are operated for flood control. In the HEC-5 simulation, the flood control pools are emptied as quickly as possible within the constraints of specified maximum discharge limits at the dams and at downstream control points. Multiple-reservoir release decisions are based on balancing the percentage of depletion of flood control storage capacity in each reservoir.

Stream gage locations are used as damage index locations as well as model control points. Frequency analyses of annual series of peak flows were performed to develop exceedance probability versus discharge relationships at each pertinent gage. Economic damage versus discharge relationships obtained from prior Corps of Engineers studies were updated using price indices. The hydrologic simulations are based on seven historical floods and several other synthetic floods, developed by scaling the historical floods using methods incorporated in HEC-5. A computational time step of one day is used.

The computations performed by HEC-5 include routing each unregulated flood through the river system, without the reservoirs, and assigning exceedance probabilities to each flood at each pertinent control point based on the frequency-discharge tables provided as input; routing each flood through the reservoir/river system, with a specified storage allocation, and determining the damages at each control point based on the damage-discharge tables provided as input; and combining the set of damages and exceedance probabilities associated with the floods to determine expected annual damages.

9.6.2 Average Annual Losses Due to Water
Supply Shortages

The water supply analysis has four components: projection of water demands, development of a water shortage versus economic loss function, development of a sequence of reservoir inflows, and simulation of the reservoir to determine average annual losses.

The economic evaluation is performed for specified alternative water demands. The water demand area is the Waco Standard Metropolitan Statistical Area, which includes the city of Waco and several smaller adjacent communities. Water demands to be supplied by Waco Reservoir were projected alternatively for the years 1990, 2000, 2010, 2020, and 2040 based on population projections, per capita water-use rates, and other socioeconomic indicators. Long-term demand management measures are reflected in the water use rates.

A relationship between water shortage, expressed as a percentage of demand, and economic losses, in dollars, was developed for various levels of demand. The estimation of economic losses due to water shortages involved the formulation of emergency demand management and supply augmentation plans, estimation of effectiveness and implementation costs for emergency measures, and estimation of other losses to water users. Development of the shortage-loss function was largely based on professional judgement utilizing information from the literature regarding experiences in responding to water shortages and an analysis of water use practices in the study area. The following emergency water management measures were selected as being representative of actions that could be taken in the event of a water shortage in the study area: (1) supply augmentation by relying on groundwater, (2) implementation of voluntary demand reduction programs, (3) modification of the water price structure, (4) implementation of mandatory demand reduction programs, and (5) importation of water by trucks or emergency pipelines. The range of water shortage severity by which each measure would be most applicable, effectiveness in reducing demands or augmenting supplies, and implementation costs were estimated. Other losses to residential and industrial water uses and the municipality were also estimated. A demand reduction of about 4.5% was estimated to result in essentially no economic losses. Past that level, the cost per unit demand reduction or supply augmentation is a rapidly increasing nonlinear function of severity of the water shortage.

Two alternative sequences of monthly inflows into Waco Reservoir were used: the 76-year historical record of gaged flows and a synthetically generated 1000-year sequence. The 76-year record of measured flows were considered to be adequate for the study. However, as a comparison, alternative simulations were performed with the 1000-year synthetic sequence. The synthetic flows were generated with the previously noted MOSS-IV Monthly Streamflow Synthesis model.

Average annual economic losses were computed for alternative reservoir storage allocations. A simulation consisted of combining a specified water demand target with a sequence of monthly reservoir inflows and evaporation rates. Monthly shortages were totalled for each year and combined with the annual shortage versus loss function to obtain economic losses, if any, for each year of the simulation. The annual losses were then averaged.

9.6.3 Comparison of Losses

Average annual flood and water shortage losses for alternative water demand levels and top of conservation pool elevations are shown in Figures 9–14 and 9–15 in terms of changes from losses associated with the original storage allocation. A

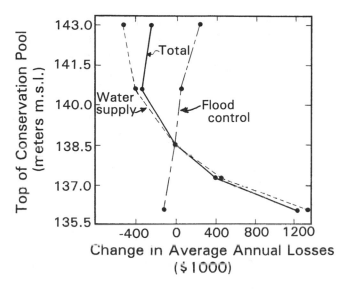

Figure 9–14   Comparison of losses for year 2000 water demand.

complete economic analysis includes comparing the change in total losses associated with a given storage reallocation plan with the discounted implementation costs, which for this project is the cost for relocating recreation facilities to be inundated by the pool raise. However, the present discussion focuses on comparing

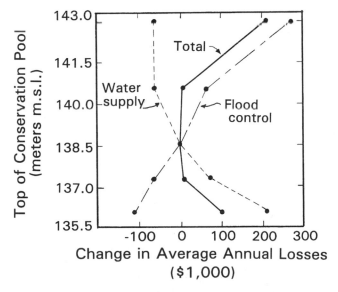

Figure 9–15   Comparison of losses for year 2020 water demand.

trade-offs between flood control and water supply. Figure 9–14 summarizes an evaluation of alternative reallocation plans based on year 2000 water demands of 2.25 m^3/s. The five alternative top of conservation pool elevations plotted in Figure 9–14 represent the original storage allocation, two pool-raising options, and two pool-lowering options. The changes in flood and water supply losses balance each other for reallocations of relatively small amounts of storage capacity. However, if the top of conservation pool is raised too much, the increase in flood losses is much greater than the decrease in water supply losses. Figure 9–15 shows the results of repeating the analysis for the year 2020 water demand of 2.94 m^3/s. With this significantly higher water demand, raising the top of conservation pool by 2.13 m results in much greater decreases in water supply losses than increases in flood control losses.

Feasibility studies of alternative storage allocations and water management strategies involve complex institutional, financial, legal, political, public opinion, and environmental considerations, as well as the hydrologic and economic factors addressed here. However, the results of hydrologic and economic modeling analyses provide an improved understanding of system performance, and meaningful quantitative information to support decision-making processes.

10

OPTIMIZATION TECHNIQUES

The terms *optimization* and *mathematical programming* are used synonymously to refer to a modeling approach in which a formal algorithm computes values for a set of decision variables that minimize or maximize an objective function subject to constraints. Mathematical programming models are formulated in a specified format for solution with available standard methods. Although other nonlinear programming techniques have also been adopted, most reservoir system optimization models involve linear or dynamic programming and extensions thereof. The fundamentals of mathematical programming are introduced in Chapter 10. A number of reservoir system optimization models that incorporate these computational algorithms are reviewed in Chapter 11.

In optimization models, the objective function and constraints are represented by mathematical expressions as a function of the decision variables. Reservoir operation models incorporate a variety of decision variables, which typically include releases and storage volumes. The objective or criterion function may be a mathematical representation of a planning or operational objective, or may be a penalty or utility function used to define operating rules based on relative priorities. Constraints reflect mass balances, storage and discharge capacities, diversion or instream flow requirements, and other aspects of the reservoir/stream system. In the multiple-stage optimization approach of dynamic programming, stages often represent time periods tied together by the state variable of storage.

10.1 LINEAR PROGRAMMING

Of the many optimization techniques, linear programming (LP) is the simplest and most widely used. Its popularity in water resources systems analysis, as well as in other operations research, management science, and systems engineering fields, is due to the following considerations. LP is applicable to a wide variety of types of problems. Efficient solution algorithms are available. Generalized computer software packages are available for applying the solution algorithms.

10.1.1 Linear Programming Format

Linear programming consists of finding values for a set of n decision variables $(x_1, x_2, ..., x_n)$ that minimize or maximize an objective or criterion function x_0 of the form:

$$x_0 = c_1 x_1 + c_2 x_2 + \cdots + c_n x_n \tag{10-1}$$

subject to a set of m constraint equations and/or inequalities of the form

$$a_{11} x_{11} + a_{12} x_{12} + \cdots + a_{1n} x_{1n} \le b_1$$
$$a_{21} x_{11} + a_{22} x_{22} + \cdots + a_{2n} x_{2n} \le b_2$$
$$\vdots \qquad\qquad \vdots \tag{10-2}$$
$$a_{m1} x_{m1} + a_{m2} x_{m2} + \cdots + a_{mn} x_{mn} \le b_m$$

and a set of constraints requiring that the decision variables be nonnegative:

$$x_j \ge 0 \quad \text{for } j = 1, 2, ..., n \tag{10-3}$$

where a_{ij}, b_i, and c_j are constants. The linear programming model is expressed in more concise notation as

minimize (or maximize) $\qquad x_0 = \displaystyle\sum_{j=1}^{n} c_j x_j \tag{10-4}$

subject to $\qquad \displaystyle\sum_{j=1}^{n} a_{ij} x_j \le b_i \quad \text{for } i = 1, 2, ..., m \tag{10-5}$

and $\qquad x_j \ge 0 \quad \text{for } j = 1, 2, ..., n \tag{10-3}$

where x_0 is the objective function, x_j are the decision variables, c_j, a_{ij}, and b_i are constants, n is the number of decision variables, and m is the number of constraints. The less than or equal sign in the constraint inequalities may be replaced by a greater than or equal or equal sign to suit the particular problem being modeled. Maximizing $-x_0$ is equivalent to minimizing x_0. The objective function and all constraints are linear functions of the decision variables. A set of values for the n variables is called a decision policy.

10.1.2 Solution of Linear Programming Problems

Considerable ingenuity and significant approximations may be required to formulate a real-world problem in the required mathematical format. However, if the problem can be properly formulated, standard LP algorithms and computer codes are available to perform the computations. The simplex algorithm, used in many linear programming computer codes, is explained in many textbooks, including [Wagner, 1975] and [Mays and Tung, 1992]. Special computationally efficient algorithms are available for certain forms of LP problems such as the network flow programming models discussed later.

Optimization models are sometimes coded from scratch in FORTRAN or other languages. Already-written FORTRAN subroutines for performing linear or nonlinear programming computations are often incorporated in model development. The same code for optimizer routines may be used in any number of different models.

Various generalized optimization computer programs are also commercially available. The user inputs values for the coefficients in the objective function and constraint equations. The optimizer program computes values for the decision variables. Linear programming capabilities are included in popular spreadsheet programs, such as Excel (marketed by Microsoft Corporation), Quattro Pro (Borland International, Inc.), and Lotus 1-2-3 (Lotus Development Corporation). However, the spreadsheet programs are designed for relatively small problems limited to several hundred decision variables and constraints. Reservoir/river system analysis models typically involve many thousands of decision variables and constraints. A number of generalized optimization programs are available for solving linear and, in some cases, nonlinear programming problems, including very large problems.

For example, the General Algebraic Modeling System (GAMS) is a general-purpose optimization package designed for solving large linear, nonlinear, and mixed-integer programming problems [Brooke *et al.*, 1992; Mays and Tung, 1992]. GAMS is a high-level language that provides data management and model formulation capabilities as well as a set of mathematical programming optimizers. GAMS was originally developed by the World Bank (International Bank for Reconstruction and Development) of the United Nations. The GAMS system consists of GAMS/MINOS and GAMS/ZOOM in addition to the basic GAMS module. GAMS/MINOS (Modular In-Core Nonlinear Optimization System) accepts GAMS formulated input to solve complex linear and nonlinear programming problems. GAMS/ZOOM (Zero/One Optimization Method) accepts GAMS-formulated input to solve mixed-integer linear programming problems. GAMS also includes a library of linear and nonlinear programming models that have been formulated for various applications. GAMS is available in microcomputer, workstation, and mainframe versions.

10.1.3 Graphical Solution of Two-Variable LP Problems

Several examples are presented to illustrate the basic concept of formulating a problem in the linear programming format. These simple examples can be solved easily with a spreadsheet program or other software. A couple of two-variable problems are presented first so that the basics can be visualized graphically. The two-variable graphical solution serves to introduce fundamental concepts incorporated in standard n-variable solution algorithms.

Example 10–1—Two-Variable Graphical Solution

The decision problem is to determine the amount of land to be planted in crop A and crop B that will maximize income within the constraints of limited land and water resources. Twenty million m^3 of water is available in storage for irrigation of the two crops. Crop A requires 9000 m^3 of water per hectare of irrigated land and produces a net income of $720 per hectare. Crop B requires 6000 m^3 per hectare and produces a profit of $1200 per hectare. Crop A is limited to 1600 hectares. Up to 2400 hectares of land is available for planting crop B. The linear programming model is formulated to determine the amount of land, in hectares, planted in crop A (x_1) and crop B (x_2) that will maximize income (x_0), in dollars.

maximize $\qquad\qquad x_0 = 720\,x_1 + 1200\,x_2$

subject to $\qquad\qquad 9000\,x_1 + 6000\,x_2 \le 20,000,000$

$$x_1 \le 1,600$$

$$x_2 \le 2,400$$

$$x_1 \text{ and } x_2 \ge 0$$

The LP problem is solved graphically in Figure 10–1. The constraint inequalities define the feasible region of the x_1 versus x_2 plane. Graphs of the objective function are parallel for alternative values of x_0. The objective function will always pass through at least one corner of the feasible region at the maximum value of x_0. Thus, the graphical solution consists of shifting the objective function (varying x_0) to find the optimum corner of the feasible region delineated by the constraints. The optimum solution determined in Figure 10–1 consists of planting 622 and 2400 hectares of crops A and B, respectively, resulting in a net income of $3,327,840.

Example 10–2—Another Two-Variable Graphical Solution

A water demand of 10,000 m^3 during a particular time period is supplied by withdrawals from two reservoirs. The total dissolved solids (TDS) concentrations in reservoirs A and B are 980 and 100 g/m^3, respectively. The maximum allowable TDS concentration is 500 g/m^3 for the water use being supplied. The capacities of the outlet structures constrain withdrawals, during the time period, to not exceed 6000 m^3 and 10,000 m^3, respectively, from reservoirs A and B. In the current time period, multiple-reservoir operating decisions are based on minimizing the amount of water withdrawn from the better-quality reservoir B, in order to maximize the amount of good-quality water remaining for future use.

The decision problem is to determine the amount of water to withdraw from reservoir

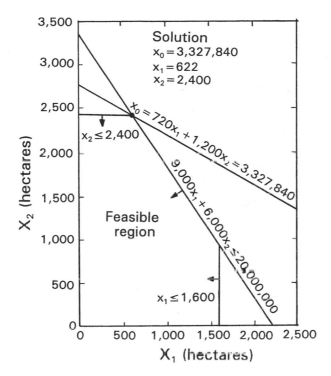

Figure 10–1   Graphical solution for Example 10–1.

A (x_1) and reservoir B (x_2) while meeting the total demand of 10,000 m³ with a concentration of 500 g/m³ or less. The amount of water withdrawn from reservoir B (x_2) is minimized.

A TDS mass balance constraint is formulated that equates the TDS load in the water supply diversion to the sum of the TDS mass withdrawn from each of the two reservoirs. The concentration of the water supply diversion is limited to 500 g/m³.

$$980\, x_1 + 100\, x_2 \leq 500\, (x_1 + x_2)$$

or

$$480\, x_1 - 400\, x_2 \leq 0$$

The linear programming model is formulated as follows, with the decision variables x_1 and x_2 expressed in units of cubic meters (m³) and concentrations in the first constraint in grams/m³.

minimize $x_0 = x_2$

subject to $480\, x_1 - 400\, x_2 \leq 0$

$$x_1 + x_2 \geq 10{,}000$$

$$x_1 \leq 6000$$

$$x_2 \leq 10{,}000$$

$$x_1 \text{ and } x_2 \geq 0$$

The problem is solved graphically in Figure 10–2 to obtain the optimum solution of withdrawing 4545 m³ and 5455 m³, respectively, from reservoirs A and B.

The plots of the constraints and objective function in Figures 10–1 and 10–2 illustrate fundamental features regarding linear programming solution algorithms. The constraints bound a feasible region. Combinations of decision variables (x_1 and x_2) within this region are feasible in that all constraints are satisfied. Values lying outside of the feasible region violate one or more constraints. The optimum solution is always on a corner of the feasible region. In the two-variable graphical solution, the objective function (x_0) can be plotted as a function of x_1 and x_2 for a specified value of x_0. Plots for alternative values of x_0 are parallel. At the maximum or minimum value of x_0, the objective function plot passes through at least one corner of the feasible region. If the objective function is parallel to a controlling constraint, an infinite number of optimum solutions are represented by two corners of the feasible region and the boundary connecting the two corners. A problem formulation may involve a set of constraints that cannot all be satisfied simultaneously, and thus there is no feasible region and no solution.

Realistic linear programming problems typically involve hundreds or thousands of decision variables. The graphical approach is applicable only to problems involving two decision variables. For problems with more than two decision variables, the objective function and constraints can be visualized as hyperplanes in n-dimensional space. Mathematical solution algorithms solve the equations at the cor-

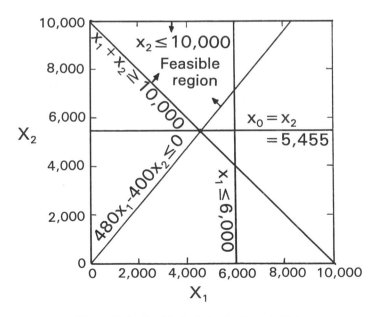

Figure 10–2   Graphical solution for Example 10–2.

ners, called *extreme points*, of the planes enclosing the feasible space, in a systematic iterative search for the optimum.

10.1.4 Linear Programming Solution Algorithms

Linear programming solution algorithms, particularly the commonly applied simplex method, are covered in detail by Wagner [1975], Mays and Tung [1992], and many other textbooks. General characteristics of linear programming are briefly noted here.

As previously discussed in conjunction with the graphical approach, the optimum set of values for the n decision variables occurs at a corner or extreme point of the feasible region in n-dimensional space. Algebraic solution algorithms are based on the following properties of extreme points:

- Only a finite number of extreme points exist for any problem.
- If an extreme point is a global optimum, its objective function value is better than all adjacent extreme points.
- If there is only one optimum solution, it is an extreme point. If there are multiple optimum solutions, at least two must be adjacent extreme points.

Solution algorithms, such as the simplex method, generally have the following features:

- The procedure begins at an extreme point of the feasible region of the n-dimensional solution space.
- Iteratively, the algorithm moves to an adjacent extreme point with a better value of the objective function. The constraint equations that intersect at the extreme point are solved.
- The iterative moves to better extreme points stop when the current extreme point is better than all adjacent extreme points.

At the beginning of the solution procedure, the inequality and equality constraints are converted to a set of m equations and n unknowns, with $n > m$. With more unknowns than equations, the system of equations is indeterminant. Thus, solution algorithms are based on letting $(n - m)$ unknowns equal zero and solving for the remaining m unknowns. In the iterative algorithm, the choice of variables to set equal to zero depends on the current extreme point.

Linear programming formulations may have no feasible solution or, in other cases, an unbounded solution. Complexities arise when the constraints define either no feasible solution space or an unbounded feasible solution space. Constraints may be formulated such that they cannot all be satisfied simultaneously. No feasible solution exists if no set of values for the decision variables satisfies all constraints. An unbounded solution occurs when the objective function can be increased or decreased infinitely. In either case, no solution is found, and the problem must be reformulated.

10.1.5 Firm Yield Analysis

Linear programming is one of several alternative approaches for performing firm yield analyses. As discussed in prior chapters, firm yield is the maximum demand that can be met continuously during a sequence of reservoir inflows representing historical hydrology. The storage capacity (C) required to meet a specified set of water demands (y_t) representing a firm yield can be computed with the following linear programming formulation.

minimize C

subject to $s_t = s_{t-1} + i_t - y_t - r_t$ for $t = 1, 2, ..., T$ (10-6)

                        $s_t \leq C$ for $t = 1, 2, ..., T$ (10-7)

                        $s_t, y_t, r_t \geq 0$ for $t = 1, 2, ..., T$ (10-8)

where

C = reservoir storage capacity
s_t = storage content at the end of period t
i_t = streamflow inflow to the reservoir during period t
y_t = demand during period t representing the firm yield
r_t = all spills and releases other than y_t during period t
T = number of time periods in the analysis

Thus, the reservoir storage capacity (C) is minimized subject to constraints, which include the reservoir mass balance, not allowing storage content (s_t) to exceed storage capacity (C), and not allowing the variables to have negative values. The constraints are repeated for each time period (t). The known firm yield withdrawals (y_t) and reservoir inflows (i_t) are provided as input data. The model computes the value of the storage capacity (C) and also values of end-of-period storage (s_t) and other releases (r_t) for each period t.

Yield analyses require an assumption regarding starting and ending storage conditions. Specification of the storage at the beginning (s_0) and end (s_T) of the overall analysis period could be added to the model formulation. Alternatively, the entire streamflow sequence representing reservoir inflows may be assumed to be repeated as necessary to achieve a repetitive cycle. This is reflected in the model by assuming that the first ($t = 1$) period of a T-period cycle follows the last period ($t = T$) of the prior cycle and specifying that the beginning ($t = 0$) and ending ($t = T$) storages be equal ($s_0 = s_T$).

With the above formulation, the reservoir storage capacity (C) is computed for a user-specified firm yield (y_t). Alternatively, a model can be formulated to determine a constant firm yield (Y) provided by a specified storage capacity, by changing the objective function to:

$$\text{maximize } Y$$

subject to the same constraints as before. With this formulation, a storage capacity is specified as input, and the model computes the firm yield (Y), along with values of s_t and r_t for each period.

Example 10–3—Storage Capacity Required for a Specified Firm Yield

Example 10–3 consists of repeating Example 7–2 using linear programming. In Example 7–2, a simulation algorithm is applied to determine the reservoir storage capacity required to provide a specified firm yield for a given sequence of inflows. The computations tabulated in Table 7–1 result in a storage capacity of 304 million m^3 to provide a specified firm yield of 50 m^3/s. In the present Example 10–3, a linear programming model is formulated to determine the storage capacity required to meet the monthly demand volumes tabulated in column 6 of Table 7–1 for the 24-month sequence of reservoir inflow volumes tabulated in column 3. The inflow and demand volumes from Table 7–1 are reproduced in the third and fourth columns of Table 10–1.

TABLE 10–1   Data and Results for the Firm Yield Model of Example 10–3

Month	t	Inflow I_t (10^6m^3)	Yield Y_t (10^6m^3)	$i_t - y_t$ (10^6m^3)	Storage s_t (10^6m^3)	Release r_t (10^6m^3)
Jan	1	123	134	−11	293	0
Feb	2	172	121	51	304	40
Mar	3	163	134	29	304	29
Apr	4	334	130	204	304	204
May	5	421	134	287	304	287
Jun	6	130	130	0	304	0
Jul	7	37	134	−97	207	0
Aug	8	19	134	−115	92	0
Sep	9	109	130	−21	71	0
Oct	10	88	134	−46	25	0
Nov	11	140	130	10	35	0
Dec	12	134	134	0	35	0
Jan	13	150	134	16	51	0
Feb	14	167	121	46	97	0
Mar	15	230	134	96	193	0
Apr	16	288	130	158	304	47
May	17	362	134	228	304	228
Jun	18	67	130	−63	241	0
Jul	19	32	134	−102	139	0
Aug	20	27	134	−107	32	0
Sep	21	98	130	−32	0	0
Oct	22	276	134	142	142	0
Nov	23	223	130	93	235	0
Dec	24	209	134	75	304	6

The linear programming problem is formulated with 49 decision variables, consisting of the reservoir storage capacity (C), end-of-period storage contents (s_t) and spills (r_t) for each of 24 months ($t = 1, 2, ..., 24$). The objective is to minimize C. Note that, in the objective function, the coefficients are 1 for the decision variable C and zero for the 48 decision variables s_t and r_t.

A set of 24 constraints are formulated to represent the reservoir mass balance for each of the 24 months of the analysis. The mass balance is expressed as

$$s_t = s_{t-1} + i_t - y_t - r_t$$

which is rearranged with the decision variables on the left of the equal sign:

$$s_t - s_{t-1} + r_t = i_t - y_t$$

The storage (s_0) at the beginning of the first month is assumed to equal the storage (s_T) at the end of the last month. Thus, for the first month ($t = 1$), the mass balance constraint is

$$s_1 - s_{24} + r_1 = i_1 - y_1$$

With values for i_1 and y_1 from Table 10–1,

$$s_1 - s_{24} + r_1 = 123 - 134 = -11$$

A set of 24 constraints specify that storage content cannot exceed capacity.

$$S_t - C \leq 0 \quad \text{for } t = 1, 2, ..., 24$$

Nonnegativity constraints specify that the 49 decision variables have values that are zero or positive numbers.

The complete formulation of the LP model is as follows:

minimize C

subject to:

$s_1 - s_{24} + r_1 = -11$	$s_{13} - s_{12} + r_{13} = 16$
$s_2 - s_1 + r_2 = 51$	$s_{14} - s_{13} + r_{14} = 46$
$s_3 - s_2 + r_3 = 29$	$s_{15} - s_{14} + r_{15} = 96$
$s_4 - s_3 + r_4 = 204$	$s_{16} - s_{15} + r_{16} = 158$
$s_5 - s_4 + r_5 = 287$	$s_{17} - s_{16} + r_{17} = 228$
$s_6 - s_5 + r_6 = 0$	$s_{18} - s_{17} + r_{18} = -63$
$s_7 - s_6 + r_7 = -97$	$s_{19} - s_{18} + r_{19} = -102$
$s_8 - s_7 + r_8 = -115$	$s_{20} - s_{19} + r_{20} = -107$
$s_9 - s_8 + r_9 = -21$	$s_{21} - s_{20} + r_{21} = -32$
$s_{10} - s_9 + r_{10} = -46$	$s_{22} - s_{21} + r_{22} = 142$
$s_{11} - s_{10} + r_{11} = 10$	$s_{23} - s_{22} + r_{23} = 93$
$s_{12} - s_{11} + r_{12} = 0$	$s_{24} - s_{23} + r_{24} = 75$

$$s_t - C \le 0 \qquad \text{for } t = 1, 2, ..., 24$$

$$s_t \ge 0 \qquad \text{for } t = 1, 2, ..., 24$$

$$r_t \ge 0 \qquad \text{for } t = 1, 2, ..., 24$$

$$C \ge 0$$

The model can be solved using any linear programming software. The results consist of a storage capacity value of

$$C = 304 \text{ million m}^3$$

and the values for s_t and r_t tabulated in Table 10–1.

10.1.6 Additional Examples of LP Formulations

Two additional simplified examples illustrate formulations of linear programming models. Example 10–4 involves determining reservoir release decisions, for each time step of the analysis period, that maximize revenues in dollars. Example 10–5 consists of allocating water between competing demands, for a given time interval, based on prespecified relative priorities. In these two examples as well as Example 10–3, the model formulations are unrealistically simplified by neglecting evaporation. Incorporation of evaporation is addressed in the subsequent section, which deals with nonlinearities in linear programming models.

Example 10–4—Determination of Reservoir Releases that Maximize Revenue

As indicated in Figure 10–3, releases from the storage reservoir flow through a hydroelectric power project located downstream, supply water for a irrigation diversion further downstream, and maintain instream flows. Flows of up to 180 million m³/month can be used to generate hydroelectric energy, flows in excess of this amount bypass the turbines. The demands shown in column 4 of Table 10–2 provide upper limits for the irrigation diversion. Irrigation diversions are limited to six months of the year. An instream flow requirement of 20 million m³ per month must be maintained. The reservoir storage capacity is 600 million m³. Revenues for supplying irrigation are $900 per million m³ of water diverted. Each million m³ of water used to generate hydroelectric power results in revenues of $400. The decision prob-

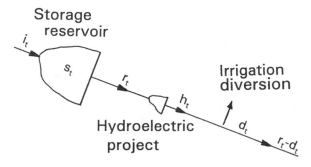

Figure 10–3 System schematic for Example 10–4.

TABLE 10–2   Data and Results for Example 10–4

(1)	(2)	(3)	(4)	(5)	(6)	(7)	(8)
					Total	Hydropower	Irrigation
		Inflow	Irrigation	Storage	Release	Release	Diversion
		i_t	Demand	s_t	r_t	h_t	d_t
Month	t	$(10^6 m^3)$	$(10^6 m^3)$	$(10^6 m^3)$	$(10^6 m^3)$	$(10^6 m^3)$	$(10^6 m^3)$
Jan	1	95	0	210	20	20	0
Feb	2	112	0	302	20	20	0
Mar	3	170	0	335	137	137	0
Apr	4	250	0	405	180	180	0
May	5	265	50	600	70	70	50
Jun	6	62	150	492	170	170	150
Jul	7	35	260	319	208	180	188
Aug	8	18	260	157	180	180	160
Sep	9	55	190	32	180	180	160
Oct	10	88	100	0	120	120	100
Nov	11	85	0	65	20	20	0
Dec	12	90	0	115	20	20	0

lem consists of determining the set of monthly reservoir releases that maximize revenues, given the inflows tabulated in column 3 of Table 10–2.

The following 42 decision variables are incorporated in the linear programming formulation, with t denoting the monthly time interval.

end-of-month storage	s_t for $t = 1, 2, ..., 12$
reservoir release	r_t for $t = 1, 2, ..., 12$
hydropower discharge	h_t for $t = 1, 2, ..., 12$
irrigation diversion	d_t for $t = 5, 6, 7, 8, 9, 10$

The objective is to maximize total annual revenues, in dollars.

$$\text{revenues} = 400h_1 + 400h_2 + 400h_3 + 400h_4 + 400h_5 + 400h_6 + 400h_7 + 400h_8 +$$
$$400h_9 + 400h_{10} + 400h_{11} + 400h_{12} + 900d_5 + 900d_6 + 900d_7 + 900d_8 +$$
$$900d_9 + 900d_{10}$$

Note that, in the objective function, the coefficients are $400 and $900 per million m^3, respectively, for the 18 decision variables h_t and d_t and zero for the 24 decision variables s_t and r_t. Constraints are as follows:

- Monthly reservoir mass balances are maintained for each of the 12 months of the analysis for the inflows (i_t) tabulated in column 3 of Table 10–2. The strorages at the beginning and end of the year are equal $(s_0 = s_{12})$

$$s_t - s_{t-1} + r_t = i_t \quad \text{for } t = 1, 2, ..., 12$$

- End-of-month storages cannot exceed the capacity of 600 million m^3.

$$s_t \le 600 \quad \text{for } t = 1, 2, ..., 12$$

- Diversions do not exceed the demands of column 4 of Table 10.2, and no more than 180 million m^3/month is used for hydroelectric power generation.

$$d_t \leq \text{demand} \quad \text{for } t = 1, 2, ..., 12$$

$$h_t \leq 180 \quad \text{for } t = 1, 2, ..., 12$$

- The instream flow requirement is 20 million m^3/month. The flow used for hydropower cannot exceed the reservoir release. The irrigation diversion cannot exceed the reservoir release.

$$r_t - d_t \geq 20 \quad \text{for } t = 1, 2, ..., 12$$

$$r_t - h_t \geq 0 \quad \text{for } t = 1, 2, ..., 12$$

- Nonnegativity constraints specify that the 49 decision variables have values that are zero or positive numbers.

$$s_t, r_t, h_t, d_t \geq 0 \quad \text{for } t = 1, 2, ..., 12$$

The complete formulation of the linear programming model is as follows:

maximize revenues $= 400h_1 + 400h_2 + 400h_3 + 400h_4 + 400h_5 + 400h_6 + 400h_7$
$+ 400h_8 + 400h_9 + 400h_{10} + 400h_{11} + 400h_{12} + 900d_5 + 900d_6$
$+ 900d_7 + 900d_8 + 900d_9 + 900d_{10}$

subject to

$$s_1 - s_{12} + r_1 = 95 \qquad s_1 \leq 600$$

$$s_2 - s_1 + r_2 = 112 \qquad s_2 \leq 600$$

$$s_3 - s_2 + r_3 = 170 \qquad s_3 \leq 600$$

$$s_4 - s_3 + r_4 = 250 \qquad s_4 \leq 600$$

$$s_5 - s_4 + r_5 = 265 \qquad s_5 \leq 600 \qquad d_5 \leq 50$$

$$s_6 - s_5 + r_6 = 62 \qquad s_6 \leq 600 \qquad d_6 \leq 150$$

$$s_7 - s_6 + r_7 = 35 \qquad s_7 \leq 600 \qquad d_7 \leq 260$$

$$s_8 - s_7 + r_8 = 18 \qquad s_8 \leq 600 \qquad d_8 \leq 260$$

$$s_9 - s_8 + r_9 = 55 \qquad s_9 \leq 600 \qquad d_9 \leq 190$$

$$s_{10} - s_9 + r_{10} = 88 \qquad s_{10} \leq 600 \qquad d_{10} \leq 100$$

$$s_{11} - s_{10} + r_{11} = 85 \qquad s_{11} \leq 600$$

$$s_{12} - s_{11} + r_{12} = 90 \qquad s_{12} \leq 600$$

$$r_1 \geq 20 \qquad\qquad r_1 - h_1 \geq 0 \qquad h_1 \leq 180$$

$$r_2 \geq 20 \qquad\qquad r_2 - h_2 \geq 0 \qquad h_2 \leq 180$$

$$r_3 \geq 20 \qquad\qquad r_3 - h_3 \geq 0 \qquad h_3 \leq 180$$

$$r_4 \geq 20 \qquad\qquad r_4 - h_4 \geq 0 \qquad h_4 \leq 180$$

$$r_5 - d_5 \geq 20 \qquad\quad r_5 - h_5 \geq 0 \qquad h_5 \leq 180$$

$$r_6 - d_6 \geq 20 \qquad\quad r_6 - h_6 \geq 0 \qquad h_6 \leq 180$$

$$r_7 - d_7 \geq 20 \qquad\quad r_7 - h_7 \geq 0 \qquad h_7 \leq 180$$

$$r_8 - d_8 \geq 20 \qquad\quad r_8 - h_8 \geq 0 \qquad h_8 \leq 180$$

$$r_9 - d_9 \geq 20 \qquad\quad r_9 - h_9 \geq 0 \qquad h_9 \leq 180$$

$$r_{10} - d_{10} \geq 20 \qquad r_{10} - h_{10} \geq 0 \quad h_{10} \leq 180$$

$$r_{11} \geq 20 \qquad\qquad r_{11} - h_{11} \geq 0 \quad h_{11} \leq 180$$

$$r_{12} \geq 20 \qquad\qquad r_{12} - h_{12} \geq 0 \quad h_{12} \leq 180$$

$$s_t \geq 0 \quad \text{for } t = 1, 2, ..., 12$$

$$r_t \geq 0 \quad \text{for } t = 1, 2, ..., 12$$

$$h_t \geq 0 \quad \text{for } t = 1, 2, ..., 12$$

$$d_t \geq 0 \quad \text{for } t = 5, 6, ..., 10$$

The model can be solved using any linear programming software. The resulting values for the decision variables are tabulated in columns 5, 6, 7, and 8 of Table 10–2. The objective function is $1,246,000 for this optimum decision policy.

As illustrated by Example 10–4, an optimization model may determine releases simultaneously for all the time steps of an analysis period. Alternatively, an optimization algorithm may be executed for each time step in turn. Example 10–5 consists of formulating an LP model to allocate limited water resources to competing users for a given time interval. Although the example addresses a single time interval, this type of LP formulation would typically be repeated for each individual time interval of a sequence of time steps, with the impacts of decisions in prior time steps being reflected in the current reservoir storage. The objective function in Example 10–5 is formulated so that water is allocated based on assigned relative priorities for different water users.

Example 10–5—Priority Based Water Allocation

A schematic of a river/reservoir system is presented in Figure 10–4. Reservoirs A and B, located at nodes 1 and 2, have storage capacities of 750×10^6 and 900×10^6 m^3, respectively. The initial storage in reservoirs A and B at the beginning of the time interval is 460×10^6 and 215×10^6 m^3, respectively. Releases are made as necessary to maintain instream flow requirements and then, to the extent possible, to meet water supply diversion targets. Requirements for minimum instream flows for the particular time period are as follows:

Reach	1–4	2–3	3–4	4–5	Below 5
Flow (10^6 m^3)	0	5	10	10	30

The supply and demand for water are shown in Table 10–3 for each of the node locations shown in Figure 10–4. The total supply at each node consists of reservoir storage at the

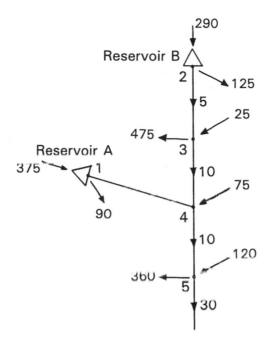

Figure 10-4   System schematic for Examples 10-5 and 10-8.

beginning of the time interval, and the local flow entering the river between the node and adjacent upstream node(s) during the time period. The demand in Table 10-3 is a target diversion at each node. If supplies are insufficient to meet all demands, allocations are based on the relative priorities tabulated in the last column. For example, with a relative priority of 5, the diversion of 360×10^6 m^3 at node 5 has the highest priority of the four diversions. Lower-priority diversions are met only to the extent that higher-priority diversions are not adversely affected.

The model incorporates the 11 decision variables defined in Table 10-4. The decision variables include instream flows in each of five river reaches (x_1, x_2, x_3, x_4, x_5), water supply diversions at four nodes (x_6, x_7, x_8, x_9), and the ending storage in the two reservoirs (x_{10}, x_{11}).

TABLE 10-3   Given Information for Examples 10-5 and 10-8

Location or Node	Initial Storage (10^6m^3)	Local Inflow (10^6m^3)	Total Supply (10^6m^3)	Demand Target (10^6m^3)	Relative Priority
1	460	375	835	90	4
2	215	290	505	125	3
3	—	25	25	475	2
4	—	75	75	—	—
5	—	120	120	360	5

TABLE 10-4   Decision Variables for Example 10-5

Decision Variable	Definition of Decision Variable	Solution $(10^6 m^3)$
X_1	Instream flow from 1 to 4	195
X_2	Instream flow from 2 to 3	380
X_3	Instream flow from 3 to 4	10
X_4	Instream flow from 4 to 5	270
X_5	Instream flow below node 5	30
X_6	Diversion at node 1	90
X_7	Diversion at node 2	125
X_8	Diversion at node 3	395
X_9	Diversion at node 5	360
X_{10}	Reservoir A ending storage	550
X_{11}	Reservoir B ending storage	0

The objective function is formulated to reflect relative priorities between water users as follows:

maximize $\qquad\qquad 4x_6 + 3x_7 + 2x_8 + 5x_9 + x_{10} + x_{11}$

The objective function coefficients are used simply to assign relative priorities to guide the allocation of the limited water resources to competing uses. The absolute values of the coefficients are arbitrary. Only the relationship of the coefficient values relative to each other affect the results. Water supply diversion x_9 is assigned the highest priority of the four diversions, as reflected by its coefficient of 5. Filling the two reservoirs (x_{10}, x_{11}) is assigned the lowest priority, as reflected by coefficients of one. The diversion targets are fully met prior to filling the reservoirs.

The instream flow requirements $(x_1, x_2, x_3, x_4, x_5)$ are assigned zero priority (coefficient values) in the objective function and are handled as constraints.

$$x_1 \geq 0 \quad x_2 \geq 5 \quad x_3 \geq 10 \quad x_4 \geq 10 \quad x_5 \geq 30$$

The constraints force the instream flow requirements to be met even if diversion targets cannot be fully met. Sufficient water is available to meet all instream flow requirements. Otherwise, there would be no feasible solution, and the model would have to be reformulated.

A volume balance constraint is written for each node. For example, at node 1, initial storage and inflows of 835 million m^3 from Table 10-3 supply downstream flows (x_1), diversions (x_6), and reservoir filling (x_{10}).

$$x_1 + x_6 + x_{10} = 835$$

The diversions are constrained to not exceed their demand targets. The reservoir storage is constrained to not exceed capacity. The complete LP model is formulated as follows:

maximize $\qquad\qquad x_0 = 4x_6 + 3x_7 + 2x_8 + 5x_9 + x_{10} + x_{11}$

subject to

$$x_1 \geq 0 \quad\quad x_2 \geq 5 \quad\quad x_3 \geq 10 \quad\quad x_4 \geq 10$$
$$x_5 \geq 30 \quad\quad x_6 \leq 90 \quad\quad x_7 \leq 125 \quad\quad x_8 \leq 475$$
$$x_9 \leq 360 \quad x_{10} \leq 750 \quad x_{12} \leq 900$$

$$x_1 + x_6 + x_{10} = 835$$
$$x_2 + x_7 + x_{11} = 505$$
$$-x_2 + x_3 + x_8 = 25$$
$$-x_1 - x_3 + x_4 = 75$$
$$-x_4 + x_5 + x_9 = 120$$

The model can be solved using any linear programming software. The resulting values for the decision variables are tabulated in the last column of Table 10–4.

10.1.7 Dealing with Nonlinearities

Nonlinearities significantly complicate linear programming models, but can be adequately handled in many cases. Nonlinearities are associated with various features of reservoir operation models, such as evaporation and hydroelectric power computations and benefit/cost functions. Successive iterative solutions of a linear programming problem are often used to handle nonlinearities. In some cases, non-linear features of a problem may be approximated by linearization techniques for incorporation into an LP model.

Examples 10–3, 10–4, and 10–5 are unrealistic because reservoir evaporation is not included in the volume-accounting computations. Evaporation volumes are typically computed as a net evaporation rate multiplied by the mean water surface area during the computational time step. Water surface area is a nonlinear function of storage. The mean storage or area during a time interval, such as a month, is typically approximated as the average of the values at the beginning and ending of the time interval. Thus, end-of-period storage is computed as a function of evaporation volume, which in turn is computed as a function of end-of-period storage.

Likewise, the nonlinearity of hydroelectric power computations complicates, but can be addressed in, LP models. As indicated by equation 6-5, hydroelectric power is a function of both head and discharge. Head is a nonlinear function of storage. End-of-period reservoir storage is computed as a function of releases required to meet hydroelectric energy requirements, which in turn depend on the available head provided by the reservoir storage.

Evaporation and hydropower are examples of nonlinear features of constraints. Objective functions may also be complicated by nonlinearities. For example, revenues and costs may be nonlinear functions of decision variables such as reservoir releases and storage.

10.1.7.1 Successive Approximations.   A common approach to dealing with nonlinear terms in linear programming models is to iteratively execute the LP model with successive approximations of terms affected by nonlinearity. For example, reservoir evaporation should be incorporated into the models of Examples 10–3, 10–4 and 10–5. Assume the computations are performed using a monthly time interval. The reservoir evaporation volume is computed by applying an evaporation rate to

the average water surface area during the month. A water surface area versus storage relationship is required as input. The average area during the month is estimated as the average of the beginning- and end-of-month water surface areas determined as a function of the corresponding storage volumes. The areas and evaporation volume are determined by the model in a routine separate from the LP algorithm. An initial estimate of evaporation volume based on the known beginning-of-month storage is input to the LP algorithm. The LP algorithm computes the end-of-month storage, which is then used to develop an improved estimate of the evaporation volume. The LP algorithm is iteratively executed with improved estimates of evaporation input until a specified stop criterion is met. The same type of iterative procedure is used to determine reservoir releases required to meet specified hydroelectric energy requirements.

10.1.7.2 Linearization of Storage-Area Curve.

An alternative approach for computing evaporation volumes involves a linear approximation of the nonlinear storage versus water surface area relationship, as illustrated by Figure 10–5. As just discussed, reservoir evaporation volume is computed as an evaporation rate multiplied by an average water surface area. The evaporation volume (E_t) during period t is computed as

$$E_t = [A_o + A_s(\frac{S_t + S_{t+1}}{2})]e_t \qquad (10\text{-}9)$$

where A_o is the water surface area for some base storage volume such as inactive storage; A_s is additional area per unit storage volume, represented by the slope of the linearized storage-area relation; e_t is the evaporation rate during period t; and s_t and s_{t+1} are storage volumes at the beginning and end of period t. Equation 10-9 is rearranged as

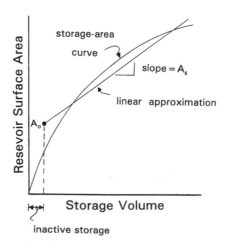

Figure 10–5   Linear approximation of the storage-area relationship.

$$E_t = A_o\, e_t + (0.5A_s\, e_t)\,(S_t + S_{t+1}) \tag{10-10}$$

Evaporation is included in the reservoir volume balance equation reflected in the constraints of linear programming models,

$$s_{t+1} = s_t + i_t - r_t - E_t \tag{10-11}$$

where i_t and r_t denote inflow and release. Equation 10-10 is substituted into equation 10-11 and rearranged to obtain

$$(1 + 0.5A_s\, e_t)s_{t+1} - (1 - 0.5A_s\, e_t)s_t + i_t - r_t - A_o\, e_t \tag{10-12}$$

where A_o, A_s, and e_t are known constants. Equation 10-12, which includes evaporation, is used to formulate volume balance constraints in linear programming models.

10.1.7.3 Linearization of Objective Functions.
Various methods are available for approximating a separable nonlinear objective function as a piece wise linear function for incorporation into the linear programming formulation [Loucks et ul., 1981]. A separable function of n variables can be written as the sum of n individual functions,

$$f(x_1, x_2, x_3, ..., x_n) = f_1(x_1) + f_2(x_2) + f_3(x_3) + \cdots + f_n(x_n)$$

each of which depends on only one of the variables. Normally, concave functions are maximized, or convex functions are minimized, using piece-wise linear approximations. Special separable linear programming techniques are available for minimizing a concave function or maximizing a convex function.

One alternative method for approximating a concave function to be maximized with piece-wise linear segments is as follows [Loucks et al., 1981]. The nonlinear function of the decision variable x is replaced with n linear functions of variables x_1, $x_2, ..., x_n$ as follows:

maximize $\qquad f(x) \simeq s_1 x_1 + s_2 x_2 + s_3 x_3 + \cdots + s_n x_n = \Sigma s_j x_j \tag{10-13}$

subject to $\qquad a_1 + x_1 + x_2 + x_3 + \cdots + x_n = x \tag{10-14}$

$\qquad\qquad x_j \le a_{j+1} - a_j \quad \text{for all segments } j \tag{10-15}$

Figure 10–6 illustrates the objective function as a nonlinear function of one of the original decision variables, x, that is to be replaced with the new variables, $x_1, x_2, ...,$ x_n, and linear functions of equations 10-13, 10-14, and 10-15. The s_j are the slopes, and the a_j are the x limits, for each linear segment. Since a concave function is being maximized, s_1 exceeds s_2, and thus x_2 will be zero unless x_1 reaches its upper limit a_2, and so forth. This approach is illustrated by Example 10–6.

Example 10–6—Nonlinear Objective Function

Economic benefits (B) are a nonlinear function of the reservoir release (r) in a time period.

$$B = 3r^{1/2}$$

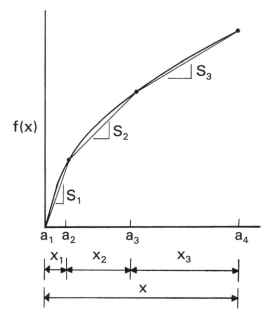

Figure 10–6   Piecewise linear approximation of a nonlinear function.

The benefits expressed as $3r^{1/2}$ are a separable term in an objective function to be maximized. The objective function may have many other terms not addressed here. In order to apply linear programming, each separable nonlinear term must be linearized.

The objective function term $3r^{1/2}$ is linearized by replacing it with three objective function terms in the form of equation 10-13 and four additional constraints in the form of equations 10-14 and 10-15. The following points, illustrated in Figure 10–6, are selected.

$$r =\quad 0 \quad 1 \quad 4 \quad 9$$

$$B = 3r^{1/2} =\quad 0 \quad 3 \quad 6 \quad 9$$

The slope (s_1) between the first two points is computed to be 3. Likewise, the slopes s_2 and s_3 are determined to be 1 and 0.6.

$$s_1 = \frac{3-0}{1-0} = 3$$

$$s_2 = \frac{6-3}{4-1} = 1 \qquad s_3 = \frac{9-6}{9-4} = 0.6$$

The objective function term $3r^{1/2}$ is replaced by the following objective function terms and constraints in the form of equations 10-13, 10-14, and 10-15:

maximize $3x_1 + x_2 + 0.6x_3$

subject to $x_1 + x_2 + x_3 - r = 0$

                                          $x_1 \leq 1 \quad x_2 \leq 3 \quad x_3 \leq 5$

10.1.8 Chance-Constrained Linear Programming

Streamflow is random. Other variables are uncertain as well. As discussed in Chapter 5, most reservoir system analysis models used by the water agencies are based on deterministic streamflow sequences reflecting historical hydrology. This book emphasizes this conventional approach. For example, sequences of monthly flows for the period January 1900 through December 1984 are incorporated in the input of the models described in Chapter 9. Likewise, the previous examples in this chapter are based on deterministic streamflow sequences. However, alternative stochastic analysis approaches are addressed extensively in the research literature. In stochastic reservoir system analysis models, streamflow is treated as a random variable represented by a probability distribution or as a stochastic process represented by a transition probability matrix or other means. Other variables, such as evaporation, may be treated as random variables as well. Stochastic models typically involve a variety of complexities, necessitating simplifying assumptions. Consequently, they are often proposed as screening models for preliminary studies, to be followed by more detailed analyses of selected alternatives using conventional approaches.

Yeh [1985] reviews many types and variations of stochastic reservoir optimization models that have been reported in the literature. Loucks et al. [1981] and Mays and Tung [1992] outline the fundamentals of stochastic analysis approaches. A variety of stochastic analysis methods can be incorporated in linear programming models to deal with uncertainties in various ways. The following discussion is limited to a relatively simple illustrative formulation of a stochastic optimization model based on the concepts of chance constraints, linear decision rule, and monthly inflows represented by a probability distribution function.

Variations of chance constraints and linear decision rules can be incorporated in a variety of linear programming models. Chance-constrained formulations are typically applied to selected constraints that limit the permissible range of reservoir storage volumes or releases. Based on probability distributions for reservoir inflows, constraints can be formulated that allow specified minimum or maximum storage or release limits to be exceeded a specified percentage of time. Other variables, in addition to streamflow, can also be treated as random variables represented by probability distributions. A set of values for a release parameter for each time interval is computed that optimizes some objective function subject to certain discharge or storage limits being violated no more than a specified percentage of time.

In a chance-constrained model, one or more of the coefficients a_{ij} and/or b_i in equation 10-2 are treated as random variables [Mays and Tung, 1992]. In our formulation, the right-hand-side coefficient b_i in certain constraints is treated as random variable B_i. The constraint is formulated in terms of the probability that it will be met. Thus, the reliability of meeting the constraint is prespecified.

The cumulative probability for a random variable X is defined as

$$F_x = P(X \le x) \tag{10-16}$$

which for our random variable B_i is written as

$$F_{B_i} = P(B_i \leq b_i) \tag{10-17}$$

The cumulative probability distribution function is the integral of the probability density function from $X = -\infty$ to $X = x$. Many continuous probability distribution functions, including the normal, can be expressed as

$$X - \mu + z\sigma \quad \text{or} \quad B_i - \mu_{B_i} + z\sigma_{B_i} \tag{10-18}$$

where μ and σ are the mean and standard deviation, and z is a frequency factor. For the normal or Gaussian probability distribution, z is the standard normal deviate tabulated in tables reproduced in statistics books.

The chance-constrained problem involves meeting particular constraints with specified reliabilities. The objective function is optimized subject to the constraints being violated no more than specified percentages of time. The constraint is met with probability α_i expressed as

$$\sum_{J=1}^{n} a_{ij} X_i \leq b_i \tag{10-19}$$

$$P[\sum_{j=1}^{n} a_{ij} x_j \leq B_i] \geq \alpha_i \tag{10-20}$$

which is equivalent to

$$P[B_i \leq \sum_{J=1}^{n} a_{ij} x_j] \leq 1 - \alpha_i$$

which can be expressed in terms of the cumulative probability distribution function of the random variable B_i where $B_i = F^{-1}(1 - \alpha_i)$ as expressed by equation 10-18.

$$F_{Bi}(\sum_{j=1}^{n} a_{ij} x_j) \leq 1 - \alpha_i$$

$$\sum_{j=1}^{n} (a_{ij} x_j) \leq F^{-1}(1 - \alpha_i)$$

Equation 10-21 is the general form of the chance constraints.

$$\sum_{j=1}^{n} a_{ij} x_j \leq \mu_{B_i} + z_{B_i, 1-\alpha_i} \; \sigma_{B_i} \tag{10-21}$$

For each constraint i, the right-hand-side constant b_i in inequality 10-2 is treated as a random variable B_i represented by the normal distribution with mean and standard deviation μ_{B_i} and σ_{B_i}. The standard normal variate z_{B_i} is read from a normal probability table as a function of α_i. The chance constraint is met with a probability of α_i.

Linear decision rules are typically associated with chance-constrained models. The decision rules simplify the model. Linear decision rules relate reservoir releases to storage, inflows, and other variables. The following rule is adopted for our formulation:

$$o_t = s_t + i_t - b_t \qquad (10\text{-}22)$$

where o_t and i_t are reservoir outflow and inflow during time period t; s_t denotes storage at the beginning of the time interval; and b_t is the decision parameter. Although for simplicity only the total outflow (o_t) is included in the model, releases, spills, and evaporation could readily be incorporated as separate components of the outflow. Comparison with the reservoir volume balance equation

$$s_{t+1} = s_t + i_t - o_t \qquad (10\text{-}23)$$

indicates that

$$b_t = s_{t+1} \qquad (10\text{-}24)$$

where s_{t+1} is the storage at the end of the current time interval t or the beginning of the next interval $t + 1$.

The chance constraint of equation 10-21 and linear decision rule of equation 10-22 are incorporated in the optimization model developed in Example 10–7. In this problem, a linear programming model is formulated to determine the reservoir storage capacity required to meet a specified outflow target each month with a specified reliability. Reservoir inflows are treated as a random variable.

Example 10–7—Chance-Constrained LP

The minimum storage capacity required to meet an outflow target of 200×10^6 m^3 per month at least 90% of the time is to be determined. In each month, the probability is 90% that the outflow will equal or exceed 200×10^6 m^3.

Reservoir inflows are assumed to be represented by the normal probability distribution function. The mean and standard deviation for each of the 12 months of the year are estimated from historical data and tabulated in Table 10–5. The monthly streamflows with a

TABLE 10–5   Data and Results for the Chance-Constrained Model of Example 10–7

Month	t	Inflow Mean (10^6m^3)	Standard Deviation (10^6m^3)	Inflow $I_{t,0.10}$ (10^6m^3)	Solution $s_{t+1} = b_t$ (10^6m^3)
Jan	1	136	38	185	290
Feb	2	169	122	325	376
Mar	3	196	107	333	376
Apr	4	311	264	649	376
May	5	391	280	750	376
Jun	6	98	27	133	309
Jul	7	34	24	65	174
Aug	8	23	9	35	9
Sep	9	103	69	191	0
Oct	10	182	145	368	168
Nov	11	181	77	280	248
Dec	12	171	67	257	305

10% exceedance probability (90% cumulative probability) are also included in Table 10–5 and were determined using equation 10.18, with a value of z of 1.282 read from the standard normal table found in any statistics or hydrology textbook. The z of 1.282 is also found in Table 7–4, recognizing that the Pearson III distribution with zero skew is equivalent to the normal distribution.

$$I_{t,90\%} = m_{It} + z_{90\%}\,\sigma_{It}$$

$$I_{t,90\%} = m_{It} + 1.282\,\sigma_{It}$$

The water balance reflected in equation 10-23 must be maintained in each month t.

$$s_{t+1} = s_t + I_t - o_t$$

The linear decision rule of equation 10-22 is adopted, with the decision parameter b_t representing the end-of-month storage. Inflow I_t is a random variable.

$$o_t = b_{t-1} + I_t - b_t$$

The chance constraint for each month is developed as follows:

$$P[o_t \geq 200] \geq 0.90$$

$$P[o_t \leq 200] \geq 0.10$$

$$P[b_{t-1} + I_t - b_t \leq 200] \leq 0.10$$

$$P[I_t \leq 200 - b_{t-1} + b_t] \leq 0.10$$

$$F_{It}(-b_{t-1} + b_t + 200) \leq 0.10$$

$$-b_{t-1} + b_t + 200 \leq I_{t,0.10}$$

$$b_t - b_{t-1} \leq I_{t,0.10} - 200$$

where $I_{t,0.10} = \mu_{I_t} + 1.282\,\sigma_{I_t}$

The $I_{t,0.10}$ is provided in Table 10–5. The 12-month sequence is assumed to cycle, with $b_0 = b_{12}$. For January ($t = 1$),

$$b_1 - b_{12} \leq 185 - 200$$

and for February ($t = 2$),

$$b_2 - b_1 \leq 325 - 200$$

Another set of constraints prevents $b_t = s_{t+1}$ from exceeding the storage capacity C.

$$b_t \leq C \quad \text{for } t = 1, 2, ..., 12$$

These are not chance constraints, since the random variable I_t is not included.

The linear programming model is formulated with 13 decision variables (b_1, b_2, ..., b_{12}, and C) as follows:

Minimize C

Subject to

$$b_1 - b_{12} \leq -15 \quad b_1 - C \leq 0$$

$$b_2 - b_1 \leq 125 \quad b_2 - C \leq 0$$

$$b_3 - b_2 \le 133 \qquad b_3 - C \le 0$$
$$b_4 - b_3 \le 449 \qquad b_4 - C \le 0$$
$$b_5 - b_4 \le 550 \qquad b_5 - C \le 0$$
$$b_6 - b_5 \le -67 \qquad b_6 - C \le 0$$
$$b_7 - b_6 \le -135 \quad b_7 - C \le 0$$
$$b_8 - b_7 \le -165 \quad b_8 - C \le 0$$
$$b_9 - b_8 \le -9 \qquad b_9 - C \le 0$$
$$b_{10} - b_9 \le 168 \qquad b_{10} - C \le 0$$
$$b_{11} - b_{10} \le 80 \qquad b_{11} - C \le 0$$
$$b_{12} - b_{11} \le 57 \qquad b_{12} - C \le 0$$

The model can be solved using any linear programming software. The resulting value for the storage capacity C is $376 \times 10^6 \ \mathrm{m}^3$. The values of b_t are tabulated in Table 10–5.

10.2 NETWORK FLOW PROGRAMMING

Network flow programming is a computationally efficient form of linear programming which can be applied to any problem that can be formulated in the required format, which involves representing the system as a network of nodes and arcs having certain characteristics. There are various recognized standard forms or classes of network flow problems and corresponding solution algorithms. Most reservoir/river system applications are formulated as a minimum-cost-capacitated network flow problem, which can be solved using the out-of-kilter algorithm as well as other linear programming algorithms. Network flow programming is addressed in detail by a number of textbooks, including [Jensen and Barnes, 1980], [Kennington and Helgason, 1980], and [Phillips and Garcia-Diaz, 1981].

10.2.1 Network Flow Programming Format

In a network flow model, the system is represented as a collection of nodes and arcs. For a reservoir/river system, the nodes are locations of reservoirs, diversions, stream tributary confluences, and other pertinent system features. Nodes are connected by arcs representing the way "flow" is conveyed. For a reservoir/river system, flow represents either a discharge rate, such as instream flows and diversions, or a change in storage per unit of time.

The general form of the network flow programming problem is as follows:

minimize	$\sum \sum c_{ij} q_{ij}$	for all arcs	(10-25)
subject to	$\sum q_{ij} - \sum q_{ji} = 0$	for all nodes	(10-26)
and	$l_{ij} \le q_{ij} \le u_{ij}$	for all arcs	(10-27)

where

q_{ij} = flow rate in the arc connecting node i to node j
c_{ij} = penalty or weighting factor for q_{ij}
l_{ij} = lower bound on q_{ij}
u_{ij} = upper bound on q_{ij}

A solution algorithm computes the values of the flows (q_{ij}) in each of n arcs (node i to node j), which minimizes an objective function consisting of the sum of the flows multiplied by corresponding weighting factors, subject to constraints including maintaining a mass balance at each node and not violating user-specified upper and lower bounds on the flows. Each arc has three parameters: a weighting, penalty, or unit cost factor (c_{ij}) associated with q_{ij}; lower bound (l_{ij}) on q_{ij}; and an upper bound (u_{ij}) on q_{ij}. The requirement for lower and upper bounds results in the term *capacitated* flow networks. Lower and upper bounds on diversions, instream flows, and reservoir storage levels are specified. The solution algorithm computes the flows and storage changes (q_{ij}).

Network flow programming provides considerable flexibility for formulating a particular river basin modeling application. The weighting factors (c_{ij}) in the objective function are defined in various ways in the models discussed in Chapter 11. The c_{ij} may be unit costs in dollars, or penalty or utility terms that provide mechanisms for expressing relative priorities in defining operating rules. A penalty weighting factor is the same as a negative utility weighting factor. Convex piecewise linear cost, penalty, or priority functions can be represented with a q_{ij}, c_{ij}, l_{ij}, and u_{ij} for each linear segment. Reservoir operating rules may be defined by user-specified values of c_{ij}, l_{ij}, and u_{ij}. For several models noted in Chapter 11, a network flow problem, as formulated in equations 10-25 through 10-27, is solved for each individual time interval in turn. For other models, a single network flow problem is solved for all time intervals of the overall period of analysis simultaneously.

10.2.2 Network Flow Programming Algorithms

Network flow programming problems can be solved using conventional linear programming methods such as the simplex algorithm. However, the network flow format facilitates the use of much more computationally efficient algorithms that save computer time and allow analysis of larger problems with numerous variables and constraints. A variety of efficient network solvers are available [Jensen and Barnes, 1980; Kennington and Helgason, 1980; Phillips and Garcia-Diaz, 1981]. Although not as computationally efficient as more specialized methods, the out-of-kilter algorithm, described in the books just cited, is the most generalized and widely used algorithm for solving capacitated network flow programming problems. It is used in several of the reservoir system analysis models cited in Chapter 11.

The out-of-kilter algorithm is based on duality theory. The structure of linear programming is characterized as having two related formulations, called *primal* and *dual*. Solution of the dual provides a solution for the primal, and vice versa. For certain types of LP problems, such as network flow programming, the relationship between the primal and dual can be used to reduce the computational burden.

10.2.3 Example of a Network Flow Programming Formulation

Several of the network flow models noted in Chapter 11 allocate water between users based on water rights priorities or other user-specified priorities. The network flow submodel allocates water for each individual month of a long multiple-year simulation period. Likewise, in the Water Rights Analysis Package (WRAP) model, discussed in Chapter 9, water is allocated between water rights holders based on the priority dates specified in their permits. In the model, junior water rights curtail use if a more senior right is adversely affected. In the Brazos River Basin simulation study discussed in Chapter 9, water is allocated to over a thousand diversion rights and storage rights in about 600 reservoirs. The original WRAP is a conventional simulation model, incorporating no mathematical programming [Wurbs *et al.*, 1993]. However, Yerramreddy and Wurbs [1995] describe a network flow programming version of the WRAP model that requires the same input and provides the same output as the original simulation model.

The c_{ij} reflect relative priorities for meeting each diversion and instream flow requirement and maintaining target reservoir storage levels. In WRAP, the c_{ij} are priority dates that reflect the relative seniority of water rights. In Example 10–8, the c_{ij} simply provide a mechanism for expressing relative priorities. The absolute magnitudes of the c_{ij} are arbitrary, with only their relative magnitudes affecting the computations. Example 10–5 is a simplified illustration of the use of linear programming for performing the water allocation for a given time interval based on specified relative priorities. Example 10–8 is a reformulation of the Example 10–5 model in the format of network flow programming.

Example 10–8—Water Allocation by Network Flow Programming

This example consists of repeating the previous Example 10–5 using the network flow format of equations 10-25, 10-26, and 10-27. Flows (q_{ij}) are computed for the arcs connecting the set of five nodes shown in Figure 10–4 along with a source node and a sink node. The source node represents the source of water entering the stream/reservoir system shown in Figure 10–4. Water leaving the system flows to the sink node. The decision variables (q_{ij}), lower and upper bounds (l_{ij} and u_{ij}) on q_{ij}, and objective function coefficients (c_{ij}) are shown in Table 10–6. The c_{ij} are the relative priorities, which for the diversions are given in Table 10–6. Five constraints in the form of equation 10-26 represent the volume balance at each of the five node locations shown in Figure 10–4. Constraints in the form of equation 10-27 place upper and lower bounds on the decision variables. In order to fit the capacitated net-

TABLE 10–6   Terms in Network Flow Formulation of Example 10–8

Decision Variable q_{ij}	Definition of q_{ij}	Nodes i	Nodes j	Lower Bound l_{ij}	Upper Bound u_{ij}	Priority c_{ij}	Solution $(10^6 m^3)$ q_{ij}
$q_{1,4}$	Instream flow from 1 to 4	1	4	0	999	0	195
$q_{2,3}$	Instream flow from 2 to 3	2	3	5	999	0	380
$q_{3,4}$	Instream flow from 3 to 4	3	4	10	999	0	10
$q_{4,5}$	Instream flow from 4 to 5	4	5	10	999	0	270
$q_{5,S}$	Instream flow below 5	5	Sink	30	999	0	30
$q_{S,1}$	Inflow + initial storage	Source	1	835	835	0	835
$q_{S,2}$	Inflow + initial storage	Source	2	505	505	0	505
$q_{S,3}$	Inflow + initial storage	Source	3	25	25	0	25
$q_{S,4}$	Inflow + initial storage	Source	4	75	75	0	75
$q_{S,5}$	Inflow + initial storage	Source	5	120	120	0	120
$q_{1,D}$	Diversion at node 1	1	Sink	0	90	4	90
$q_{2,D}$	Diversion at node 2	2	Sink	0	125	3	125
$q_{3,D}$	Diversion at node 3	3	Sink	0	475	2	395
$q_{5,D}$	Diversion at node 5	5	Sink	0	360	5	360
$q_{1,S}$	Reservoir A storage	1	Sink	0	750	1	550
$q_{2,S}$	Reservoir B storage	2	Sink	0	900	1	0

work flow format, the given initial storage and inflows are treated as decision variables with both lower and upper bounds set at the values specified in Table 10–6.

The network flow model is formulated as follows:

maximize

$$4q_{1,D} + 3q_{2,D} + 2q_{3,D} + 5q_{5,D} + q_{1,S} + q_{2,S}$$

subject to

$$q_{S,1} - q_{1,D} - q_{1,4} - q_{1,S} = 0$$

$$q_{S,2} - q_{2,D} - q_{2,3} - q_{2,S} = 0$$

$$q_{S,3} + q_{2,3} - q_{3,D} - q_{3,4} = 0$$

$$q_{S,4} + q_{1,4} - q_{4,5} = 0$$

$$q_{S,5} + q_{4,5} - q_{5,D} - q_{5,S} = 0$$

$$0 \le q_{1,4} \le 999 \quad 835 \le q_{S,1} \le 835 \quad 0 \le q_{1,D} \le 90$$

$$5 \le q_{2,3} \le 999 \quad 505 \le q_{S,2} \le 505 \quad 0 \le q_{2,D} \le 125$$

$$10 \le q_{3,4} \le 999 \quad 25 \le q_{S,3} \le 25 \quad 0 \le q_{3,D} \le 475$$

$$10 \le q_{4,5} \le 999 \quad 75 \le q_{S,4} \le 75 \quad 0 \le q_{5,D} \le 360$$

$$30 \le q_{5,S} \le 999 \quad 120 \le q_{S,5} \le 120 \quad 0 \le q_{1,S} \le 750$$

$$0 \le q_{2,S} \le 900$$

The problem can be solved using any linear programming software. Since the model is formulated in the format of equations 10-25, 10-26, and 10-27, network flow programming algorithms can be used. The solution is tabulated in the last column of Table 10–6.

10.3 DYNAMIC PROGRAMMING

The general approach of dynamic programming (DP) is introduced in the following section. DP was originally invented by Bellman [1957]. DP theory and its application in various fields are covered in depth in books by Dreyfus and Law [1977], Cooper and Cooper [1981], and Denardo [1982]. Loucks *et al.* [1981], Esogbue [1989], and Mays and Tung [1992] describe the fundamentals of DP from the perspective of water resources planning and management. Numerous problem formulations, solution techniques, expansions, and extensions of the general DP strategy have been developed. Yeh [1985] and Esogbue [1989] review the various types and variations of DP formulations of reservoir operation models.

Unlike linear programming, for which many general-purpose software packages are available, the availability of generalized dynamic programming codes is limited. Most DP computer programs have been developed for specific applications. Labadie [1990] describes a generalized microcomputer DP package called CSUDP, which is one of the few available general-purpose DP computer programs. CSUDP, developed at Colorado State University (CSU), has been used for a broad range of water resources planning and management applications, including reservoir operation. The basic DP computational algorithms are coded in CSUDP, with the model user furnishing FORTRAN subroutines reflecting the objective function and state and constraint equations for the particular application.

10.3.1 DP Strategy

Dynamic programming is not a precisely structured algorithm like linear programming, but rather a general approach to solving optimization problems. DP involves decomposing a complex problem into a series of simpler subproblems which are solved sequentially, while transmitting essential information from one stage of the computations to the next using state concepts. DP models have the following characteristics:

- The problem is divided into a sequence of stages with a decision required at each stage. The stages may represent different points in time (as in determining reservoir releases for each time interval), different points in space (for example, releases from different reservoirs), or different activities (such as releases for different water users).
- Each stage of the problem must have a finite number of states associated with it. The states describe the possible conditions in which the system might be at that stage. The amount of water in storage is an example of a typical state variable.
- The effect of a decision at each stage of the problem is to transform the current state of the system into a state associated with the next stage. If the decision variable is how much water to release from the reservoir during the cur-

rent time period, this decision will transform the amount of water stored in the reservoir (state variable) from the current amount to a new amount for the next stage (time period).

- A return function indicating the utility or cost of the transformation is associated with each potential state transformation. The return function allows the objective function to be represented by stages.
- The optimality of the decision required at the current stage is judged in terms of its impact on the return function for the current stage and all subsequent stages.

The last two characteristics are reflected in a recursive relationship that defines the objective function in terms of stages,

$$f_i(s_i, x_i) = \text{maximize or minimize } [r_i(x_i) + f_{i+1}(s_{i+1})] \qquad (10\text{-}28)$$

where i denotes the current stage in the DP procedure. Equation 10-28 is written in the format of backward DP, in which the computations proceed from stage $i + 1$ to stage i. For forward DP, the term $f_{i+1}(s_{i+1})$ is replaced with $f_{i-1}(s_{i-1})$ in equation 10-28, and the computations proceed from stage $i - 1$ to stage i. The objective function $(f_i(s_i, x_i))$ is a function of the decision (x_i) and state $(s_i$ and $s_{i+1})$ variables. The return (r_i) is dependent on the decision (x_i) made in stage i. At the completion of the DP computations, the recursive stage-by-stage $f_i(s_i, x_i)$ provides the final value of the objective function.

The dynamic programming strategy involves the following general steps:

1. Formulate the decision variables, recursive objective function, return function, constraints, stages, and state variables.
2. Solve a single-stage optimization problem.
3. Solve a two-stage optimization problem.
4. Solve a three-stage optimization problem and so forth until all stages are included.

Dynamic programming is based on the following fundamental principle of optimality:

- No matter in what state of which stage one may be, in order for a policy to be optimal, one must proceed from that stage and state in an optimal manner.
- No matter in what state of which stage one may be, in order for a policy to be optimal, one had to get to that stage and state in an optimal manner.

DP is a general methodology for applying state concepts to the formulation and solution of problems which can be viewed as optimizing a multiple-stage decision process. DP is much more efficient than the brute-force approach to a multiple-stage decision problem: an exhaustive enumeration of all combinations of decisions at all stages. DP minimizes the number of combinations of decisions that must be considered to find the optimum. The computational effort of exhaustive enumeration tends to increase geometrically with the number of stages. With DP, the solution effort tends to increase only linearly with an increasing number of stages.

10.3.2 DP versus LP

In some cases, the same problem can be solved by alternative dynamic programming formulations or by either DP or LP. In general, linear programming has the advantage over DP of being more precisely defined and easier to understand. The degree of generalization and availability of generalized computer codes is much more limited for DP than for LP. However, the strict linear form of the LP formulation can be a significant hindrance. Nonlinear properties of a problem can readily be reflected in a DP formulation. Functional relationships in the objective function and constraints can be nonlinear, nonconvex, and discontinuous in DP. However, various assumptions, including a separable objective function, limit the range of application, and require ingenuity and understanding by the modeler in applying DP. The so-called curse of dimensionality is a major consideration in DP, meaning that increasing the number of state variables greatly increases the computational burden. For example, since reservoir storage is typically a state variable, the number of reservoirs that can be included in a DP model may be limited.

10.3.3 Deterministic versus Stochastic DP

Reservoir system optimization models are classified as deterministic or stochastic depending primarily on the manner in which the streamflow inflows are handled. Variables other than streamflow can be treated as being stochastic as well. In a deterministic model, the inflows for each time interval are given. In a stochastic model, the inflows are treated as a stochastic process rather than as known values. The typical approach is to model inflows as a Markov process represented by transition probability matrices. The inflow in a given time interval can be various discrete values with probabilities conditioned on the flow in the previous time period.

The DP computational algorithm is similar for either deterministic or stochastic inflows. In deterministic DP, the objective function is determined for a specified decision based on the known inflow. The difference in stochastic DP is that the expected value of the objective function that reflects the full range of possible inflows with their associated probabilities is determined. For simplicity, the inflows are typically treated as falling in discrete ranges. The expected value $E[X]$ of a discrete random variable is expressed as

$$E[X] = \Sigma x_i P(x_i) \qquad (10\text{-}29)$$

where $P(x_i)$ is the probability of the random variable X taking the value x_i. In stochastic DP, the conditional inflow probabilities are provided by a transition probability matrix, as discussed in Chapter 5.

Thus, the recursive objective function of equation 10-28, expressed for a deterministic problem, is rewritten in terms of expected value for a stochastic problem.

$$f_i(s_i, x_i) = \text{maximize or minimize } \{r_i(x_i) + \Sigma[P_{ij}f_{i+1}(s_{i+1})]\} \qquad (10\text{-}30)$$

P_{ij} is the discrete probability that the inflow falls within the range represented by i given that it is in range j in the subsequent or previous period. Example 10–11 is a stochastic dynamic programming problem.

10.3.4 Illustrative Examples

The basic concepts of dynamic programming are illustrated with three simple examples.

Example 10–9—Reservoir Optimization with DP

A reservoir supplies water during a three-month dry season during which there are no inflows. The net benefits to be derived from supplying various amounts of water during each of the three months are shown in Table 10–7. The decision problem is to allocate the water available in storage at the beginning of the dry season over the three months, based on maximizing total benefits. The allocation is repeated assuming different amounts of water are available in storage at the beginning of the dry season.

The stages i are the three months. The state variables s_1, s_2, and s_3 are the volume of water available in storage at the beginning of each month. The decision variables x_1, x_2, and x_3 are the amounts of water supplied in each of three months. The state and decision variables are specified as a discrete value of either 0, 1, 2, or 3 million m³. The amount of water supplied during the first month (x_1) determines the state or amount of water available (s_2) for allocation (x_2 and x_3) between the remaining two months. The stage return $r(x_i)$ is the net benefits, in dollars, associated with the optimum water allocation for each month. The objective function $f_i(s_i, x_i)$ is expressed in the recursive format of equation 10-28.

In the backward DP solution, the first step is simply to tabulate the amount of water supplied in month 3 for given amounts in storage at the beginning of month 3. The amount of water available in month 3 depends on how water is supplied during months 1 and 2. The next task is to allocate the water (s_2) available at the beginning of month 2 between months 2 and 3 (x_2 and x_3). Then the water available at the beginning of the first month (s_1) is allocated among all three months. Discrete amounts of either 0, 1, 2, or 3 units of water are assumed to be available at the beginning of each month.

TABLE 10–7   Benefit Relationships for Example 10–9

Water Allocated (10⁶m³)	Net Benefits ($100,000)		
	First Month	Second Month	Third Month
0	0	0	0
1	3	1	4
2	6	3	5
3	8	9	6

TABLE 10–8   **Stage 3 DP Allocation for Month 3 of Example 10–9**

Water Available S_3 $(10^6 m^3)$	Water Supplied X_3 $(10^6 m^3)$	Net Benefits B_3 ($100,000)
0	0	0
1	1	4
2	2	5
3	3	6

Results for the three stages of the backward DP procedure are summarized in Tables 10–8, 10–9, and 10–10. The stage 3 allocation is tabulated in Table 10–8. The total available (s_3) is the amount supplied (x_3), with the resulting net benefits (B_3).

Stage 2 consists of allocating available water between months 2 and 3. The computations are presented in Table 10–9. If $1 \times 10^6 \ m^3$ of water is available for use in months 2 and 3, the optimum allocation is zero and $1 \times 10^6 \ m^3$, respectively, for x_2 and x_3, which results in benefits (B_{total}) of $400,000. If 3 units are available for use in months 2 and 3, the optimum allocation is $x_2 = 3$ and $x_3 = 0$, with a resulting $B_{total} = \$900,000$.

Stage 1 consists of allocating the total amount of water available between month 1 and the combination of months 2 and 3. The computations are shown in Table 10–10. For any allocation to the combined months 2 and 3 $(s_2 = x_1 + x_2)$, the optimum allocation between x_1 and x_2 is available from the stage 2 results shown in Table 10–9.

The final solution to the water allocation problem is determined by backtracking through Tables 10–10 and 10–9. The optimum allocation for the alternative discrete levels of water in storage are tabulated in Table 10–11. If three units of water are available in storage for use during the three-month dry season, the optimum allocation is to use 2, 0, and 1 units in months one, two, and three, respectively, which results in benefits of $1,000,000. Optimum allocations are also tabulated in Table 10.11, alternatively assuming the total available storage is either one or two units.

TABLE 10–9   **Stage 2 DP Allocation Between Months 2 and 3 of Example 10–9**

$S_2 = X_2 + X_3$	Supplied $(10^6 m^3)$		Net Benefits ($100,000)			Optimum
	X_2	X_3	B_2	B_3	B_{total}	
1	1	0	1	0	1	
	0	1	0	4	4	*
2	2	0	3	0	3	
	1	1	1	4	5	*
	0	2	0	5	5	*
3	3	0	9	0	9	*
	2	1	3	4	7	
	1	2	1	5	6	
	0	3	0	6	6	

TABLE 10–10   Stage 1 DP Allocation Between Month 1 Versus Months 2 and 3
             of Example 10–9

$S_2 =$ $X_1 + X_2 + X_3$	Supplied (10^6m³)		Net Benefits ($100,000)			
	X_1	$X_2 + X_3$	B_1	$B_2 + B_3$	B_{total}	Optimum
1	1	0	3	0	3	
	0	1	0	4	4	*
2	2	0	6	0	6	
	1	1	3	4	7	*
	0	2	0	5	5	
3	3	0	8	0	8	
	2	1	6	4	10	*
	1	2	3	5	8	
	0	3	0	9	9	

Example 10–10—Shortest Route Using DP

Figure 10–7 shows all possible routes a water conveyance canal could take from start to end. Flow is from left to right in the figure. The length of each link, in kilometers, is shown. (Alternatively, the problem could be formulated in terms of cost rather than distance.) The optimization problem is to determine the route from the start node to the end node of the minimum total distance. Dynamic programming is used to find the shortest route.

The DP stages are segments of the overall route. Stage 1 consists of connecting the start node with node A, B, or C. Stage 2 consists of linking nodes A, B, and C with nodes D, E, and F. Stage 3 is getting from node D, E, or F to node G, H, or I, and so forth. The state variable is the node location at the beginning and end of each stage. For example, at the end of stage 1 or the beginning of stage 2, the possible states are nodes A, B, and C. At each stage, for each state, the decision variable is which link to select. For example, with forward DP, for stage 2, if the beginning state is at node B, the decision is selecting of link B-D, B-E, or B-F. The objective function to be minimized is the total distance from start to end. The return is the lengths of the individual links between the nodes. For example, the return associated with connecting nodes A and D is 6 km, the length between the two node locations. The recursive representation of the objective function is the distance from either the start or end node to the node currently being considered. Constraints consist of limiting the nodes and the choice of connecting links to those shown in Figure 10–7.

Alternative solutions are presented in Tables 10–12 and 10–13, respectively, based on

TABLE 10–11   Solution for Example 10–9

Water Available (10^5m³)	Allocation (10^6m³)			Benefits B_{total} ($100,000)
	Month 1 X_1	Month 2 X_2	Month 3 X_3	
0	0	0	0	0
1	0	0	1	4
2	1	0	1	7
3	2	0	1	10

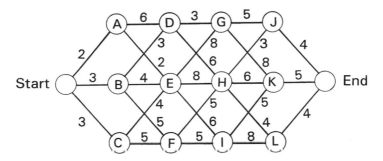

Figure 10–7   Schematic for DP Example 10–10.

forward DP and backward DP. The forward DP procedure begins at the start node and works forward. The backward DP solution consists of beginning the algorithm at the end node and working backwards to reach the start node. Either approach results in the same solution. For this problem, either approach is equally convenient: in other applications, either forward or backward computations may be advantageous.

The solution is that the shortest route from the start to end nodes passes through nodes B, D, G, and J. This optimum path has a total length of 18 kilometers.

Example 10–11—Reservoir Optimization with Stochastic DP

Reservoir releases (R_t) are determined for each time period (t) which represent the dry and wet seasons of successive years. Storage and diversion targets are 6 and 8 million m³, respectively. The objective function to be minimized consists of the sum for all time periods of the squared deviations from the target storage and diversion.

TABLE 10–12   Forward DP Solution of Example 10–10

Stage	State	Decision	Return
	Ending node	Optimum link	Kilometers from start
1	A	start-A	2
	B	start-B	3
	C	start-C	3
2	D	B-D	$3 + 3 = 6$
	E	A-E	$2 + 2 = 4$
	F	B-F or C-F	$3 + 5 = 8$
3	G	D-G	$6 + 3 = 9$
	H	D-H or E-H	$6 + 6 = 12$ or $4 + 8 = 12$
	I	E-I	$4 + 6 = 10$
4	J	G-J	$9 + 5 = 14$
	K	I-K	$10 + 5 = 15$
	L	H-L	$12 + 4 = 16$
5	End	J-end	$14 + 4 = 18$

The shortest route is start-B-D-G-J-end, which has a length of 18 km.

TABLE 10–13   Backward DP Solution of Example 10–10

Stage	State Ending node	Decision Optimum link	Return Kilometers from start
5	J	J-end	4
	K	K-end	5
	L	L-end	4
4	G	G-J	$4 + 5 = 9$
	H	H-J	$4 + 3 = 7$
	I	K-I	$5 + 5 = 10$
3	D	D-G	$9 + 3 = 12$
	E	E-H	$7 + 8 = 15$
	F	F-H	$7 + 5 = 12$
2	A	A-E	$15 + 2 = 17$
	B	B-D	$12 + 3 = 15$
	C	C-F	$12 + 5 = 17$
1	start	start-B	$15 + 3 = 18$

The shortest route is start-B-D-G-J-end, which has a length of 18 km.

$$\text{Minimize } \Sigma \, [(S_t - 6)^2 + (R_t - 8)^2]$$

Releases (R_t) are determined as a function of beginning-of-period storage (S_t), such that the objective function is minimized. Backward stochastic DP is used, with the objective function expressed in the following recursive format:

$$f_t(S_t, I_t, S_{t+1}) = \text{minimize } \{r_t(R_t) + \Sigma[P_{ij}f_{t+1}(S_{t+1})]\}$$

where

$$r(R_t) = (S_t - 6)^2 + (R_t - 8)^2$$

Reservoir inflows are treated stochastically. Since $f_{t+1}(S_{t+1})$ depends on inflows, the expected value $\Sigma[P_{ij}f_{t+1}(S_{t+1})]$ is incorporated in the recursive objective function. The probability P_{ij} reflects the correlation between streamflows in successive time periods i and j. Transition probability matrices for streamflows are provided in Table 10–14. As an example of interpreting Table 10–14, if the preceding dry season inflow is 2 million m^3, the probabil-

TABLE 10–14   Streamflow Transition Probability Matrices for Example 10–11

Flow in Dry Season (Q_t)	Flow in Wet Season (Q_{t+1}) 6 Mm^3	 8 Mm^3	Flow in Wet Season (Q_t)	Flow in Dry Season (Q_{t+1}) 2 Mm^3	 4 Mm^3
	$P_{ij} = P(Q_{t+1}|Q_t)$			$P_{ji} = P(Q_{t+1}|Q_t)$	
2 Mm^3	0.6	0.4	6 Mm^3	0.8	0.2
4 Mm^3	0.3	0.7	8 Mm^3	0.1	0.9

ity P_{ij} of the wet season flow being 6 or 8 million m³ is 0.6 and 0.4, respectively. If the preceding wet season flow is 8 units, the probability of the dry season flow being either 2 or 4 units is 0.1 and 0.9, respectively.

All possible combinations of inflows, releases, and beginning- and end-of-period storages are tabulated in Table 10–15, in units of 10^6 m³, for each of the two seasons of the year. The variables are limited to these discrete values to simply the DP computations. Storage is assumed to fall within discrete ranges represented by values of 2, 4, or 6 million m³. Likewise, inflows and releases are assumed to fall within discrete ranges represented by values of 2, 4, 6, or 8 million m³. For each possible combination of these variables, Table 10–15 includes the computed squared deviations from the storage and release targets $[(S_t - 6)^2 + (R_t - 8)^2]$

The backward stochastic dynamic programming computations are summarized in Table 10–16. The first time period considered is a wet season. The optimum end-of-period storage (S_{t+1}) and corresponding release (R_t) are determined for each possible combination of beginning storage (S_t) and inflow (I_t) based on minimizing the return function $r(R_t)$.

$$f_t(S_t) = r(R_t) = (S_t - 6)^2 + (R_t - 8)^2$$

For each S_t and I_t, the optimum S_{t+1} and R_t are tabulated in the last two columns of Table 10–16.

The second period considered in the DP algorithm is the previous dry season. The optimum S_{t+1} is determined for each possible combination of S_t and I_t based on the recursive objective function

TABLE 10–15 Discrete Sets of Variable Values for Example 10–11

Initial Storage (10^6m³)	Inflow (10^6m³)	Final Storage (10^6m³)	Release (10^6m³)	Squared Deviations		
				Storage	Release	Total
			Dry Season			
4	2	2	4	16	16	32
4	2	4	2	4	36	40
4	4	2	6	16	4	20
4	4	4	4	4	16	20
6	2	2	6	16	4	20
6	2	4	4	4	16	20
6	4	2	8	16	0	16
6	4	4	6	4	4	8
			Wet Season			
2	6	4	4	4	16	20
2	6	6	2	0	36	36
2	8	4	6	4	4	8
2	8	6	4	0	16	16
4	6	4	6	4	4	8
4	6	6	4	0	16	16
4	8	4	8	4	0	4
4	8	6	6	0	4	4

TABLE 10–16   Dynamic Programming Computations for Example 10–11

Storage S_t	Inflow	Squared Deviations		$f_t(S_t)$	Optimum S_{t+1}	Optimum Release
		S_{t+1} = 4 or 6		Period 1: Wet Season		
2	6	20	36	20	4	4
2	8	8	16	8	4	4
4	6	8	16	8	4	4
4	8	4	4	4	4&6	6&8
		S_{t+1} = 2 or 4		Period 2: Dry Season		
4	2	47.2	46.4	46.4	4	2
4	4	31.6	25.2	25.2	4	4
6	2	35.2	26.4	26.4	4	4
6	4	27.6	13.2	13.2	4	6
		S_{t+1} = 4 or 6		Period 3: Wet Season		
2	6	62.16	59.76	59.76	6	2
2	8	35.32	30.32	30.32	6	4
4	6	50.16	38.16	38.16	6	4
4	8	31.32	18.32	18.32	6	6
		S_{t+1} = 2 or 4		Period 4: Dry Season		
4	2	79.98	70.22	70.22	4	2
4	4	58.67	44.27	44.27	4	4
6	2	67.02	50.22	50.22	4	4
6	4	54.67	32.27	32.27	4	6
		S_{t+1} = 4 or 6		Period 5: Wet Season		
2	6	85.03	82.63	82.63	6	2
2	8	54.87	50.06	50.06	6	4
4	6	73.03	62.63	62.63	6	4
4	8	50.87	38.07	38.07	6	6
		S_{t+1} = 2 or 4		Period 6: Dry Season		
4	2	106.60	92.81	92.81	4	2
4	4	79.83	65.44	65.44	4	4
6	2	89.60	72.81	72.81	4	4
6	4	75.83	53.44	53.44	4	6

$$f_t(S_t, I_t, S_{t+1}) = \text{minimize} \ \{r_t(R_t) + \Sigma[P_{ij} f_{t+1}(S_{t+1})]\}$$

where

$$r(R_t) = (S_t - 6)^2 + (R_t - 8)^2$$

from Table 10–15.

The Σ squared deviations tabulated in Table 10–16 for period 2 are computed as follows:

$$f(4,2,2) = 32 + 0.6(20) + 0.4(8) = 47.2$$

$$f(4,2,4) = 40 + 0.6(8) + 0.4(4) = 46.4$$

$$f(4,4,2) = 20 + 0.3(20) + 0.7(8) = 31.6$$

$$f(4,4,4) = 20 + 0.3(8) + 0.7(4) = 25.2$$

$$f(6,2,2) = 20 + 0.6(20) + 0.4(8) = 35.2$$

$$f(6,2,4) = 20 + 0.6(8) + 0.4(4) = 26.4$$

$$f(6,4,2) = 16 + 0.3(20) + 0.7(8) = 27.6$$

$$f(6,4,4) = 8 + 0.3(8) + 0.7(4) = 13.2$$

The third period is the wet season of the previous year. The Σ squared deviations tabulated in Table 10–16 for period 3 are computed as follows:

$$f(2,6,4) = 20 + 0.8(46.4) + 0.2(25.2) = 62.16$$

$$f(2,6,6) = 36 + 0.8(26.4) + 0.2(13.2) = 59.76$$

$$f(2,8,4) = 8 + 0.1(46.4) + 0.9(25.2) = 35.32$$

$$f(2,8,6) = 16 + 0.1(24.4) + 0.9(13.2) = 30.32$$

$$f(4,6,4) = 8 + 0.8(46.4) + 0.2(25.2) = 50.16$$

$$f(4,6,6) = 16 + 0.8(24.4) + 0.2(13.2) = 38.16$$

$$f(4,8,4) = 4 + 0.1(46.4) + 0.9(25.2) = 31.32$$

$$f(4,8,6) = 4 + 0.1(24.4) + 0.9(13.2) = 18.32$$

The computations are repeated for each season of each year, working backward in time. A steady state is reached after the first two periods, in which the release decisions remain constant each year. Optimum releases are shown in Table 10–17, as a function of beginning-of-

TABLE 10–17   Final Solution for Example 10–11

Initial Storage (10^6m³)	Inflow (10^6m³)	Optimum Release (10^6m³)	Final Storage (10^6m³)
		Wet Season	
2	6	2	6
2	8	4	6
4	6	4	6
4	8	6	6
		Dry Season	
4	2	2	4
4	4	4	4
6	2	4	4
6	4	6	4

period storage and inflows, with all variables limited to the discrete values considered in the DP formulation.

10.4 UNIVARIATE GRADIENT SEARCH

A broad range of nonlinear optimization techniques are classified as search methods. Search techniques involve iteratively changing estimated values of the decision variables in such a way as to move closer to the optimum value of the objective function. Search techniques provide more flexibility than other nonlinear programming methods that are limited by strict mathematical format requirements. Search techniques have been combined with LP and DP and also with simulation models.

An approach is outlined here that links a search algorithm with a simulation model to solve optimization problems that cannot be formulated in the format required by other linear and nonlinear programming methods. The optimization modeling strategy combines the following:

- The univariate gradient search procedure
- The Newton-Raphson formula to determine an improved value of a decision variable during each iteration of the search
- Finite difference formulas to approximate the derivative terms in the Newton-Raphson formula
- A simulation model to provide the objective function evaluations required for the finite difference approximations of the derivative terms in the Newton-Raphson formula

10.4.1 Univariate Gradient Search Procedure

The univariate gradient search approach, which is also called a *cyclic coordinate search*, consists of optimizing one variable at a time, with the other variables temporarily held constant. The computations are repeated iteratively to optimize each individual variable in turn. The cyclic algorithm continues until a specified stop criterion is met. The stop criterion is based on no further changes in the decision variables and no additional improvement in the objective function having occurred in the latest iterations. Thus, the search procedure consists of the following steps:

1. Start the search with an initial estimate of the values of the decision variables.
2. Using the Newton-Raphson method, optimize one decision variable while all others are held constant.
3. Check whether the stop criterion has been satisfied. If not, repeat step 2 by optimizing the next decision variable.

For each univariate search iteration, the Newton Raphson formula

$$x^* = x - \frac{f'(x)}{f''(x)} \tag{10-31}$$

is used to determine an improved value x^* given the current value of x and the first and second derivatives of the objective function, $f'(x)$ and $f''(x)$. The derivation of the Newton Raphson formula is based on approximating the function $f(x)$ with the truncated Taylor series. From calculus, the derivative of the truncated Taylor series is set equal to zero to derive equation 10-31. Example 10–12 illustrates the general concept of a cyclic coordinate search with the Newton Raphson formula applied at each step.

Example 10–12—Univariate Gradient Search

The problem consists of determining values for the variables x_1 and x_2 that result in a minimum value for the following function:

$$f(x_1, x_2) = x_1^2 + 2x_2^2 - 4x_1 - 2x_1 x_2$$

This simple problem can be solved analytically by setting the derivative of the function with respect to each variable equal to zero and solving the resulting two equations to obtain $x_1 = 4.0$ and $x_2 = 2.0$. However, the problem is solved here using the univariate gradient search method. The solution obtained can be compared with the true analytical solution of 4.0 and 2.0.

The first and second derivatives of the function with respect to each variable are as follows:

$$f'(x_1) = 2x_1 - 4 - 2x_2$$
$$f''(x_1) = 2$$
$$f'(x_2) = 4x_2 - 2x_1$$
$$f''(x_2) = 4$$

The algorithm starts by arbitrarily assuming initial estimates of $x_1 = 1$ and $x_2 = 1$. Equation 10-31 is applied at each iteration to determine improved values for the variables. The iterative computations are summarized in Table 10–18. The first cycle consists of first determining an improved value x^* of 3.0 for x_1 and then determining an improved value x^* of 1.5 for x_2. The second cycle consists of applying equation 10-31 twice to obtain improved values for x_1 and x_2 of 3.5 and 1.75, respectively. At each step, the most recently computed values for x_1 and x_2 are used in the computation of improved values. The iterative computations are stopped when the variables are no longer significantly changing. At the completion of the sixth cycle of computed improved values for the two variables, the solution is as follows:

$$x_1 = 4.0 \quad \text{and} \quad x_2 = 2.0$$

10.4.2 Combining the Search Algorithm with a Simulation Model

For typical applications, the $f'(x)$ and $f''(x)$ terms in equation 10-31 cannot be determined analytically and thus must be determined numerically using finite-difference formulas, such as the following:

TABLE 10–18   Univariate Gradient Search of Example 10–12

Cycle	Optimize	x_1	x_2	$f'(x)$	$f''(x)$	x^*
1	x_1	1	1	−4	2	3.0
	x_2	3	1	−2	4	1.5
2	x_1	3	1.5	−1	2	3.5
	x_2	3.5	1.5	−1	4	1.75
3	x_1	3.5	1.75	−0.5	2	3.75
	x_2	3.75	1.75	−0.5	4	1.875
4	x_1	3.75	1.875	−0.25	4	3.875
	x_2	3.875	1.875	−0.25	4	1.938
5	x_1	3.875	1.938	−0.125	2	3.938
	x_2	3.938	1.938	−0.125	4	1.969
6	x_1	3.938	1.969	−0.0625	2	3.969
	x_2	3.969	1.969	−0.0625	4	1.984

$$f'(x) = \frac{f(x + \Delta x) - f(x - \Delta x)}{2\Delta x} \tag{10-32}$$

$$f''(x) = \frac{f(x + \Delta x) - 2f(x) + f(x - \Delta x)}{\Delta x^2} \tag{10-33}$$

A finite-difference approximation of the second derivative requires at least three objective function evaluations. Thus, if the search algorithm is linked with a simulation model that determines values for the objective function, the simulation model is executed three times each time equation 10-31 is applied in the cyclic search.

The general idea of developing an optimization model by linking a search algorithm and a simulation model has broad applicability. The simulation model may be very complex. It must simply be able to compute values for an objective function for a specified set of values for the decision variables. The search algorithm repeatedly executes the simulation model with different values for the decision variables in a systematic automated search for the optimum. For example, this optimization approach is incorporated in the HEC-1 Flood Hydrograph Package [Hydrologic Engineering Center, 1990] for two different optional tasks, parameter calibration and flood damage reduction system optimization.

The HEC-1 search routine automates the calibration of watershed (precipitation-runoff) and flood-routing parameters. As illustrated by Figure 10–8, values of parameters are adjusted based on matching a computed hydrograph to an observed hydrograph. Loss rate and unit hydrograph parameters are computed that best reproduce the observed hydrograph resulting from a rainfall event. Parameters for the Muskingum and other routing methods are determined based on observed upstream and downstream hydrographs.

The objective function is

$$\text{Minimize } \Sigma \left((QO_i - QC_i)^2 \, (WT_i/n) \right)^{1/2} \tag{10-34}$$

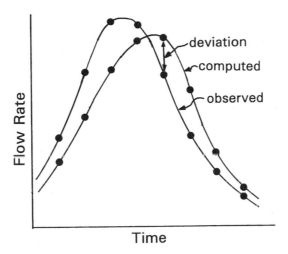

Figure 10-8   The HEC-1 parameter calibration objective function is based on minimizing the sum of squared deviations between observed and computed hydrograph ordinates.

where

$$WT_i = (QO_i + Q_a) / (2*Q_a)$$

The subscript i denotes hydrograph ordinates. As shown in Figure 10–8 and equation 10-34, the sum of the squared deviations between the observed flows (QO_i) and computed flows (QC_i) are minimized. The weighting factor WT_i gives a greater weight to high flows. Q_a is the average of the observed flows QO_i. The observed hydrograph is provided as input to the model. The computed hydrograph is developed by either the precipitation-runoff or streamflow routing feature of the simulation model. The loss rate and unit hydrograph parameters or routing parameters are the decision variables. The univariate gradient search algorithm adjusts one parameter at a time. In order to develop finite-difference estimates of the derivative terms in the Newton-Raphson formula (equation 10-31), each time a parameter is adjusted, the simulation is executed three times, with incrementally different values of the parameter, to compute the three hydrographs with ordinates QC_i needed to determine the three objective function values ($f(x), f(x - \Delta x), f(x + \Delta x)$).

The same cyclic coordinate search algorithm is used in the HEC-1 flood damage reduction system optimization option. The decision variables are the capacity of each component of a proposed flood control system. The decision variables include storage capacities for reservoirs and flow capacities for channels, diversions, and pumping plants. The objective function is to minimize total system economic cost in dollars discounted to annual worth,

$$\text{Minimize cost} = \Sigma\ C_i + \Sigma\ EAA_j \qquad (10\text{-}34)$$

where C_i are the costs of each reservoir, channel improvement, or other system component i and EAA_j are the expected annual damages for each damage index location j. The simulation model computes expected annual flood damages as described in Chapter 7. Thus, a complex simulation model is executed for each of the numerous evaluations of the objective function performed during the iterative search procedure.

10.5 OTHER OPTIMIZATION METHODS

Optimization methods are covered in numerous operations research and mathematics textbooks. Optimization techniques are classified as unconstrained and constrained. Unconstrained optimization problems are typically solved using classical techniques of calculus. Search techniques are also employed.

Various nonlinear programming (NLP) algorithms are available for solving problems of the form of maximizing or minimizing an objective function subject to constraints. Certain NLP methods are associated with special mathematical formulations that are similar to linear programming. Separable programming involves a separable nonlinear objective function and linear constraints. A separable objective function can be written as the sum of n functions, each of which is a function of only one variable, and each of which may be nonlinear.

$$f(x_1, x_2, ..., x_n) = f(x_1) + f(x_2) + ... + f(x_n) \qquad (10\text{-}35)$$

Quadratic programming involves linear constraints and a nonlinear objective function which is the summation of linear and quadratic terms. Integer programming is linear programming with decision variables constrained to integer values. Zero-one integer programming is linear programming with decision variables limited to values of zero or one. NLP techniques have been applied relatively little, compared to LP and DP, to problems of optimizing reservoir operations. The significant advancements in computer technology in recent years have removed computational constraints, which could result in greater use of NLP in the future.

11

RESERVOIR SYSTEM OPTIMIZATION MODELS

During World War II, the Allies organized interdisciplinary teams to solve complex scheduling and allocation problems involved in military operations. Mathematical optimization models were found to be very useful in this work. After the war, the evolving discipline of operations research continued to rely heavily on optimization models for solving a broad range of problems in private industry. The same mathematical programming techniques also became important tools in the various systems engineering disciplines, including water resources systems analysis. Reservoir operations have been viewed as an area of water resources planning and management which has particularly great potential for beneficial application of optimization models. Developing reservoir models based on mathematical programming techniques has been a major focus of the water resources research literature since the 1960s.

The literature related to optimization models in general and their application to reservoir operations in particular is extensive. Mathematical programming is covered by numerous operations research and mathematics textbooks as well as by water resources systems analysis books, including those by Hall and Dracup [1970], Loucks *et al.* [1981], and Mays and Tung [1992]. Yeh [1982, 1985] presents a comprehensive state-of-the-art review of reservoir operation models, with a strong emphasis on mathematical programming. Literally thousands of journal and conference papers have been published on reservoir system optimization models. University research projects involving case studies account for most of the applications of optimization techniques to reservoir operations. Of the various optimization methods, linear programming is most often adopted for reservoir system analysis models. Variations and extensions of dynamic programming represent the next largest group of reservoir system optimization models reported in the research literature. Various other nonlinear programming methods have been used to a lesser extent.

Measures of system performance incorporated in optimization models are discussed in Chapter 4. The objective or criterion function is the heart of an optimization model. The objective function, which is to be minimized or maximized subject to constraints, is expressed in the proper mathematical format as a function of the decision variables. The objective function may be an expression of relative priorities used to define operating rules, a penalty or utility function, or a mathematical expression of a planning or operational objective.

The basics of mathematical programming are introduced in Chapter 10. The present chapter addresses the use of mathematical programming algorithms in reservoir system analysis models. Several models representative of the numerous models reported in the literature are cited. General categories and characteristics of reservoir system optimization models are discussed.

11.1 NETWORK FLOW PROGRAMMING MODELS

Network flow programming has a proven record of practical application in reservoir system analysis, as compared with many more mathematically sophisticated models which have been addressed extensively in academic research. The network flow models discussed later are notable for combining advantageous features of simulation with a formal mathematical programming algorithm. Most of the models cited are designed such that the user provides the required input data, with the network flow formulation being essentially transparent. Since the network flow formulation is built into the computer code of the generalized model, various reservoir/river systems can be modeled without redeveloping the general network flow problem formulation and solution algorithm. This is in contrast to most other optimization models reported in the literature, which were developed for specific reservoir systems.

Network flow programming provides considerable flexibility for formulating a particular river basin modeling application. For the majority of the models cited in the following paragraphs, a network flow problem, formulated in the format of equations 10-25 through 10-27, is solved for each individual time interval in turn. For other models, such as the Hydrologic Engineering Center Prescriptive Reservoir Model (HEC-PRM), discussed later, a single network flow problem is solved for all time intervals of the overall period of analysis simultaneously. In several of the models, the objective function coefficients c_{ij} of equation 10-25 are penalty or utility terms which provide a mechanism for expressing relative priorities for use in defining operating rules. In the HEC-PRM, the c_{ij} are unit costs in dollars. Convex piecewise linear penalty functions are represented with a q_{ij}, c_{ij}, l_{ij}, and u_{ij} for each linear segment. Successive iterative algorithms are used in several models to handle nonlinearities such as those associated with reservoir evaporation and hydroelectric power computations. The out-of-kilter algorithm, developed by Barr *et al.* [1974], is used in several of the models. However, more computationally efficient algorithms have been developed to replace the out-of-kilter algorithm in some network flow models, such as MODSIM and HEC-PRM.

A number of reservoir system analysis models based on network flow programming have been reported in the literature. Several representative models are briefly described in the following paragraphs. The first group of models are descriptive simulation models. The descriptive models do not automatically find an optimal set of reservoir release and storage values, but do show the releases and storages which would result from a particular operating plan. The descriptive models have the advantage of allowing the user to more precisely define the operating plan. Next, a prescriptive model is discussed. Prescriptive models provide the advantage of determining the sequences of operating decisions which optimize a specified criterion function.

11.1.1 Generalized Reservoir/River System Simulation Models Based on Network Flow Programming

The Texas Water Development Board began development of a series of models in the late 1960s in conjunction with the formulation of the Texas Water Plan. Since that time, several generalized models, reflecting pioneering applications of network flow programming, have evolved through various versions. SIMYLD-II, AL-V, and SIM-V incorporate a capacitated network flow formulation solved with the out-of-kilter linear programming algorithm.

SIMYLD-II provides capabilities for analyzing water storage and transfer within a multireservoir or multibasin system with the objective of meeting a set of specified demands in a given order of priority [Texas Water Development Board, 1972]. If sufficient water is not available to meet competing demands during a particular time interval, the shortage is assigned to the lowest-priority demand node. SIMYLD-II also determines the firm yield of a single reservoir within a multiple-reservoir system. An iterative procedure is used to adjust the demands at a reservoir in order to converge on its firm yield.

The Surface Water Resources Allocation Model (AL-V) and Multireservoir Simulation and Optimization Model (SIM-V) simulate and optimize the operation of an interconnected system of reservoirs, hydroelectric power plants, pump canals, pipelines, and river reaches [Martin, 1981, 1982, 1983]. SIM-V is used to analyze short-term reservoir operations. AL-V is for long-term operations. Hydroelectric benefits, which are complicated by nonlinearity, are incorporated by solving successive network flow problems, where flow bounds and unit costs are modified between successive iterations to reflect first-order changes in hydroelectric power generation with flow release rates and reservoir storage.

MODSIM was developed at Colorado State University and was originally based on modifying and updating the SIMYLD-II model. Various versions of MODSIM are described by Labadie et al. [1984], Frevert et al. [1994], and Labadie *et al.* [1994]. MODSIM is a generalized river basin network simulation model for hydrologic and water rights analyses of complex water management systems. Water is allocated based on user-specified priorities. The user assigns relative priorities for meeting diversion, instream flow, and storage targets, as well as lower and upper bounds on flows and storages. The model computes values for all flows and storages. A network flow programming problem is solved for each individual time interval. Thus, release decisions are not affected by future inflows and release decisions. Monthly, weekly, or daily time intervals may be used. Recent versions of MODSIM include capabilities for analyzing water quality and conjunctive use of surface and groundwater. The out-of-kilter algorithm incorporated in earlier versions of MODSIM has been replaced with a more efficient algorithm based on a Lagrangean relaxation strategy. MODSIM has been applied by university researchers and water management agencies in studies of a number of reservoir/river systems.

Brendecke *et al.*, [1989] describe the Central Resource Allocation Model (CRAM) developed by WBLA, Inc., for use in preparing a water supply master plan for the city of Boulder, Colorado. The model was used to compute the yields which could be achieved with various system operation plans. MODSIM served as the basis for the development of CRAM, with various improvements pertinent to the particular application being added to CRAM.

Acres International Corporation developed a generalized model, called the Acres Reservoir Simulation Program (ARSP), which has been applied to a number of river basins. Sigvaldason [1976] describes pioneering work in applying network flow programming to reservoir system management that later evolved into the current ARSP model. The original model was developed to assess alternative operation policies for a 48-reservoir multiple-purpose water supply, hydropower, and flood control system in the Trent River Basin in Ontario, Canada. The model was originally developed for planning, but has also been used for real-time operation. In the model, each reservoir was subdivided into five storage zones, and time-based rule curves were specified. The combined rule curve and storage zone representation is similar to the Hydrologic Engineering Center's HEC-5 Simulation of Flood Control and Conservation Systems model, discussed in Chapters 8 and 9. However, the Acres model was formulated as a network flow programming problem. Penalty coefficients were assigned to those variables which represented deviations from ideal conditions. Different operating policies were simulated by altering relative values of these coefficients. The out-of-kilter algorithm used to solve the network flow problem is similar to the Texas Water Development Board models cited earlier. Bridgeman *et al.* [1988] describe more recent applications of a later version of the network flow model designed to forecast inflows, simulate operations, and post-process results. The ACRES consulting firm continues to apply the ARSP model to various reservoir/river systems for various clients.

The Water Assignment Simulation Package (WASP) was developed to analyze the water supply system of the city of Melbourne, Australia, which includes nine reservoirs and a complex conveyance and distribution system, but is generalized for application to other systems as well [Kuczera and Diment, 1988]. WASP allocates water according to the following criteria in order of decreasing priority:

1. satisfy all demands,
2. satisfy instream requirements,
3. minimize spills,
4. ensure that water assignments are consistent with user-defined operating rules, and
5. minimize operating cost.

The network programming solution is based on minimizing a weighted penalty function, with a hierarchy of penalties based on the above priorities.

The California Department of Water Resources Simulation Model (DWR-SIM) was developed to simulate the combined operation of the State Water Project and Central Valley Project [Chung et al., 1989]. The model continues to be applied by various water management agencies in California. The original DWRSIM was a conventional simulation model, with no mathematical programming, developed based on modifying the Hydrologic Engineering Center's model HEC-3 Reservoir System Analysis for Conservation. However, the DWRSIM was later revised to incorporate the out-of-kilter network flow programming algorithm. The versions of DWRSIM with and without the network flow programming algorithm are used for the same types of analyses and have essentially the same input and output formats. The network flow formulation was incorporated into DWRSIM to enhance capabilities for analyzing the consequences of different operational scenarios. The most significant model improvements provided by the network flow algorithm are capabilities to provide a better balance among the reservoirs in the system, assign different relative priorities to different demand points, assign different relative priorities to the different components that make up a demand point, and allocate storage within a reservoir to specific demands.

11.1.2 Prescriptive Optimization of Reservoir System Operations

The Hydrologic Engineering Center Prescriptive Reservoir Model (HEC-PRM) was developed in conjunction with studies of major reservoir systems in the Missouri and Columbia River Basins [Hydrologic Engineering Center, 1992, 1993]. However, the HEC-PRM is generalized for application to any reservoir system. HEC-PRM is a network flow programming model designed for prescriptively oriented applications. Improved network flow computational algorithms have been developed in conjunction with the model. HEC-PRM is used in combination with the HEC-DSS Data Storage System [Hydrologic Engineering Center, 1994], which provides input data preparation and output analysis capabilities.

The model minimizes a cost-based objective function. Reservoir release decisions are made based on minimizing costs associated with convex piecewise linear penalty functions (dollars versus storage or flow] associated with various purposes, including hydroelectric power, recreation, water supply, navigation, and flood control. However, schemes have also been devised to include noneconomic components in the basically economic HEC-PRM objective function. User-specified lower and upper bounds on flows and storages are reflected in the constraint equations.

The first application of the HEC-PRM was to the system of six main-stem Missouri River reservoirs described in Chapter 2. The competing interests addressed by the analysis included lake recreation, hydroelectric power generation, flood control, water supply, and downstream navigation and environmental concerns [Hydrologic Engineering Center, 1992]. The major environmental concern is maintenance of steady flows for sandbar-nesting birds. Both the Missouri and Columbia River studies involved evaluations of the operations of existing reservoir systems motivated by water shortages during recent droughts which exacerbated competition among water users.

The second HEC-PRM application was in conjunction with the Columbia River System Operation Review, involving 14 reservoirs, conducted by the U.S. Army Corps of Engineers, Bureau of Reclamation, and Bonneville Power Administration [Hydrologic Engineering Center, 1993; Hayes *et al.*, 1993]. For the Columbia River Basin study, the HEC-PRM objective function reflects cost-based piecewise linear convex penalty functions representing hydropower, flood control, navigation, anadromous fish, water supply, and recreation. The primary environmental concern is maintenance of seasonal flows to aid in the migration of salmon and steelhead. The hydroelectric power penalty function is expressed in terms of dollars versus both flow and storage. The penalty functions for each of the other uses are expressed in terms of dollars per unit of monthly flows. The penalty functions vary monthly to reflect seasonal characteristics. The objective is to minimize total system costs. Basin hydrology is represented by gaged monthly streamflows for the period from 1928 to 1978, adjusted to represent 1980 conditions of basin development. Various alternative system operation scenarios were evaluated. The analyses were performed on an 80486 MS-DOS-based microcomputer.

HEC-PRM applications to date have used a monthly time interval with historic period-of-record streamflows. Unlike the previously discussed descriptive simulation models, the HEC-PRM performs the computations simultaneously for all the time intervals. Thus, the HEC-PRM results show a set of reservoir storages and releases that would minimize cost (as defined by the user-input penalty functions) for the given inflow sequences, assuming all future flows known as release decisions are made during each period. Since in the real world future streamflows are not actually known when a release decision is made, the model provides an upper limit or best possible scenario on what can be achieved. Although the model provides only one set of decision variable values, combinations of a range of values for each variable may result in the same value of the objective function. Various strategies are adopted for using HEC-PRM results to develop alternative reservoir system operating plans and then evaluate the plans in more detail using a descriptive simulation model [Hydrologic Engineering Center, 1994, 1995].

11.2 OTHER LINEAR PROGRAMMING MODELS

Network flow programming, as represented by equations 10-25 through 10-27, is a special form of linear programming, as represented more generally by equations 10-1 through 10-3. Numerous other reservoir system analysis models have also been developed based on linear programming (LP). Several models are cited below as a representative sampling of the variety of ways in which LP models have been applied.

Dorfman [1962] illustrated the use of linear programming with three versions of a model, each with increasing complexity, in which values of decision variables (reservoir storage capacities and release targets) were computed to maximize an economic objective function. The following alternative approaches for representing inflows are reflected in the three versions of the model: average seasonal flows, critical period flows, and treating flows stochastically.

A number of researchers have developed stochastic LP models with random serially correlated inflows represented by a Markov chain with transition probabilities estimated from historical streamflows. Loucks [1968] developed a stochastic LP model for a single reservoir, which determined release rates that minimized an objective function consisting of the sum of the expected squared deviations from target reservoir volumes and discharges. Streamflow input data consisted of an inflow transition probability matrix. Houck and Cohon [1978] reported a multiple-reservoir system model with streamflows represented by a discrete Markov structure. A nonlinear formulation was approximated by solving two LP problems.

A number of papers in the research literature have focused on chance-constrained formulations and associated linear decision rules. Revelle et al. [1969] published one of the key early papers on these techniques. Other researchers, such as Loucks and Dorfman [1975] and Houck et al. [1980], built upon and extended the basic concepts. As discussed in Chapter 10, in chance-constrained LP formulations, probability characteristics of inflows and other random variables are reflected in the constraints. Certain constraints are violated a specified percentage of the time. A linear decision rule provides a mechanism for reflecting reservoir operating rules in an LP model, which also simplifies solution of the model. Inflows, storage, and releases are related to a decision parameter. For example, for an LP formulation with chance constraints and a linear decision rule, a set of values for a release parameter for each month (or other time period) of the year may be computed which minimizes some specified objective function subject to certain discharge or storage limits being violated no more than a specified percentage of time.

Windsor [1973] developed an LP model for analyzing multiple-reservoir flood control operations. Release schedules were determined which minimized the total damage cost at pertinent locations for a design storm. Reservoir and channel routing equations are incorporated in the model.

Reznicek and Simonovic [1990] developed a successive linear programming approach for analyzing hydroelectric power operations. The model determines releases that maximize system revenue and minimize the cost of satisfying energy demands described by a given load duration curve for a given set of stream inflows. A multiple-reservoir system operated by Manitoba Hydro was used as a case study to test the modeling approach.

Palmer *et al.* [1982] used LP to determine firm yields for single reservoirs and a multiple-reservoir system in the Potomac River Basin. Trade-off analyses were performed to determine the impact of instream flow requirement constraints on system yields.

Palmer and Holmes [1988] describe the Seattle Water Department integrated drought management expert system. An LP model is incorporated in this decision support system to determine optimal operating policies and system yield. The LP model is based on the two objectives of maximizing yield and minimizing the economic loss associated with deficits from a specified target.

Randall, Houck, and Wright [1990] developed an LP model to study the operation, during drought, of a metropolitan water system consisting of multiple reservoirs, groundwater, treatment plants, and distribution facilities. Four objectives were incorporated in the modeling study: (1) maximize net revenues, which were the difference between revenues for selling water and electrical pumping costs; (2) maximize reliability, expressed as the minimum of the ratios of consumption to demand for each water use district; (3) maximize reservoir storage at the end of the optimization horizon; and (4) maximize the minimum flow in the streams. Alternative versions of the model were formulated with one objective being optimized as the objective function and the others incorporated as constraints at user-specified levels. Trade-off curves were developed to show the trade-offs between the four alternative objectives.

Martin [1987] describes the MONITOR-I model developed by the Texas Water Development Board to analyze complex surface water storage and conveyance systems operated for hydroelectric power, water supply, and low-flow augmentation. Unlike the site-specific models previously cited, MONITOR-I is generalized for application to any system. The LP model uses an iterative successive LP algorithm to handle nonlinearities associated with hydroelectric power and other features of the model. The decision variables are daily reservoir releases, water diversions, and pipeline and canal flows. The objective function to be maximized is an expression of net economic benefits.

Martin [1995] combined LP and simulation to analyze the operations of a system of six reservoirs operated by the Lower Colorado River Authority in Texas. Capabilities for meeting winter peak power requirements, without adversely impacting reservoir levels and water supply capabilities, were evaluated. An hourly-time-step LP model determined the maximum hydroelectric power that could be generated by plants at four dams without violating a set of constraints which included requirements for maintaining specified storage levels in each of six reservoirs. Since linear approximations of power generation-discharge curves were used, manual adjustments were performed between iterative executions of the LP model. The LP computations were performed using the general-purpose LP software package MILP88, which is marketed by Eastern Software Products of Alexandria, Virginia. A daily-time-step simulation model was used to simulate daily operation of the multiple-purpose reservoir system during the winter season, with the power generation capacity determined with the LP model. The daily streamflows and power demands are treated as probability distributions in the simulation. The computations required for the simulation are performed with LOTUS 1-2-3, a popular spreadsheet package marketed by Lotus Development Corporation of Cambridge, Massachusetts.

The Tennessee Valley Authority has particularly notable experience in practical application of optimization techniques. Shane and Gilbert [1982] and Gilbert and Shane [1982] describe a model, called HYDROSIM, used to simulate the 42-reservoir Tennessee Valley Authority system based on an established set of operating priorities. HYDROSIM has been used for various purposes, including:

- Evaluation of the operation of new operating requirements on established objectives
- Continual checking of current reservoir system status to warn of possible future problems
- Development of long-range operating guides
- Forecasting of reservoir system operation in terms of possible and likely pool-level and discharge variations, constraint violations, and hydroelectric generation characteristics anticipated in the next one to 52 weeks of operation

A database includes weekly streamflows at all pertinent locations for the period since 1903. The HYDROSIM model uses LP to compute reservoir storages, releases, and hydroelectric power generation for each week of a 52-week period beginning at the present based on alternative sequences of historical streamflows from the database.

As mandated by the legislative act creating the TVA, the order of priority for operating the system is as follows: flood prevention, navigation, water supply, power generation (energy and capacity assurance), water quality, drawdown rates, recreation, minimization of power production costs, and balancing of reservoirs. The HYDROSIM model is based on these priorities. A series of operating constraints are formulated to represent these objectives. The model sequentially minimizes the violation of these constraints in their order of priority. The violation of each constraint is minimized subject to the condition that the violation of no higher-priority constraint is increased. This general approach has been used elsewhere and is called *preemptive goal programming*. An LP algorithm is used to perform the computations. Finally, a nonlinear hydropower cost function is minimized subject to the condition that no constraint violation is increased. The cost function is in the form of current power cost plus expected future power cost. Cost for the current week is the total cost (thermal, purchase, and peak sharing) of meeting the load for the current week. The expected future cost is the expected cost of meeting the power load for the remainder of the planning horizon. The nonlinear hydropower cost function is minimized, subject to the priority constraints, by a search procedure which involves iteratively solving a sequence of linear programming problems.

11.3 DYNAMIC PROGRAMMING

As discussed in Chapter 10, dynamic programming (DP) is a general approach to optimization in which a problem is decomposed into stages, with a decision required at each stage. The objective function and constraints may be highly nonlinear. Yeh [1985] and Esogbue [1989] outline the variations and extensions of DP that have been used to develop reservoir system analysis models. Several examples of DP models are cited in the following paragraphs.

Buras [1966] describes early applications of DP in water resources development. Hall *et al.* [1968] used DP to determine releases over time for a single reservoir which maximized revenues from the sale of water and energy. Liu and Tedrow [1973] combined dynamic programming and a multivariable pattern search technique to determine seasonal rule curves for flood control and conservation operation of a five-reservoir system in the Oswego River Basin in New York. Collins [1977] developed a dynamic programming model to determine least-cost withdrawal and release schedules for a four-reservoir water supply system operated by the city of Dallas. The objective function consisted of electricity costs for operating pumps in the water distribution system and a water loss penalty function related to evaporation losses. Trezos and Yeh [1987] developed a DP methodology for improving the operation of systems of multiple hydroelectric power projects. Giles and Wunderlich [1981] describe a model developed by the TVA based on DP that is similar to the previously discussed HYDROSIM model, which uses LP.

Karamouz and Houck [1987] compare the use of stochastic versus deterministic DP in developing single-reservoir operating rules. Tejada-Guibert *et al.* [1995] investigated the use of alternative hydrologic state variables in applying stochastic dynamic programming to a multiple-reservoir system.

Allen and Bridgeman [1986] applied DP to three case studies involving hydroelectric power scheduling, which included optimal instantaneous scheduling of hydropower units with different generating characteristics to maximize overall plant efficiency; optimal hourly scheduling of hydropower generation between two hydrologically linked power plants to maximize overall daily/weekly system efficiency; and optimal monthly scheduling of hydropower generation to minimize the purchase cost of imported power supply subject to a time-of-day rate structure.

Martin [1987] incorporated a DP algorithm in a modeling procedure for determining an optimal expansion plan for a water supply system. The optimization procedure determines the least costly sizing, sequencing, and operation of storage and conveyance facilities over a specified set of staging periods. A Texas Water Development Board DP-based model, called DPSIM-I, is combined with the previously noted TWDB models Al-V and SIM-V.

Chung and Helweg [1985] combined DP with HEC-3 in an analysis of operating policies for Lake Oroville and San Luis Reservoir, which are components of the California State Water Project. The HEC-3 reservoir system simulation model, discussed in Chapters 8 and 9, was used to determine the amount of excess water still available for export after all system commitments were met. A DP model was then used to determine how the reservoirs should be operated to maximize the net benefits of exporting the excess water. The DP decision variables were reservoir releases in each time period, and the objective function was an expression of revenues from selling the water. Since approximations were necessary for formulating the DP model, HEC-3 was used to check and refine the release schedules determined with the DP model.

11.4 COMBINATIONS OF LINEAR AND DYNAMIC PROGRAMMING

A real-time optimization procedure, involving combined use of DP and LP, was developed to determine multiple-reservoir release schedules for hydroelectric power generation in the operation of the California Central Valley Project [Yeh, 1981]. The overall procedure optimizes, in turn, a monthly model over a period of one year, a daily model over a period of up to one month, and an hourly model for 24 hours. Output from one model (monthly ending storages or daily releases] are used as input to the next-echelon model. The monthly model is a combined LP-DP formulation which computes releases and storages based on the objective of minimizing the loss of stored potential energy. Given end-of-month storage levels, the daily model uses LP to determine the daily releases for each power plant which minimizes loss of stored potential energy in the system. The hourly model uses a combination of LP and DP to determine the hourly releases for each plant which maximize total daily system power output.

Simonovic [1992] describes an intelligent decision support system, called REZES, which includes a library of 11 models for performing various analyses for a single reservoir. The models utilize various simulation and optimization techniques, including LP, DP, and nonlinear programming. REZES is an expert system that facilitates the selection and application of alternative reservoir system analysis methods. The system is structured based on the four phases of the reservoir analysis process: (1) problem identification and formulation, (2) model selection, (3) data preparation and computation, and (4) presentation and evaluation of results. The major components of the expert system are the user interface, knowledge base, model library, input data preparation module, algorithmic routines, and output data analysis module. The model library contains the following single-reservoir models:

- RESER is a simulation-optimization (search algorithm) model for determining the minimum storage capacity required to meet specified demands and to evaluate reliability.
- CYIELD is an LP-based model for minimizing total storage capacity.
- AYIELD is an LP-based reservoir sizing model for within-year analysis.
- ILP is an iterative LP model for planning hydroelectric power operations.
- EMSLP is a successive LP optimization model for long-term planning of an interconnected hydro utility.
- DP is a deterministic dynamic programming model for long-term planning of a multiple-purpose reservoir.
- CCCP is a chance-constrained LP model for long-term planning of a multiple-purpose reservoir.
- RPORC is a reliability programming model for long-term planning of a multiple-purpose reservoir.
- SDP is a predictive stochastic dynamic programming model for long-term multipurpose planning.
- PROFEXI combines LP with Kalman filtering and multi-objective compromise programming to analyze short-term multipurpose operations.
- FCCP is a fuzzy chance-constrained model for long-term planning that accepts both quantitative and qualitative input information.

11.5 SEARCH METHODS

Duren and Beard [1972] incorporated a univariate gradient search algorithm, with the Newton-Raphson convergence technique, into a reservoir simulation model to develop a method for determining the economically optimum flood control diagram for a single multipurpose reservoir. The model was applied to Folsom Reservoir in California. As discussed in Chapter 10, the general approach of incorporating the univariate gradient search algorithm into a simulation model was also adopted for the parameter calibration and flood damage reduction system optimization options of the HEC-1 Flood Hydrograph Package.

Gagnon et al. [1974] optimized the operation of a large hydroelectric system using a method called *elimination by affine transformation* which incorporated the Fletcher-Reeves gradient search method. Chu and Yeh [1978] developed a gradient projection model for optimizing the hourly operation of a hydropower reservoir. Simonovic and Marino [1980] applied the gradient projection method with a two-dimensional Fibonacci search to solve a reliability programming problem for a single multipurpose reservoir. Rosenthal [1981] applied a reduced gradient method and integer programming to maximize the benefits in a hydroelectric power system.

Ford, Garland, and Sullivan [1981] combined a reservoir yield simulation model and the Box-Complex search algorithm [Box 1965] to analyze the operation of the multipurpose conservation pool of Sam Rayburn Reservoir in Texas. The combined simulation-optimization approach for selecting an optimal operation policy was as follows. The simulation model is used to simulate a given operating policy, satisfying all demands when possible and allocating the available water according to specific priorities when conflicts occur. The simulation model is linked to the search algorithm, which automatically selects the optimal operation policy given data generated by the simulation model and a user-specified objective function. The operation policy identified by the optimization model is then smoothed using engineering judgment based on experience with operation of the system. The system response with the smoothed operating policy is then simulated with the simulation model, and adjustments in the operating policy are made as necessary.

The optimization problem was formulated with the decision variables being the allocation of the fixed conservation storage capacity to four zones. The reservoir operating rules were based on specifying hydroelectric power requirements, water supply demands, and downstream releases to prevent saltwater intrusion, as functions of the zone within which the water surface elevation happened to be at a particular time. An objective function was formulated as the sum of ten weighted indices. The relative weights assigned to each index was user-specified and could be varied in alternative runs of the model to facilitate various trade-off analyses. The ten indices the objective function comprised are as follows:

- Energy shortage index computed as the sum of the squares of the annual shortage ratios multiplied by 100/number of years of analysis, where the shortage ratio is the annual shortage divided by the annual requirement
- Downstream discharge shortage index computed similar to the above energy shortage index
- Number of times a downstream saltwater barrier is installed in the period of analysis
- Number of times a saltwater barrier fails in the period of analysis
- Average annual energy shortage
- Average annual downstream discharge shortage
- Average monthly conservation pool elevation fluctuation
- Average annual energy
- Number of times the conservation pool is emptied
- Number of times downstream discharge shortage occurs

11.6 CHARACTERIZATION OF OPTIMIZATION MODELS

The aforementioned models illustrate the wide variety of ways in which reservoir system optimization models have been formulated and applied. The remainder of the chapter outlines general characteristics and categories of models.

11.6.1 Descriptive Simulation versus Prescriptive Optimization

Reservoir system analysis models can be categorized as:

- Descriptive simulation models that use no mathematical programming algorithms
- Descriptive simulation models based on mathematical programming
- Prescriptive optimization models

Descriptive versus prescriptive refers to a general modeling orientation rather than a precise categorization of models. In general, descriptive models demonstrate what will happen if a specified plan is adopted. Prescriptive models determine the plan that should be adopted to satisfy specified decision criteria.

A descriptive simulation model is a representation of a system used to predict its behavior under a given set of conditions. A simulation study consists of constructing a model and investigating alternative management strategies, scenarios of conditions affecting system performance, and modeling premises, based on multiple runs of the model. Prescriptive optimization strategies may consist of iterative trial-and-error runs of a simulation model, which may be automated to various degrees.

Optimization models incorporate mathematical programming algorithms that automatically search for an optimum set of decision variable values. Optimization models may be either descriptively or prescriptively oriented.

A number of conventional simulation models that incorporate no formal mathematical programming algorithms are described in Chapter 8. A number of simulation models based on network flow programming and other mathematical programming techniques are noted earlier in the present chapter. The objective functions and constraints in these models are used to specify water use requirements and relative priorities, and to define reservoir operating rules. Simulation models developed with and without incorporation of mathematical programming are often applied in essentially the same manner. Other optimization models described in this chapter are prescriptively oriented. Reservoir releases are determined that optimize an economic or other type of planning or operating objective.

Conventional simulation has the advantage of generally permitting a more detailed and realistic representation of the complex characteristics of a reservoir/river system. Greater modeling flexibility is provided since the computational algorithms are not restricted to a particular mathematical format. Mathematical programming methods provide useful capabilities for analyzing problems characterized by a need to consider an extremely large number of combinations of values for decision variables. The advantages of mathematical programming also include facilitating a more prescriptive analysis in some cases, and providing more systematic and efficient computational algorithms. Many different models, representing diverse applications in engineering, science, and business, can be developed based on the same standard linear and nonlinear programming algorithms.

Capabilities for formulating objectives and assessing performance in meeting these objectives are a driving consideration in applying prescriptive optimization models. Significant complexities in developing prescriptive models for optimizing reservoir operations also derive from the facts that in actual operations, future streamflows are unknown at the time a release is made; and general operating rules are typically needed, rather than computed releases corresponding to specified streamflow sequences.

Reservoir/river system modeling studies are usually based on deterministic streamflow sequences representing historical hydrology. Optimization techniques tend to naturally fit the format of computing releases, which minimize or maximize a specified objective function for given sequences of streamflows. Prescriptive optimization models typically make all release decisions simultaneously, considering all streamflows covering the entire hydrologic period of analysis. Simulation models perform computations period by period in such a way that future streamflows are not reflected in release decisions, except for some models which include features for limited short-term forecasts. Many descriptive simulation models which incorporate mathematical programming algorithms make period-by-period sequential release decisions just like conventional simulation models.

Many optimization models compute the releases that optimize an objective function without directly using detailed operating rules. Simulation models, including both those built on mathematical programming and those not, generally provide mechanisms for the user to define the operating rules in greater detail. Reservoir/river system management strategies are reflected in the formulation of both the objective function and constraints.

Prescriptive optimization models and descriptive simulation models can be used in combination. For example, a typical reservoir system analysis problem consists of establishing operating rules that best achieve certain water management objectives. A prescriptive optimization model may be used to determine sequences of reservoir release decisions that maximize or minimize a criterion function which provides a measure of performance in meeting the water management objectives. Professional judgment and various analyses are then used to develop operating rules that appear to be consistent with the sequences of release decisions reflected in the optimization model results. These rules are then tested using a descriptive simulation model. In various other types of applications as well, preliminary screening of numerous alternatives using a prescriptive optimization model may be followed by a more detailed evaluation of selected plans using a descriptive simulation model.

11.6.2 Prescriptive Criterion Functions

Reservoir-system analysis models can also be categorized by the types of analyses or measures of system performance used. For example, many simulation and optimization models perform yield/reliability analyses for water supply and/or hydroelectric energy. Economic analysis models compute benefits and/or costs associated with alternative operating plans. Typically, in an economic analysis, the computer model simply assigns dollars as a function of computed storages, diversions, and instream flows. The difficult, time-consuming part of the study is developing a conceptual basis and supporting field data for assigning benefits and costs that can be incorporated in the model input data.

Examples of system performance measures that have been reflected in the objective functions of optimization models have been cited in both Chapters 4 and 11. These same types of criteria are important in simulation studies as well. A key philosophical question is how completely and accurately the criterion function incorporated in a model needs to reflect actual societal objectives in order to provide meaningful information for use in the decision-making process. This is a basic issue in assessing the practical utility of systems analysis tools, particularly optimization models, in general, as well as a key consideration in formulating a modeling approach for a particular application. The usefulness of a model depends on how meaningfully the complex real world can be represented by a set of mathematical equations. Necessary simplifications and approximations severely limit the utility of models. Even if planning objectives can be precisely articulated, which is typically not the case, it will likely not be possible to incorporate a criterion function into a model that captures the total essence of the planning objectives. However, models still provide valuable analysis tools. A model can significantly contribute to the evaluation process even though it can never tell the whole story. The criterion function can be a simple index of the relative utility of alternative operating plans that provide significant information regarding which alternative plan best meets the planning objectives. Modeling exercises with alternative decision criteria help address different aspects of the overall story.

An optimization model can normally incorporate only one objective function. However, several different objectives will typically be of concern in a particular reservoir system analysis study. Multiple objectives can be combined in a single function if they can be expressed in commensurate units, such as dollars. However, the different objectives of concern are typically not quantified in commensurate units. Various approaches are adopted in considering multiple objectives. One approach is to execute the optimization model with one selected objective reflected in the objective function and the other objectives treated as constraints at fixed user-specified levels. Alternative runs of the model are made to examine the trade-offs between optimizing for each objective. Another approach is to prioritize the objectives and optimize each objective subject to not adversely impact any higher-priority objective.

An alternative approach for analyzing trade-offs between incommensurate objectives involves treating each objective as a weighted component of the objective function. The objective function is the sum of each component multiplied by a weighing factor reflecting the relative importance of that objective. The weighing factors can be arbitrary, with no physical significance other than to reflect relative weights assigned to the alternative objectives included in the objective function. The model can be executed iteratively, with different sets of weighing factor values to analyze the trade-offs between the objectives with alternative operating plans.

11.6.3 Deterministic Versus Stochastic Models

Reservoir operation models can be classified as either deterministic or stochastic, based on the manner in which streamflow inflows are represented. Deterministic models use specific sequences of streamflows, which may be either historical or synthetically generated. Stochastic models are based on representing the streamflow inflows as probability distributions or stochastic processes in various formats that capture the probabilistic characteristics of the adjusted historical data.

Modeling studies are commonly based on historical gaged-streamflow data adjusted to represent flow conditions at pertinent locations for a specified past, present, or future condition of river-basin development. The data can be expressed in either stochastic or deterministic formats.

If flows in sequential time periods are statistically independent, a stochastic model may be developed by expressing the flows as a probability distribution. However, monthly and shorter-time-interval streamflows typically exhibit a high degree of autocorrelation. The probability distribution of flow must be conditioned on previous flows. The approach often adopted in stochastic reservoir operation models is to represent inflows by a transition probability matrix that describes the discrete probability of a certain inflow conditioned on the previous period inflow. Other more complex approaches are also reported in the literature. Meaningful representation of floods and droughts is difficult in a stochastic model. Interpretation and communication of model results are also typically more difficult for a stochastic model as compared to a deterministic model.

Deterministic reservoir system analysis models are based on input streamflow sequences consisting of either adjusted historical period-of-record streamflows; adjusted historical streamflows during a critical period or other selected subperiod of the period of record; synthetically generated streamflow sequences which preserve selected statistical characteristics of the adjusted historical data; or flows synthesized using a precipitation-runoff model.

Analyses conducted by reservoir management agencies, involving conservation storage operations, are typically based on sequences of adjusted historical period-of-record or critical-period streamflows. Other stochastic hydrology approaches, based on either synthetically generated streamflow sequences or transition probability matrices, have been used extensively in studies reported in the literature, most typically in university research projects. Studies involving flood-control operations include flood hydrographs computed using precipitation-runoff models as well as adjusted historical gaged streamflows. Conventional simulation models, rather than optimization models, are normally used in evaluating flood control operations.

11.6.4 System Representation

Models vary significantly in the mechanisms adopted to represent the water management and use system. Some models provide more flexibility than others in regard to realistically representing reservoir-operating rules, water use requirements and priorities, and complex system configurations. Most of the previously cited models are applicable to multiple-reservoir systems, but some are limited to analysis of a single reservoir. Simulation models typically include mechanisms for detailed specification of operating rules. Optimization models often compute the releases that optimize an objective function without directly addressing the finer details of operating rules. Simulation models based on network flow programming tend to provide greater flexibility than more conventional simulation models for specifying relative priorities between competing water uses and users.

11.6.5 Generalized versus System-Specific Models

Generalized versus system-specific represents another way in which models differ. A model may be developed for a specific reservoir system or generalized for application to essentially any reservoir system in any river basin. With system-specific models, unique features of the particular reservoir system are built into the computer code. Generalized models are designed for application to a range of problems dealing with systems of various configurations and locations, rather than to analyze one specific reservoir system. With a generalized model, the unique features and information reflecting the particular reservoir system of concern are provided in the input data.

Most reservoir optimization models reported in the literature were developed to address specific types of problems for specific reservoir systems. The generalized models cited in this chapter are notable exceptions in this regard.

11.6.6 Mathematical Programming Techniques

Linear programming (LP) dominates as the most widely applied of the numerous optimization methods. Numerous reservoir operation models based on dynamic programming (DP) and search algorithms have been developed in university research projects. Various other nonlinear programming methods have been used less frequently. Two or more techniques are often applied in combination.

LP has the advantage over other optimization methods of being a well-defined, easy-to-understand, readily available algorithm. Generalized computer codes are available for solving LP problems, including very efficient algorithms applicable to particular formulations such as network flow problems. Many reservoir-operation problems can be represented realistically by a linear objective function and a set of linear constraints. Various linearization techniques have been used to deal with nonlinearities such as evaporation and hydropower computations. However, the strict linear form of LP does limit its applicability. Nonlinear properties of a problem can readily be reflected in a DP formulation. DP is not a precise algorithm like LP, but rather is a general approach to solving optimization problems. DP is applicable to problems that can be formulated by optimizing a multiple-stage decision process. Numerous variations and extensions to the general DP approach have been developed specifically for reservoir-system analysis problems. Search algorithms have the advantage of being readily combined with a complex simulation model. The simulation model captures the complexities of the real-world reservoir system operation problem. The search algorithm provides a mechanism to systematize and automate the series of iterative executions of the simulation model required to find a near-optimum decision policy.

12

WATER QUALITY MODELS

Water quality models provide a means of predicting the impacts of natural processes and human activities on the physical, chemical, and biological characteristics of water in a reservoir/river system. Alternative reservoir operating plans are evaluated from the perspective of the effects of releases on in-pool and downstream water quality. Various other reservoir management strategies, in addition to release policies, can be evaluated as well. Models are widely used to evaluate the impacts of pollutant loads from various point and nonpoint sources. Models can be used in conjunction with water quality monitoring activities to interpolate or extrapolate sampled data to other locations and times. Models are also used as research tools to develop an understanding of the processes and interactions affecting water quality.

Water quality modeling dates back to development of the Streeter and Phelps oxygen sag equation in conjunction with a study of the Ohio River in the 1920s. One of the first computer models of water quality was developed for a study of the Delaware Estuary [Thomann, 1963]. The Delaware Estuary model extended the Streeter and Phelps equation to a multiple-segment system. A number of models were developed and applied during the 1960s and early 1970s. During that period, water quality modeling focused on temperature, dissolved oxygen, and biochemical oxygen demand. By the mid-1970s, the need was apparent for developing capabilities for analyzing a more comprehensive range of water quality parameters. The literature since the 1970s is extensive.

Books covering the fundamentals of water quality modeling include [Tchobanoglous and Schroeder, 1985] and [Thomann and Mueller, 1987]. Chapra and Reckhow [1983] treat water quality of lakes. McCutcheon [1989] provides in-depth coverage of transport and surface exchange in rivers. Orlob [1984, 1992],

Dortch and Martin [1989], and Stefan, Ambrose, and Dortch [1989] provide state-of-the-art reviews of surface water quality modeling.

12.1 WATER QUALITY ASPECTS OF RESERVOIR MANAGEMENT

Water quality concerns addressed by reservoir managers range from meeting state and federal water quality standards to dealing with water supply treatment problems to aesthetics to enhancing fisheries. Reservoir management strategies affect the quality of both downstream flows and the water in the reservoir.

12.1.1 Water Quality Problems

Problems with water quality of the streamflow below dams, called *tailwaters*, are often associated with seasonal thermal stratification of the reservoir [Dortch et al., 1992]. Release temperatures that are too warm, too cool, or fluctuate too much or too rapidly are common problems addressed in reservoir operation. Stratification is often accompanied by depletion of dissolved oxygen (DO) in the bottom waters or the hypolimnion illustrated in Figure 12-1. Oxygen depletion and the establishment of reducing conditions in the hypolimnion increase mobilization from the sediments of dissolved nutrients (i.e., ammonium and inorganic substances), sulfide, reduced metals (e.g., iron and manganese), and organic substances (i.e., simple organic acids and methane). These substances can accumulate in the hypolimnion, impacting in-pool and release water quality. Reservoir releases that are low in dissolved oxygen and high in reduced substances can adversely impact downstream ecosystems, cause water supply treatment problems, and be obnoxious to downstream recreational users.

A variety of in-pool water quality problems are common. Turbidity and sus-

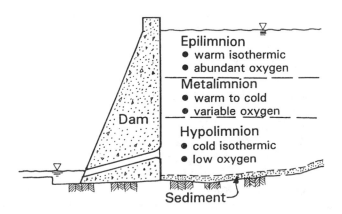

Figure 12-1   Thermally stratified reservoir (USACE Waterways Experiment Station).

pended solids and associated impacts on fisheries, algae, and water quality are typical concerns. Pollution from watershed activities such as acid mine drainage, oil field operations, agricultural activities, and municipal and industrial wastewater effluents are problems in many areas.

In-pool problems are often related to eutrophication. Eutrophication is the process of excessive addition of organic matter, plant nutrients, and silt to reservoirs at rates sufficient to cause increased production of algae and rooted plants. Symptoms of eutrophication include algae blooms, weed-choked shallow areas, low dissolved oxygen, and accumulation of bottom sediments. Resulting problems include elimination of reservoir fisheries, adverse impacts on downstream ecosystems, degradation of water supplies, and reduced storage capacity.

Turbidity is an optical condition caused by suspended solids. Many types of materials such as silt, clay, finely divided organic matter, algae, and other microscopic organisms cause light to be scattered and absorbed. The abiogenic and biogenic substances can be divided into settling and nonsettling suspended matter. There is a continuum of particulate material in reservoirs throughout the year, entering in various amounts and rates. In quiescent, stratified water bodies, particles settle through the epilimnion at fairly uniform rates until they reach the metalimnion. A thermocline plane can abruptly reduce the settling rate. Settling through the isothermal hypolimnion is usually at a uniform, slow rate because the cold water has greater viscosity. This is counteracted to some extent by the increased specific gravity of the particle with decreases in temperature. Turbidity in reservoirs is dependent on the hydrodynamics of flow patterns, which can be very complex.

Temperature and dissolved oxygen concentration are commonly considered the two most important parameters affecting reservoir water quality [Cassidy, 1989]. Due to the variation of water density with temperature, the physical characteristics associated with water temperature have major effects on the hydrodynamic distribution of water being impounded. Temperature also has a major effect on the metabolic rates and physiological responses of aquatic biota and on the rates of chemical, biochemical, and biogeochemical reactions in a reservoir. The dissolved oxygen concentrations present in reservoir waters affect respiration and other physiological responses of aquatic biota, including their distribution, abundance, and behavior. Dissolved oxygen levels also affect the amount and rate of chemical and biochemical nutrients released from the sediment into the water column.

12.1.2 Reservoir Management Measures

Cassidy [1989], Cooke and Kennedy [1989], and Price and Meyer [1992] summarize a variety of reservoir management strategies for dealing with water quality problems. Selective withdrawal through multiple-level outlet works intake structures provides a means for partially controlling temperature and other characteristics of releases. Releasing from selected elevations in the reservoir can be used to manage the in-pool water quality as well as the quality of downstream releases. The operation of a multilevel intake structure requires the consideration of numer-

ous project conditions and constraints, the most important of which is thermal stratification. The multilevel intake structure illustrated in Figure 12–2 allows releases
to consist of mixtures of water taken from selected vertical levels. Various other
outlet structure configurations allow releases to be made from selected elevations in
the reservoir. Some reservoirs have outlet structures specifically designed for selective withdrawal. Other projects have some limited capabilities for releasing through
outlets at different levels even though selective withdrawal was not specifically
considered in their design. Cassidy [1989] estimates that the U.S. Army Corps of
Engineers operates more than 70 reservoirs with selective withdrawal capabilities,
the Bureau of Reclamation about 15 such reservoirs, and the Tennessee Valley Authority four.

Other alternative techniques, in addition to selective withdrawal, are applied
to reduce problems associated with dissolved oxygen. Artificial aeration techniques
include artificial circulation, hypolimnetic aeration, and aeration/oxygenation of
reservoir releases. Artificial circulation methods that mix and destratify the water in
the reservoir include air-lift systems, mechanical pumps, and water jets. Hypolim
netic aeration consists of air injection and other techniques for increasing the dissolved oxygen content of bottom waters without disruption of the normal stratification pattern. Aeration/oxygenation of releases from outlet structures involve gas
transfer manipulations in the hydraulic structures to maximize dissolved oxygen uptake. Dissolved oxygen problems in tailwaters are often associated with hydroelectric power operations. Cassidy [1989] describes structural mechanisms for aerating
hydropower releases, including hub baffle installation, draft tube air aspiration,
draft tube forced-air systems, vacuum breaker modifications, small-pore diffusers,
and epilimnetic pump installation. The oxygenation of flood control and other non-
hydropower releases also can be enhanced by manipulation of the flow through
conduits and over spillways.

Cooke and Kennedy [1989] review methods used by reservoir managers to

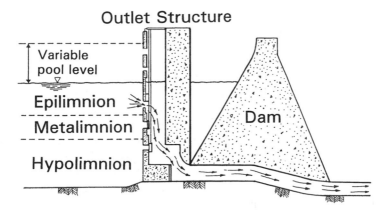

Figure 12–2   Selective withdrawal release structure (USACE Waterways Experiment
Station).

deal with eutrophication-related problems. Alternative approaches involve siltation basins, wetlands, and other means for removing nutrients and particulate matter from reservoir inflows; phosphorus inactivation through addition of aluminum sulfate or sodium aluminate to the reservoir; dilution and flushing; sediment removal; hypolimnetic aeration; artificial circulation; water-level drawdown; harvesting of aquatic plants; biological control of nuisance macrophytes and algae; sediment covers to eliminate nuisance rooted plant growth; and herbicides and algicides.

Reservoir release strategies may incorporate various other mechanisms for managing water quality. For example, poor-quality inflows during certain seasons of the year may be considered in developing seasonal rule curves. Pool levels are drawn down and inflows passed through the reservoir during the periods of poor-quality inflows. A similar strategy involves inflow routing. Reservoir release decisions are based on passing undesirable inflows through the reservoir to minimize adverse impacts on the quality of the water stored in the reservoir [Price and Meyer, 1992].

12.2 CHARACTERISTICS OF WATER QUALITY MODELS

Water quality models simulate hydrodynamic as well as water quality conditions. Hydrodynamics and water quality are often reflected in separate interfacing submodels. The models incorporate equations for transport or conservation of mass for water and equations for transport and transformation of materials in the water. The hydrodynamic equations deal with water volumes, flow rates, velocities, and depths. The materials transport and transformation equations include expressions for energy transfer and expressions for chemical equilibrium or chemical and biological kinetics. Classical formulations for nutrient uptake, growth, photosynthesis, predation, and microbial decomposition have been used in many models.

Water quality constituents can be categorized as organic, inorganic, radiological, thermal, and biological. Pollutants may be classified by specific forms, such as biochemical oxygen demand, nitrogen, phosphorus, bacteria, or specific toxic substances. Unstable pollutants, which decay with time, are termed nonconservative. Many inorganic pollutants are treated as being conservative. Pollutants may be loaded into a watercourse from either point or nonpoint sources. Water quality parameters that have been modeled include temperature, total dissolved solids or inorganic salts, inorganic suspended sediments, dissolved oxygen, biochemical nutrients (phosphorus, nitrogen, silicon), inorganic carbon (carbon, pH), biomass and food chains (chlorophyll α, zooplankton, etc.), metals (lead, mercury, cadmium), synthetic organic chemicals (polychlorinated biphenyls), radioactive materials (radium, plutonium), and herbicides and insecticides [Dieldrin, DDT).

Parameter values for water quality models often must be developed through field and laboratory studies. Initial estimates can sometimes be approximated from information available in the literature. Sets of parameter values are required for model calibration and verification. Parameter values are also required as model input to define boundary and initial conditions. Boundary conditions typically include specification of flows and loads entering the river/reservoir system. Boundary conditions also include specification of meteorological conditions that govern the calculation of the energy flux through the water surface and chemical or biological characteristics that govern the exchange of material, such as dissolved oxygen, between the bed and water column. Flow models may also require specification of flow or depth at the downstream end of the model domain to take into account backwater effects.

Initial conditions are required for dynamic models to define parameter values at the beginning of the simulation period. Initial conditions are typically defined by initial values of flow, depth, and all water quality parameters included in the model.

12.2.1 Model Categorization

Standing bodies of water (lakes and reservoirs) are somewhat different than flowing streams and rivers. A model may include capabilities for simulation of rivers or reservoirs or both. Capabilities for modeling the vertical distribution of temperature and chemical and biological parameters in reservoirs are a particularly important feature included in some models. Models may be pertinent to estuaries and coastal waters as well as lakes, ponds, reservoirs, streams, and rivers.

Models may be categorized as zero-dimensional, one dimensional, two-dimensional, or three-dimensional. Although two-dimensional river models are not uncommon, rivers are typically treated as one-dimensional, with values for water quality and flow parameters changing only in the longitudinal direction. Zero-dimensional and multiple-dimensional modeling is more commonly associated with reservoirs. Reservoirs are also often treated as one-dimensional, with gradients in the vertical direction. Thus, one-dimensional models typically reflect longitudinal gradients for rivers and vertical gradients for reservoirs. Zero-dimensional input-output models contain no information on hydrodynamics other than the assumption that the water body in well mixed. Three-dimensional models include vertical, lateral, and longitudinal changes in water quality and flow parameters.

In steady-state models, parameter values do not vary with time. Dynamic or unsteady models allow parameters to vary with time. Steady versus unsteady refers to both flow parameters (discharge, velocity, depth) and water quality parameters. In a fully dynamic model, both flow and water quality parameters are unsteady. In some models, some parameters are allowed to vary with time while other parameters are assumed to be steady-state.

For purposes of the following review of available models, water quality models are grouped as follows:

- One-dimensional (vertical) models for reservoirs and lakes
- Multidimensional models for reservoirs and lakes
- Steady-state models for riverine systems
- Dynamic models for riverine systems

12.3 REVIEW OF AVAILABLE MODELS

The water quality modeling literature is extensive. The following review focuses on generalized models that can be applied to any reservoir/river system rather than being developed for a specific site. Several models representative of the state of the art are highlighted. These are well-documented public-domain software packages, with proven records of practical applications. A majority of the models noted were developed and are maintained by the following agencies:

- Center for Exposure Assessment Modeling, Environmental Research Laboratory, U.S. Environmental Protection Agency, 960 College Station Road, Athens, Georgia 30613-0801
- Waterways Experiment Station, U.S. Army Corps of Engineers, 3909 Halls Ferry Road, Vicksburg, Mississippi 39180-6199
- Hydrologic Engineering Center, U.S. Army Corps of Engineers, 609 Second Street, Davis, California 95616

The generalized simulation models with supporting documentation and related publications can be obtained from the agencies. Executable versions of the models are available for MS-DOS-based microcomputers. The FORTRAN programs have been compiled and executed on various other computer systems as well.

The steady-state and dynamic riverine system water quality models cited in the remainder of the chapter provide capabilities for simulating reservoirs as components of the overall aquatic system. Models that deal specifically with vertical stratification in reservoirs and lakes are grouped together and presented first. Of the river/reservoir system models discussed later, some reflect only one-dimensional longitudinal gradients. However, others such as WASP, WQRRS, and HEC-5 include features for modeling vertical stratification in reservoirs along with a range of other simulation capabilities.

12.3.1 One-Dimensional Models for Reservoirs and Lakes

A number of models provide capabilities for simulating the vertical distribution of thermal energy and chemical and biological materials in a reservoir through time. These models include LAKECO [Chen *et al.*, 1975], WRMMS [Tennessee Valley Authority, 1976], DYRESM [Imberger *et al.*, 1978], RESTEMP [Brown and Shiao, 1981], the USGS model [House, 1981], MS CLEAN [Park *et al.*, 1981], RESQUALII [Stefan *et al.*, 1982], and MINLAKE [Riley and Stefan, 1987] as well as the SELECT and CE-QUAL-R1 models, described below.

12.3.1.1 SELECT.
The USACE Waterways Experiment Station model SELECT [Davis *et al.*, 1987] predicts the vertical extent and distribution of withdrawal from a reservoir of known density and quality distribution for a given dis-

charge from a specified location. Using this prediction for the withdrawal zone, SELECT computes the quality of the release for parameters (such as temperature, dissolved oxygen, and iron) treated as conservative substances. SELECT can be used as a stand-alone program, but has also been incorporated, in subroutine form, into other models including CE-QUAL-R1.

12.3.1.2 CE-QUAL-R1. CE-QUAL-R1 [Waterways Experiment Station, 1986] determines values for water quality parameters as a function of vertical location and time. A reservoir is conceptualized as a vertical sequence of horizontal layers with thermal energy and materials uniformly distributed in each layer. The distribution of inflows among the horizontal layers is based on density differences. Vertical transport of thermal energy and materials occurs through entrainment and turbulent diffusion. The primary physical processes modeled include surface heat transfer, shortwave and longwave radiation and penetration, convective mixing, wind and flow-induced mixing, entrainment of ambient water by pumped-storage inflows, inflow density current placement, selective withdrawal, and density stratification as impacted by temperature and dissolved and suspended solids. Chemical and biological processes simulated by CE-QUAL-R1 include the effects on dissolved oxygen of atmospheric exchange, photosynthesis, respiration, organic matter decomposition, nitrification, and chemical oxidation of reduced substances; uptake, excretion, and regeneration of phosphorus and nitrogen and nitrification-denitrification under aerobic and anaerobic conditions; carbon cycling and alkalinity-pH-CO_2 interactions; trophic relationships for phytoplankton and macrophytes; transfers through higher trophic levels; accumulation and decomposition of detritus and organic sediment; coliform bacteria mortality; and accumulation and reoxidation of manganese, iron, and sulfide when anaerobic conditions prevail. Reservoir outflows may be based optionally on a user-specified schedule of port releases or model-selected port releases to meet user-specified release amounts and temperature.

12.3.2 Multidimensional Models for Reservoirs and Lakes

Two-dimensional models are significantly more complex than one-dimensional models and have been used less frequently. However, in long deep reservoirs, both vertical and longitudinal water quality gradients may be important. Three-dimensional water quality models have been used much less frequently than two-dimensional models. The three-dimensional models reported in the literature are complex, and are not operational in the sense of the one- and two-dimensional models cited in the following paragraphs. Examples of three-dimensional models include those reported by Simons [1973], Leendertse and Liu [1975], Funkquist and Gidhapen [1984] and Thomann et al. [1979]. Several two-dimensional models are cited below.

The Box Exchange Transport Temperature and Ecology of Reservoirs

(BETTER] model has been applied to Tennessee Valley Authority Reservoirs [Brown, 1985]. BETTER incorporates a modeling approach in which the reservoir is segmented into an array of volume elements or boxes. The flow patterns of the reservoir are modeled as longitudinal and vertical flow transfers between the array of volume elements.

The Computation of Reservoir Stratification (COORS) model [Waldrop *et al.*, 1980; Tennessee Valley Authority, 1986] and the Laterally Averaged Reservoir Model (LARM) program [Edinger and Buchak, 1983; Buchak and Edinger, 1984] solve advection/diffusion equations in a vertical-longitudinal plane through a reservoir. COORS and LARM both provide capabilities for predicting the temperature structure of deep reservoirs throughout the annual stratification cycle. The models also develop the temporal and spatial hydrodynamics of reservoirs to provide advective components for water quality models. CE-QUAL-W2 was developed by expanding LARM to include 20 water quality constituents.

12.3.2.1 CE-QUAL-W2. The Waterways Experiment Station [1986] model CE-QUAL-W2 is a numerical, two-dimensional, laterally averaged model of hydrodynamics and water quality. The model was developed primarily for reservoirs but can also be applied to rivers and estuaries. The model simulates the vertical and longitudinal distributions of thermal energy and selected biological and chemical materials in a water body through time. The model provides capabilities for assessing the impact of reservoir design and operations on the water quality variables. The model determines in-pool water volumes, surface elevations, densities, vertical and longitudinal velocities, temperatures, and constituent concentrations as well as downstream release concentrations.

CE-QUAL-W2 consists of directly coupled hydrodynamic and water quality transport models. The unsteady flow hydrodynamics are influenced by variable water density caused by temperature, salinity, and dissolved and suspended solids. The hydrodynamic computations reflect the variable density effects on the flow field. The water quality model simulates the dynamics of up to 20 constituents in addition to temperatures and circulation patterns. The model simulates the interaction of physical factors (such as flow and temperature), chemical factors (such as nutrients), and an algal assemblage. The constituents are arranged in four levels of optional modeling complexity, permitting flexibility in model application. The first level includes materials that are conservative and noninteractive. The second level includes the interactive dynamics of oxygen-phytoplankton-nutrients. The third level allows simulation of pH and carbonate species. The fourth level allows simulation of total iron.

The physical, chemical, and biological processes of CE-QUAL-W2 are similar to those of the previously discussed CE-QUAL-R1, with the following exceptions. CE-QUAL-W2 does not include transfer to higher trophic levels of zooplankton and fish. It does not account for substances accumulated in the sediments other than organic matter. It contains only one algal group rather than three. It does not include macrophytes. It does not allow the release and oxidation of sulfur and manganese when anaerobic conditions prevail, although it does allow specification, as a boundary condition, of flux from the sediments of iron, ammonia nitrogen, and phosphate phosphorus during anaerobic conditions.

12.3.3 Steady-State Water Quality Models for Riverine Systems

12.3.3.1 QUAL2E.   The Enhanced Stream Water Quality Model (QUAL2E) is a basically steady-state riverine quality model with some dynamic features [Brown and Barnwell, 1987]. Steady-state hydraulics (flows) are combined with water quality parameters that can optionally be steady-state or reflect daily variations. QUAL2E is maintained by the EPA Center for Exposure Assessment Modeling. The model is widely used by many entities for waste load allocations, discharge permit evaluations, and other studies throughout the United States and in other countries as well. It is an accepted standard, particularly for waste load allocation studies of stream systems.

QUAL2E and its variations stem from early models developed by the Texas Water Development Board (TWDB). QUAL I [TWDB, 1971] was developed by expanding the earlier DOSAG model, which solves the steady-state oxygen sag problem for a multiple-segment river reach. QUAL II [Roesner et al., 1973] was developed for the U.S. Environmental Protection Agency by expanding and improving QUAL I.

QUAL2E is a one-dimensional (longitudinal) model for simulating well-mixed streams and lakes [Brown and Barnwell, 1987]. A watercourse is represented as a series of piecewise segments or reaches of steady nonuniform flow. Flows are constant with time and uniform in each reach, but can vary from reach to reach. QUAL2E allows simulation of point and nonpoint loadings, withdrawals, branching tributaries, and in-stream hydraulic structures. The model allows simulation of 15 water quality constituents, including dissolved oxygen, biochemical oxygen demand, temperature, algae as chlorophyll α, organic nitrogen, ammonia nitrogen, nitrate nitrogen, organic phosphorus, inorganic phosphorus, coliforms, an arbitrary nonconservative constituent, and three arbitrary conservative constituents.

QUAL2E has optional features for analyzing the effects on water quality, primarily dissolved oxygen and temperature, caused by diurnal variations in meteorological data. Diurnal dissolved oxygen variations caused by algal growth and respiration can also be modeled. QUAL2E also has an option for determining flow augmentation required to meet any prespecified dissolved oxygen level.

QUAL2E-UNCAS is an enhanced version of QUAL2E which provides capabilities for uncertainty analysis. The uncertainty analysis capabilities include sensitivity analysis with an option for factorially designed combinations of input variable perturbations; first-order error analysis with output consisting of a normalized sensitivity coefficient matrix, and a components of variance matrix; and Monte Carlo simulation with summary statistics and frequency distributions of the output variables.

12.3.3.2 Other Steady-State Models. Simpler one-dimensional steady-state models for analyzing water quality in stream networks include the Waterways Experiment Station STEADY model [Martin, 1986] and U.S. Geological Survey (USGS) Streeter-Phelps model [Bauer *et al.,* 1979]. STEADY computes temperature, dissolved oxygen, and biochemical oxygen demand. The USGS model simulates dissolved oxygen, nitrogenous oxygen demand, carbonaceous biochemical oxygen demand, orthophosphate phosphorus, total and fecal coliform bacteria, and three conservative substances.

Two-dimensional steady-state riverine water quality models have not been widely applied. Examples of two-dimensional models include SARAH [Ambrose and Vandergrift, 1986], RIVMIX [Krishnappan and Lau, 1985], Ontario Ministry of the Environment models [Gowda, 1984], and the USGS model [Bauer and Yotsukuru, 1974].

12.3.4 Dynamic Models for Riverine Systems

With unsteady or dynamic models, the flow and water quality variables vary as a function of time. The WASP, CE-QUAL-RIV1, WQRRS, and HEC-5 models are dynamic models for simulating river/reservoir systems.

12.3.4.1 WASP. The Water Quality Analysis Simulation Program (WASP), maintained by the EPA Center for Exposure Assessment Modeling, is a generalized modeling framework for simulating contaminant fate and transport in aquatic systems including rivers, reservoirs, estuaries, and coastal waters. The various versions of the model evolved from the original WASP [DiToro *et al.,* 1983] which incorporated concepts from a number of earlier models. Version 4, WASP4 [Ambrose *et al.,* 1988], was replaced by the newer version, WASP5 [Ambrose *et al.,* 1993].

WASP is designed to provide a flexible modeling system. The time-varying processes of advection, dispersion, point and diffuse mass loading, and boundary exchange are represented in the basic program. Water quality processes are modeled in special kinetic subroutines that are either selected from a library or supplied by the user. WASP is structured to permit easy substitution of kinetic subroutines into an overall package to form problem-specific models. A compartment modeling approach represents the aqueous system as segments which can be arranged in one, two, or three dimensions.

WASP consists of two stand-alone computer programs, DYNHYD and WASP, that can be run in conjunction or separately. The hydrodynamics program DYNHYD simulates the movement of water. The water quality program WASP models the movement and interaction of pollutants within the water. EUTRO and TOXI are submodels which can be incorporated into the water quality program. EUTRO is used to analyze conventional pollution involving dissolved oxygen, biochemical oxygen demand, nutrients and eutrophication. TOXI simulates toxic pollution involving organic chemicals, metals, and sediment.

12.3.4.2 CE-QUAL-RIV1.

The Waterways Experiment Station model CE-QUAL-RIV1 is a fully dynamic one-dimensional flow and water quality simulation model for streams [Waterways Experiment Station, 1990]. The original version of the model was developed at Ohio State University for the Environmental Protection Agency [Bedford et al., 1983], primarily for predicting water quality associated with stormwater runoff. The present CE-QUAL-RIV1 has been modified to handle control structures. The model is designed for analyzing highly unsteady streamflow conditions, such as that associated with peaking hydropower tailwaters. The model also allows simulation of branched river systems with multiple control structures such as reregulation dams and navigation locks and dams.

The CE-QUAL-RIV1 package includes two stand-alone programs, RIV1H and RIV1Q, which can be interfaced or used separately. RIV1H performs hydraulic routing based on a numerical solution of the full St. Venant equations. RIV1Q is the water quality program. The model is similar to QUAL2E in that it simulates temperature, dissolved oxygen, biochemical oxygen demand, and nutrient kinetics. However, the dynamic CE-QUAL-RIV1 can model sharp flow and water quality gradients.

12.3.4.3 WQRRS.

The Water Quality for River-Reservoir Systems (WQRRS) is a package of dynamic water quality and hydrodynamic models [Hydrologic Engineering Center, 1985]. The WQRRS package includes the models SHP, WQRRSQ, and WQRRSR, which interface with each other. The Stream Hydraulics Package (SHP) and Stream Water Quality (WQRRSQ) programs simulate flow and quality conditions for stream networks, which can include branching channels and islands. The Reservoir Water Quality (WQRRSR) program is a one-dimensional model used to evaluate the vertical stratification of physical, chemical, and biological parameters in a reservoir. The SHP provides a range of optional methods for computing discharges, velocities, and depths as a function of time and location in a stream system. The hydraulic computations can be performed optionally using input stage-discharge relationships, hydrologic routing, kinematic routing, steady flow equations, or the full unsteady-flow St. Venant equations. The WQRRSR and WQRRSQ programs provide capabilities for analyzing up to 18 constituents, including chemical and physical constituents (dissolved oxygen, total dissolved solids), nutrients (phosphate, ammonia, nitrite, and nitrate), carbon budget (alkalinity, total carbon), biological constituents (two types of phytoplankton, benthic algae, zooplankton, benthic animals, three types of fish), organic constituents (detritus, organic sediment), and coliform bacteria.

12.3.4.4 HEC-5Q. The HEC-5 Simulation of Flood Control and Conservation Systems model is described in Chapter 8. The quantity-only HEC-5 has been applied in many reservoir system studies. A water quality version, HEC-5Q [Hydrologic Engineering Center, 1986], is also available but has not been as widely used. The HEC-5 model discussed in Chapter 8 provides the flow simulation module for the water quality version HEC-5Q. Additional subroutines provide the water quality simulation module. The water quality module accepts system flows generated by the flow module and computes the vertical distribution of temperature and other constituents in the reservoirs and the water quality in the associated downstream reaches. The water quality module also includes an option for selecting the gate openings for reservoir selective withdrawal structures to meet user-specified water quality objectives at downstream control points. If the downstream quality objectives cannot be satisfied by selective withdrawal, the model will determine if the objectives can be met by an increase in flow amounts.

The water quality model can be applied in three alternative modes: calibration, annual simulation, and long-term mode. In the calibration mode, values of parameters such as decay rates and dispersion coefficients are computed based on input historical flow, water quality, and reservoir operation data. In the annual simulation mode, the model uses a daily computational time interval in determining the effects of reservoir operations on the water quality in the reservoirs and downstream river reaches. The long-term mode is similar to the annual mode, except the time steps are longer (generally 30 days), so the effects of reservoir operations on water quality can be examined over longer planning horizons of several or many years.

Water quality constituents vary with two alternative simulation options. With the first option, the following constituents can be included in the model: water temperature (always required), up to three conservative constituents, up to three nonconservative constituents, and dissolved oxygen. The other option, referred to as the *phytoplankton option*, requires the following eight constituents: water temperature, total dissolved solids, nitrate nitrogen, phosphate phosphorus, phytoplankton, carbonaceous BOD, ammonia nitrogen, and dissolved oxygen.

12.3.4.5 HEC Utility Programs. Several utility computer programs are available from the USACE Hydrologic Engineering Center, including GEDA, HEATX, and WEATHER, which are intended for use with the WQRRS and HEC-5Q packages. The Geometric Elements from Cross Section Coordinates (GEDA) program serves as a preprocessor for WQRRS (SHP) and HEC-5Q, which prepares tables of hydraulic elements from HEC-2 (Chapter 6) cross-sections. The Heat Exchange Program (HEATX) is used to analyze day-to-day variations in meteorologic variables, and to compute equilibrium temperatures and coefficients of surface heat exchange between a water surface and the atmosphere. HEATX outputs coefficients required for the WQRRS and HEC-5Q models. Program WEATHER was developed to assist users of WQRRS and HEC-5Q with the preparation of required weather input data. WEATHER reads an NOAA National Climatic Center weather data file and outputs a file in the proper input format for either WQRRS or HEC-5Q.

12.3.5 Model Selection

The selection and use of models for investigating river and reservoir water quality problems are governed by the goals and objectives of the study. The choice of model depends on the questions to be answered by the simulation exercises. The appropriateness of modeling assumptions is also an important consideration in selecting a model for a particular application. The ability to collect the proper data to define boundary and initial conditions, and to calibrate and verify the model is particularly important in water quality modeling. Limited data collection resources can constrain the choice of model.

REFERENCES

ALLEN, R. B., and S. G. BRIDGEMAN, "Dynamic Programming in Hydropower Scheduling," *Journal of Water Resources Planning and Management,* American Society of Civil Engineers (ASCE), Vol. 112, No.3, July 1986.

AMBROSE, R. B., T. A. WOOL, J. L. MARTIN, J. P. CONNOLLY, and R. W. SCHANZ, "WASP5.x, A Hydrodynamic and Water Quality Model-Model Theory, User's Manual, and Programmer's Guide," Environment Research Laboratory, U.S. Environmental Protection Agency, Athens, GA, 1993.

AMBROSE, R. B., "WASP4, A Hydrodynamic and Water Quality Model: Model Theory, User's Manual, and Programmer's Guide," EPA-600-3-86-058, Environmental Research Laboratory, U.S. Environmental Protection Agency, Athens, GA, 1988.

AMBROSE, R. B., JR. and S. VANDERGRIFT, "SARAH, A Surface Water Assessment Model for Back Calculating Reductions in Abiotic HazardoU.S. Wastes," Report EPA/600/3-86/058, Environmental Research Laboratory, U.S. Environmental Protection Agency, Athens, GA, 1986.

BARNES, G. W., and F. I. CHUNG, "Operational Planning for California Water System," *Journal of Water Resources Planning and Management*, ASCE, Vol. 112, No. 1, January 1986.

BAUER, D. P. and N. YOTSUKURA, "Two-Dimensional Excess Temperature Model for a Thermally Loaded Stream," Report WRD-74-044, U.S. Geological Survey, Water Resources Division, Gulf Coast Hydroscience Center, Bay St. Louis, MS, 1974.

BAUER, D. P., M. E. JENNINGS, and J. E. MILLER, "One-Dimensional Steady-State Stream Water Quality Model," Water Resources Investigation 79-45, U.S. Geological Survey, Bay St. Louis, MS, 1979.

Beard, L. R., "Transfer of Streamflow Data Within Texas," Report No. 104, Center for Research in Water Resources, University of Texas at Austin, August 1973.

BECKER, L., and W. W-G. YEH, "Optimization of Real Time Operation of Multiple Reservoir System", *Water Resources Research*, American Geophysical Union, Vol. 10, No. 6, December 1974.

BEDFORD, K. W., R. M. SYKES, and C. LIBICKI, "Dynamic Advective Water Quality Model for Rivers," *Journal of Environmental Engineering Division*, ASCE, Vol. 109, No. 3, pp. 535–554, 1983.

BEDIENT, P. S., and W. C. HUBER, *Hydrology and Floodplain Analysis*, 2nd ed., Reading, MA: Addison-Wesley, 1992.

BELLMAN, R., *Dynamic Programming*, Princeton, NJ: Princeton University Press, 1957.

Borland, W. M., and C. R. Miller, "Distribution of Sediment in Large Reservoirs," *Journal of the Hydraulics Division,* ASCE, Vol. 84, No. HY2, April 1958.

BOX, M. J., "A New Method of Constrained Optimization and a Comparison with Other Methods," *The Computer Journal*, Vol. 8, No. 1, 1965.

BOYER, J. M., "Addressing Central Valley Project Policy Issues Using a General-Purpose Model," *Water Policy and Management: Solving the Problems* (D. G. Fontane and H. N. Tuvel, eds.), ASCE, 1994.

BRAS, R. L., and I. RODRIGUEZ-ITURBE, *Random Functions and Hydrology*, New York, NY: Addison-Wesley, 1985.

BRENDECKE, C. M., W. B. DEOREO, E. A. PAYTON, and L. T. ROZAKLIS, "Network Models of Water Rights and System Operations," *Journal of Water Resources Planning and Management*, ASCE, Vol.115, No. 5, September 1989.

BRIDGEMAN, S. G., D. J. W. NORRIE, H. J. COOK, and B. KITCHEN, "Computerized Decision-Guidance System for Management of the Trent River Multireservoir System," *Computerized Decision Support Systems for Water Managers* (Labadie, Brazil, Corbu, and Johnson, eds.), ASCE, 1988.

BROOKE, A. , D. KENDRICK, and A. MEERAUS, *Release 2.25 GAMS, A User's Guide*, South San Francisco, CA: Scientific Press, 1992.

BROWDER, L. E., "RESOP-II Reservoir Operating and Quality Routing Program, Program Documentation and User's Manual," UM-20, Texas Department of Water Resources, Austin, TX, 1978.

BROWN, R. T., and T. O. BARNWELL, "The Enhanced Stream Water Quality Models QUAL2E and QUAL2E-UNCAS: Documentation and User Manual," EPA/600/3-87/007, U.S. Environmental Protection Agency, Environmental Research Laboratory, Athens, GA, May 1987.

BROWN, R. T., and M. C. SHIAO, "Reservoir Temperature Modeling: Case Study of Normandy Reservoir," Report WR28-1-86-101, Tennessee Valley Authority, Water Systems Development Branch, 1981.

BROWN, R. T. "BETTER, A Two-Dimensional Reservoir Water Quality Model: Model User's Guide," Center for the Management, Utilization, and Protection of Water Resources, Tennessee Technological University, Cookeville, TN, 1985.

BROWN, B. W., and R. A. SHELTON, "TVA's Use of Computers in Water Resource Management," *Journal of Water Resources Planning and Management*, ASCE, Vol. 112, No. 3, July 1986.

BUCHAK, E. M., and J. E. EDINGER, "Generalized, Longitudinal-Vertical Hydrodynamics and

Transport: Development, Programming and Applications," Document No. 84-18-R, U.S. Army Corps of Engineers, Vicksburg, MS, 1984.

BURAS, N., "Dynamic Programming and Water Resources Development," *Advances in Hydroscience*, Vol. 3, New York: Academic Press, 1966.

Bureau of Reclamation, *Design of Small Dams*, U.S. Government Printing Office, Washington, DC, 1987.

Bureau of Reclamation, *PROSIM 2.0 User's Manual,* U.S. Department of the Interior, Denver, CO, November 1990.

Bureau of Reclamation, "Inventory of Hydrologic Models," Global Climate Change Response Program, U.S. Department of the Interior, Denver, CO, August 1991.

Bureau of Reclamation, "Statistical Compilation of Engineering Features on Bureau of Reclamation Projects," U.S. Department of the Interior, Denver, CO, 1992.

Bureau of Reclamation, "The CALIDAD Framework, User's Manual," U.S. Department of Interior, Denver, CO, February 1994.

Bureau of Reclamation, "CALIDAD Framework, Programmer's Manual," U.S. Department of Interior, Denver, CO, Draft, May 1994.

CASSIDY, R. A., "Water Temperature, Dissolved Oxygen, and Turbidity Control in Reservoir Releases," *Alternatives in Regulated River Management*, Boca Raton, FL: CRC Press, 1989.

Center for Advanced Decision Support for Water and Environmental Systems, "River Simulation System Documentation," University of Colorado at Boulder, October 1992.

CHAPRA, S. C., and K. H. RECKHOW, *Engineering Approaches for Lake Management*, two volumes, Boston, MA: Butterworth Publishers, 1983.

CHAUDRY, M. II., *Open-Channel Flow*, Englewood Cliffs, NJ: Prentice Hall, 1993.

CHEN, C. W., M. LORENZEN, and D. J. SMITH, "A Comprehensive Water Quality-Ecologic Model for Lake Ontario," Great Lakes Environmental Research Laboratory, National Oceanic and Atmospheric Administration, Washington, DC, 1975.

CHOW, V. T., *Open-Channel Hydraulics*, McGraw-Hill, 1959.

CHU, W. S., and W. W-G. YEH, "A Nonlinear Programming Algorithm for Real Time Hourly Reservoir Operations, " *Water Resources Bulletin*, American Water Resources Association (AWRA), Vol. 14, No. 5, 1978.

CHUNG, F. I., and O. HELWEG, "Modeling the California State Water Project," *Journal of Water Resources Planning and Management*, ASCE, Vol. 111, No. 1, January 1985.

CHUNG, F. I., M. C. ARCHER, and J. J. DEVRIES, "Network Flow Algorithm Applied to California Aqueduct Simulation," *Journal of Water Resources Planning and Management*, ASCE, Vol. 115, No. 2, March 1989.

CIESLIK, P. E., and R. F. MCALLISTER, "Missouri River Master Manual Review and Update," *Water Policy and Management: Solving the Problems* (D. G. Fontane and H. N. Tuvel, eds.), ASCE, 1994.

COHON, J. L., *Multiobjective Programming and Planning*, San Diego, CA: Academic Press, 1978.

COLLINS, M. A., "Implementation of an Optimization Model for Operation of a Metropolitan Reservoir System," *Water Resources Bulletin*, AWRA, Vol. 13, No. 1, 1977.

COLON, R., and G. F. MCMAHON, "BRASS Model: Application to Savannah River System

Reservoirs," *Journal of Water Resources Planning and Management*, ASCE, Vol. 113, No. 2, March 1987.

Columbia Water Management Group, "Columbia River Water Management Report, Water Year 1993," April 1994.

COOKE, G. DENNIS, and ROBERT H. KENNEDY, "Water Quality Management for Reservoirs and Tailwaters; Report 1," Technical Report E-89-1, U.S. Army Corps of Engineers, Vicksburg, MS, January 1989.

COOPER, L. L. and M. W. COOPER, *Introduction to Dynamic Programming*, Elmsford, NY: Pergamon Press, 1981.

DAVIS, C. V., and K. E. SORENSEN, eds., *Handbook of Applied Hydraulics*, 3rd ed., New York: McGraw-Hill, 1984.

DAVIS, S. A., B. B. JOHNSON, W. J. HANSEN, J. WARREN, F. R. REYNOLD, C. O. FOLEY, and R. L. FULTON, "National Economic Development Procedures Manual-Urban Flood Damage," IWR Report 88-R-2, USACE Institute for Water Resources, March 1988.

DAVIS, J. E., J. P. HOLLAND, M. L. SCHNEIDER, and S. C. WILHELMS, "SELECT: A Numerical One-Dimensional Model for Selective Withdrawal," Instruction Report E-87-2, U.S. Army Engineer Waterways Experiment Station, Vicksburg, MS, 1987.

DENARDO, E. V., *Dynamic Programming Theory and Applications*, Englewood Cliffs, NJ: Prentice Hall, 1982.

DIAZ, G. E., and D. G. FONTANE, "Hydropower Optimization Via Sequential Quadratic Programming," *Journal of Water Resources Planning and Management*, ASCE, Vol. 115, No. 6, November 1989.

DITORO, D. M., *et al.*, "Documentation for Water Quality Analysis Simulation Program (WASP) and Model Verification Program (MVP)," EPA-600/3-81-044, U.S. Environmental Protection Agency, Duluth, MN, 1983.

DORFMAN, R., "Mathematical Models: The Multi-Structure Approach," *Design of Water Resources Systems* (A. Maass, ed.), Cambridge, MA: Harvard University Press, 1962.

DORTCH, M. S., D. H. TILLMAN, and B. W. BUNCH, "Modeling Water Quality of Reservoir Tailwaters," Technical Report W-92-1, Environmental Laboratory, USACE Waterways Experiment Station, Vicksburg, MS, May 1992.

DORTCH, M. S., and J. L. MARTIN, "Water Quality Modeling of Regulated Streams," *Alternatives in Regulated River Management* (J. A. Gore and G. E. Petts, eds.), Boca Raton, FL: CRC Press, 1989.

DREYFUS, S. and A. LAW, *The Art and Theory of Dynamic Programming*, New York: Academic Press, 1977.

DUREN, F. K., and L. R. BEARD, "Optimization of Flood Control Allocation for a Multipurpose Reservoir," *Water Resources Bulletin*, AWRA, Vol. 8, No. 4, August 1972.

DZIEGIELEWSKI, B., and J. E. CREWS, "Minimizing the Cost of Coping with Droughts: Springfield, Illinois," *Journal of Water Resources Planning and Management*, ASCE, Vol. 112, No. 4, October 1986.

EDINGER, J. E., and E. M. BUCHAK, "Developments in LARM2: A Longitudinal-Vertical, Time-Varying Hydrodynamic Reservoir Model," Technical Report E-83-1, USACE Waterways Experiment Station, Vicksburg, MS, 1983.

ESOGBUE, A. O., Editor, *Dynamic Programming for Optimal Water Resources Systems Analysis*, Englewood Cliffs, NJ: Prentice Hall, 1989.

ESOGBUE, A. O., "A Taxonomic Treatment of Dynamic Programming Models of Water Resources Systems," *Dynamic Programming for Optimal Water Resources Systems Analysis* (A. O. Esogbue, ed.), Englewood Cliffs, NJ: Prentice Hall, 1989.

FARNSWORTH, R. K., and E. S. THOMPSON, "Mean Monthly, Seasonal, and Annual Pan Evaporation for the United States," NOAA Technical Report NWS 34, National Oceanic and Atmospheric Administration, Washington, DC, December 1982.

FARNSWORTH, R. K., E. S. THOMPSON, and E. L. PECK, "Evaporation Atlas for the Contiguous United States," NOAA Technical Report NWS 33, National Oceanic and Atmospheric Administration, Washington, DC, June 1982.

FELDMAN, A. D., "HEC Models for Water Resources System Simulation: Theory and Experience," *Advances in Hydroscience* (V. T. Chow, ed.), Volume 12, New York: Academic Press, 1981.

FITCH, W. N., P. H. KING, and G. K. YOUNG, "The Optimization of the Operation of Multipurpose Water Resource System," *Water Resources Bulletin*, AWRA, Vol. 6, No. 4, 1970.

FORD, D. T., "Reservoir Storage Reallocation Analysis with PC," *Journal of Water Resources Planning and Management*, ASCE, Vol. 116, No. 3, June 1990.

FORD, D. T., R. GARLAND, and C. SULLIVAN, "Operation Policy Analysis: Sam Rayburn Reservoir," *Journal of the Water Resources Planning and Management Division*, ASCE, Vol. 107, No. WR2, October 1981.

FREAD, D. L., "Chapter 10: Flood Routing," *Handbook of Hydrology* (D. R. Maidment, ed.), New York: McGraw-Hill, 1993.

FREAD, D. L., "National Weather Service Operational Dynamic Wave Model," Hydrologic Research Laboratory, National Weather Service, Silver Spring, MD, April 1978, reprinted April 1987.

FREAD, D. L., "The NWS DAMBRK Model, Theoretical Background and User Documentation," HDR-256, Hydrologic Research Laboratory, National Weather Service, Silver Spring, MD, June 1988.

FREAD, D. L., and J. M. LEWIS, "FLDWAV: A Generalized Flood Routing Model," *Proceedings of the 1988 National Conference on Hydraulic Engineering* (S. R. Abt and J. Gessler, eds.), ASCE, 1988.

FREVERT, D. K., J. W. LABADIE, ROGER K. LARSON, and N. L. PARKER, "Integration of Water Rights and Network Flow Modeling in the Upper Snake River Basin," *Water Policy and Management: Solving the Problems* (D. G. Fontane and H. N. Tuvel, eds.), ASCE, 1994.

FROEHLICH, D. C., "Finite Element Surface-Water Modeling System: Two-Dimensional Flow in a Horizontal Plane, User's Manual," FHWA-RD-88-177, Federal Highway Administration, Turner-Fairbank Highway Research Center, McLean, VA, April 1989.

FUNKQUIST, L., and L. GIDHAPEN, "A Model for Pollution Studies in the Baltic Sea," Report RHO 39, Swedish Meteorological and Hydrological Institute, Norkoping, Sweden, 1984.

GAGNON, C. R., R. H. HICKS, S. L. S. JACOBY, and J. S. KOWALIK, "A Nonlinear Programming Approach to a Very Large Hydroelectric Systems Optimization", *Mathematical Programming*, Vol. 6, Amsterdam: North-Holland Publishing Company, 1974.

GILBERT, K. C., and R. M. SHANE, "TVA Hydro Scheduling Model: Theoretical Aspects," *Journal of Water Resources Planning and Management*, ASCE, Vol. 108, No. 1, March 1982.

GILES, J. E., and W. O. WUNDERLICH, "Weekly Multipurpose Planning Model for TVA Reservoir System," *Journal of Water Resources Planning and Management*, ASCE, Vol. 107, No. 2, October 1981.

GOICOECHEA, A., D. R. HANSEN, and L. DUCKSTEIN, *Multiobjective Decision Analysis with Engineering and Business Applications*, New York: John Wiley & Sons, 1982.

GOLZE, A. R., Editor, *Handbook of Dam Engineering*, New York: Van Nostrand Reinhold Company, 1977.

GOODMAN, A. S., *Principles of Water Resources Planning*, Englewood Cliffs, NJ: Prentice Hall, 1984.

GOULD, B. W., "Statistical Methods for Estimating the Design Capacity of Dams," *Journal of the Institution of Engineers*, Australia, Vol. 33, No. 12, 1961.

GOWDA, T. P., "Water Quality Prediction in Mixing Zones of Rivers," *Journal of Environmental Engineering*, Vol. 110, No. 4, 751, 1984.

HALL, W. A., W. S. BUTCHER, and A. ESOGBUE, "Optimization of the Operations of a Multi-Purpose Reservoir by Dynamic Programming," *Water Resources Research*, AGU, Vol. 4, No. 3, June 1968.

HALL, W. A., and J. A. DRACUP, *Water Resources Systems Engineering*, New York: McGraw-Hill, 1970.

HANSEN, W. J., "National Economic Development Procedures Manual—Agricultural Flood Damage," IWR Report 87-R-10, USACE Institute for Water Resources, October 1987.

HASHIMOTO, T., J. R. STEDINGER, and D. P. LOUCKS, "Reliability, Resiliency, and Vulnerability Criteria for Water Resources System Performance Evaluation," *Water Resources Research*, Vol. 18, No. 1, 1982.

High Performance Systems, Inc., "STELLA II Technical Documentation," Hanover, New Hampshire, 1990.

HOGGAN, D. H., *Computer-Assisted Hydrology & Hydraulics Featuring the U.S. Army Corps of Engineers HEC-1 and HEC-2 Software System*, New York: McGraw-Hill, 1989.

HOUCK, M. H., J. L. COHON, and C. REVELLE, "Linear Decision Rule in Reservoir Design and Management, 6, Incorporation of Economic Efficiency Benefits and Hydroelectric Energy Generation," *Water Resources Research*, AGU, Vol. 16, No. 1, 1980.

HOUSE, L. B. "One-Dimensional Reservoir-Lake Temperature and Dissolved Oxygen Model," U.S. Geological Survey, Water-Resources Investigations, No. 82-5, Gulf Coast Hydroscience Center, NSTL Station, MS, 1981.

HUFSCHMIDT, M. F., and M. B. FIERING, *Simulation Techniques for Design of Water Resources Systems*, Cambridge, MA: Harvard University Press, 1966.

HULA, R. L., "Southwestern Division Reservoir Regulation Simulation Model," *Proceedings of the National Workshop on Reservoir Systems Operations* (G. H. Toebes and A. A. Sheppard, Eds.), ASCE, New York, NY, 1981.

Hydrologic Engineering Center, "HEC-4 Monthly Streamflow Simulation, Users Manual," U.S. Army Corps of Engineers, Davis, CA, February 1971.

Hydrologic Engineering Center, "HEC-3 Reservoir System Analysis for Conservation, User's Manual," U.S. Army Corps of Engineers, Davis, California, March 1981.

Hydrologic Engineering Center, "HEC-5 Simulation of Flood Control and Conservation Systems, User's Manual," U.S. Army Corps of Engineers, Davis, CA, April 1982.

Hydrologic Engineering Center, "Water Quality for River-Reservoir Systems (WQRRS), User's Manual," U.S. Army Corps of Engineers, CPD-8, Davis, CA., October 1978, revised February 1985.

Hydrologic Engineering Center, "HEC-5 Simulation of Flood Control and Conservation Systems, Appendix on Water Quality Analysis," USACE, Davis, CA, Draft, September 1986.

Hydrologic Engineering Center, "HEC-5 Simulation of Flood Control and Conservation Systems, Exhibit 8, Input Description," USACE, Davis, CA, January 1989.

Hydrologic Engineering Center, "HEC-1 Flood Hydrograph Package, User's Manual," USACE, Davis, CA, September 1990.

Hydrologic Engineering Center, "A Preliminary Assessment of Corps of Engineer Reservoirs, Their Purposes and Susceptibility to Drought," Research Document No. 33, USACE, Davis, CA, December 1990.

Hydrologic Engineering Center, "Optimization of Multiple-Purpose Reservoir System Operations: A Review of Modeling and Analysis Approaches," Research Document No. 34, USACE, Davis, CA, January 1991.

Hydrologic Engineering Center, "HEC-2 Water Surface Profiles, User's Manual," Davis, CA, September 1990 (revised February 1991).

Hydrologic Engineering Center, "Missouri River Reservoir System Analysis Model: Phase II," PR-17, USACE, Davis, CA, January 1992.

Hydrologic Engineering Center, "Flood Frequency Analysis (HEC-FFA), User's Manual," CPD-13 USACE, Davis, CA, May 1992.

Hydrologic Engineering Center, "Columbia River System Model-Phase II," PR-21, USACE, Davis, CA, December 1993.

Hydrologic Engineering Center, "HEC-6 Scour and Deposition in Rivers and Reservoirs, User's Manual," Davis, CA, August 1993.

Hydrologic Engineering Center, "HEC Reservoir-Database Network, Installation and User's Guide," TD-35, USACE, Davis, CA, September 1994.

Hydrologic Engineering Center, "HECDSS User's Guide and Utility Program Manuals," CPD-45, USACE, Davis, CA, October 1994.

Hydrologic Engineering Center, "UNET One Dimensional Unsteady Flow Through a Full Network of Open Channels, Draft User's Manual," CPD-66, USACE, Davis, CA, May 1995.

Hydrologic Engineering Center, "Preliminary Operating Rules for the Columbia River System from HEC-PRM Results," PR-26, USACE, Davis, CA, June 1995.

Hydrologic Engineering Center, "River Analysis System, Users Manual," USACE, Davis, CA, August 1995.

IMBERGER, J., J. PATTERSON, B. HERBERT, and L. LOH, "Dynamics of Reservoirs of Medium Size," *Journal of Hydraulic Engineering*, ASCE, Vol. 104, No. 4, 1978.

Institute for Water Resources, "Catalog of Residential Depth-Damage Functions Used by the

Army Corps of Engineers in Flood Damage Estimation," Report 92-R-3, Alexandria, VA, May 1992.

Interagency Advisory Committee on Water Data, "Guidelines for Determining Flood Flow Frequency," Bulletin 17B of the Hydrology Subcommittee, U.S. Geological Survey, Reston, VA, 1982.

International Commission on Large Dams, *World Register of Dams,* Paris, France, 1984.

ISERI, K. T., and W. B. LANGBEIN, "Large Rivers of the United States," Circular 686, U.S. Geological Survey, 1974.

JAMES, L. D., and R. R. LEE, *Economics of Water Resources Planning*, New York: McGraw-Hill, 1971.

JANSEN, R. B., Editor, *Advanced Dam Engineering for Design, Construction, and Rehabilitation*, New York: Van Nostrand Reinhold, 1988.

JENSEN, P. A., and J. W. BARNES, *Network Flow Programming*, New York: John Wiley and Sons, 1980.

JOHNSON, W. K., R. A. WURBS, and J. E. BEEGLE, "Opportunities for Reservoir Storage Reallocation," *Journal of Water Resources Planning and Management*, ASCE, Vol. 116, No. 4, July 1990.

KACZMAREK, Z., and J. KINDLER, eds., *The Operation of Multiple Reservoir Systems*, Laxenburg, Austria: International Institute for Applied Systems Analysis, 1982.

KANE, J. K., "Monthly Reservoir Evaporation Rates for Texas, 1950 Through 1965," Report 64, Texas Water Development Board, October 1967.

KARAMOUZ, M., and M. H. HOUCK, "Comparison of Stochastic and Deterministic Dynamic Programming for Reservoir Operating Rule Generation," *Water Resources Bulletin*, AWRA, Vol. 23, No. 1, February 1987.

KARPACK, L. M., and R. N. PALMER, "The Use of Simulation Modeling to Evaluate the Effect of Regionalization on Water Supply Performance," Water Resources Series Technical Report No. 134, University of Washington, Seattle, WA, March 1992.

KENNINGTON, J. L., and R. V. HELGASON, *Algorithms for Network Programming*, New York: Wiley-Interscience, 1980.

KEYES, A. M., and R. N. PALMER, "The Role of Object Oriented Simulation Models in the Drought Preparedness Studies," *Water Management in the '90s, a Time for Innovation, Proceedings of 20th Annual WRPMD Conference*, ASCE, New York, NY, 1993.

KLEMES, V., "Applied Stochastic Theory of Storage in Evolution," *Advances in Hydroscience*, Volume 12 (V. T. Chow, ed.), New York: Academic Press, 1981.

KLEMES, V., "The Essence of Mathematical Models of Reservoir Storage," *Canadian Journal of Civil Engineering*, Vol. 9, No. 4, 1982.

KOLARS, J. F., and W. A. MITCHELL, *The Euphrates River and the Southeast Anatolia Development Project*, Carbondale, IL: Southern Illinois University Press, 1991.

KRISHNAPPAN, B. G. and Y. L. LAU, "User's Manual: Prediction of Transverse Mixing in Natural Streams Model-RIVMIX MK2," Environmental Hydraulics Section, Hydraulics Division, National Water Research Institute Canada Centre for Inland Waters, Ontario, Canada, 1985.

KUCZERA, G., and G. DIMENT, "General Water Supply System Simulation Model: Wasp,"

Journal of Water Resources Planning and Management, ASCE, Vol. 114, No. 4, July 1988.

LABADIE, J. W., "Dynamic Programming with the Microcomputer," *Encyclopedia of Microcomputers, Volume 5* (A. Kent, J. G. Williams, R. Kent, and C. M. Hall, eds.), New York: Marcel Dekker, 1990.

LABADIE, J. W., A. M. PINEDA, and D. A. BODE, "Network Analysis of Raw Supplies Under Complex Water Rights and Exchanges: Documentation for Program MODSIM3," Colorado Water Resources Institute, Fort Collins, CO, March 1984.

LABADIE, J. W., D. G. FONTANE, and T. DAI, "Integration of Water Quantity and Quality in River Basin Network Flow Modeling," *Water Policy and Management: Solving the Problems* (D. G. Fontane and H. N. Tuvel, eds.), ASCE, 1994.

LANE, W. L., and D. K. FREVERT, "Applied Stochastic Techniques (LAST, A Set of Generally Applicable Computer Programs), User Manual," Engineering and Research Center, U.S. Bureau of Reclamation, Denver, CO, 1985.

LATKOVICH, V. J., and G. H. LEAVESLEY, "Chapter 25: Automated Data Acquisition and Transmission," *Handbook of Hydrology* (D. R. Maidment, ed.), New York: McGraw Hill, 1993.

LEENDERTSE, J. J., and S. K LIU, "A Three-Dimensional Model for Estuaries and Coastal Seas: Aspects of Computation," R-1764-OWRT, U.S. Department of Interior, Washington, DC, 1975.

LINSLEY, R. K., M. A. KOHLER, and J. L. H. PAULHUS, *Hydrology for Engineers*, 3rd ed., New York: McGraw-Hill, 1982.

LINSLEY, R. K., J. B. FRANZINI, D. L. FREYBERG, and G. TCHOBANOGLOUS, *Water-Resources Engineering*, 4th ed., New York: McGraw-Hill, 1992.

LIU, C. S. and A. C. TEDROW, "Multi-lake River System Operating Rules," *Journal of the Hydraulics Division*, ASCE, Vol. 99, No. HY9, 1973.

LOUCKS, D. P., "Computer Models for Reservoir Regulations," *Journal of the Sanitary Engineering Division*, ASCE, Vol. 94, No. SA4, August 1968.

LOUCKS, D. P., and P. J. DORFMAN, "An Evaluation of Some Linear Decision Rules in Chance-Constrained Models for Reservoir Planning and Operation," *Water Resources Research*, AGU, Vol. 11, No. 6, December 1975.

LOUCKS, D. P., K. A. SALEWICZ, and M. R. TAYLOR, "IRIS: An Interactive River System Simulation Model, General Introduction and Description," Cornell University, Ithaca, NY, and International Institute for Applied Systems Analysis, Laxenburg, Austria, November 1989.

LOUCKS, D. P., J. R. STEDINGER, and D. A. HAITH, *Water Resource Systems Planning and Analysis*, Englewood Cliffs, NJ: Prentice Hall, 1981.

LOUCKS, D. P., P. N. FRENCH, and M. R. TAYLOR, "IRAS: Interactive River-Aquifer Simulation, Program Description and Operation," Cornell University and Resource Planning Associates, Ithaca, NY, June 1995.

LOUCKS, D. P., K. A. SALEWICZ, and M. R. TAYLOR, "IRIS: An Interactive River System Simulation Model, User's Manual," Cornell University, Ithaca, NY, and International Institute for Applied Systems Analysis, Laxenburg, Austria, January 1990.

MAASS, A., M. M. HUFSCHMIDT, R. DORFMAN, H. A. THOMAS, S. A. MARGLIN, and G. M.

FAIR, *Design of Water-Resource Systems,* Cambridge, MA: Harvard University Press, 1966.

MANZER, D. F., and M. P. BARNETT, "Analysis by High-Speed Digital Computer," *Design of Water-Resource Systems*, Cambridge, MA: Harvard University Press, 1966.

MARTIN, J. L., "Simplified, Steady-State Temperature and Dissolved Oxygen Model: User's Guide," Instruction Report E-86-4, USACE Waterways Experiment Station, Vicksburg, MS, 1986.

MARTIN, Q. W., "Optimal Daily Operation of Surface Water Systems," *Journal of Water Resources Planning and Management*, ASCE, Vol. 113, No. 4, July 1987.

MARTIN, Q. W., "Optimal Operation of Multiple Reservoir Systems," *Journal of Water Resources Planning and Management*, ASCE, Vol. 109, No. 1, January 1983.

MARTIN, Q. W., "Surface Water Resources Allocation Model (AL-V), Program Documentation and User's Manual," UM-35, Texas Department of Water Resources, Austin, TX, October 1981.

MARTIN, Q. W., "Multivariate Simulation and Optimization Model (SIM-V), Program Documentation and User's Manual," UM-38, Texas Department of Water Resources, Austin, TX, March 1982.

MARTIN, Q. W., "Optimal Daily Operation of Surface-Water System," *Journal of Water Resources Planning and Management*, ASCE, Vol. 113, No. 4, July 1987.

MARTIN, Q. W., "Optimal Reservoir Control for Hydropower on Colorado River, Texas," *Journal of Water Resources Planning and Management*, ASCE, Vol. 121, No. 6, November/December 1995.

MAYS, L. W., and Y-K. TUNG, *Hydrosystems Engineering and Management*, New York: McGraw-Hill, 1992.

MCCRORY, J. A., "Natural Salt Pollution Control, Brazos River Basin, Texas," *Salinity in Watercourses and Reservoirs* (R. H. French, ed.), Boston, MA: Butterworth Publishers, 1984.

MCCUEN, R. H., *Hydrologic Analysis and Design*, Englewood Cliffs, NJ: Prentice Hall, 1989.

MCCUTCHEON, S. C. *Water Quality Modeling, Volume I Transport and Surface Exchange in Rivers*, Boca Raton, FL: CRC Press, 1989.

MCMAHON, G. F., and R. G. MEIN, *River and Reservoir Yield*, Fort Collins, CO: Water Resources Publications, 1986.

MCMAHON, G. F., R. FITZGERALD, and B. MCCARTHY, "BRASS Model: Practical Aspects," *Journal of Water Resources Planning and Management*, ASCE, Vol. 110, No. 1, January 1984.

MORAN, P. A. P., *The Theory of Storage*, London: Methuen, 1959.

MOY, W. S., J. L. COHON, and C. S. REVELLE, "A Programming Model for Analysis of the Reliability, Resilience, and Vulnerability of a Water Supply Reservoir," *Water Resources Research*, AGU, Vol. 22, No. 4, 1986.

National Weather Service, "Application of Probable Maximum Precipitation Estimates, United States East of the 105th Meridian," Hydrometeorological Report No. 52, National Oceanic and Atmospheric Administration, Silver Spring, MD, 1992.

OPRICOBIC, S., and B. DJORDJIVIC, "Optimal Long-Term Control of a Multi-Purpose Reservoir with Indirect Users," *Water Resources Research*, AGU, Vol. 12, No. 6, 1976.

ORLOB, G. T., ED., " Water Quality Modeling, Streams, Lakes, and Reservoirs," IIASA State of the Art Series, London: Wiley-Interscience, 1984.

ORLOB, G. T., "Water-Quality Modeling for Decision Making," *Journal of Water Resources Planning and Management*, ASCE, Vol. 118, No. 3, May 1992.

PALMER, R. N., J. A. SMITH, J. L. COHON, and C. S. REVELLE, "Reservoir Management in Potomac River Basin," *Journal of Water Resources Planning and Management*, ASCE, Vol. 108, No WR1, March 1982.

PALMER, R. N., and K. J. HOLMES, "Operational Guidance During Droughts: Expert System Approach," *Journal of Water Resources Planning and Management*, ASCE, Vol. 114, No. 6, November 1988.

PALMER, R. N., J. A. SMITH, J. L. COHON, and C. S. REVELLE, "Reservoir Management in Potomac River Basin," *Journal of Water Resources Planning and Management*, ASCE, Vol. 108, No. 1, March 1982.

PALMER, R. N., J. R. WRIGHT, J. A. SMITH, J. L. COHON, and C. S. REVELLE, "Policy Analysis of Reservoir Operation in the Potomac River Basin, Volume I, Executive Summary," John Hopkins University, Department of Geography and Environmental Engineering, 1980.

PARK, R. A., C. D. COLLINS, D. K. LEUNG, C. W. BOYDEN, J. ALBANESE, P. DECAPPARIIS, and H. FORSTNER, "The Aquatic Ecosystem Model MS CLEANER." *Proceedings of the First International Conference on State of the Art of Ecological Modeling* (S. E. Jorgensen, ed.), Copenhagen. Elsevier 1981.

PATENODE, G. A., and K. L. WILSON, "Development and Application of Long Range Study (LRS) Model for Missouri River System," *Water Policy and Management: Solving the Problems* (D. G. Fontane and H. N. Tuvel, eds.), ASCE, 1994.

PHILLIPS, D. T., and A. GARCIA-DIAZ, *Fundamentals of Network Flow Programming*, Englewood Cliffs, NJ: Prentice Hall, 1981.

PRENDERGAST, J., "Fishy Business," *Civil Engineering*, ASCE, Vol. 64, No. 11, November 1994.

PRICE, RICHARD E., and EDWARD B. MEYER, "Water Quality Management for Reservoirs and Tailwaters; Report 2," Technical Report E-89-1, USACE, Vicksburg, MS, July 1992.

RANDALL, D., M. H. HOUCK, and J. R. WRIGHT, "Drought Management of Existing Water Supply System," *Journal of Water Resources Planning and Management*, ASCE, Vol. 116, No. 1, January 1990.

REVELLE, C., E. JOERES, and W. KIRBY, "The Linear Decision Rule in Reservoir Management and Design, 1, Development of the Stochastic Model," *Water Resources Research*, AGU, Vol. 5, No. 4, August 1969.

REZNICEK, K. K., and S. P. SIMONOVIC, "An Improved Algorithm for Hydropower Optimization," *Water Resources Research*, Vol. 26, No. 2, February 1990.

RILEY, M. J., and H. G. STEFAN, "Dynamic Lake Water Quality Simulation Model MIN-LAKE," Report 263, St. Anthony Falls Hydraulic Laboratory, University of Minnesota, Minneapolis, MN, 1987.

RIPPL, W. "The Capacity of Storage Reservoirs for Water Supply," *Proceedings of the Institute of Civil Engineers,* Vol. 71, 1883.

ROESNER, L. A., J. R. MONSER, and D. E. EVENSON, "Computer Program Documentation for the Stream Quality Model, QUAL-II," U.S. Environmental Protection Agency, Washington, DC, 1973.

ROSENTHAL, R. E., "A Nonlinear Network Flow Algorithm for Maximization of Benefits in a Hydroelectric Power System," *Operations Research,* Vol. 29, No. 4, 1981.

SALAS, J. D., "Chapter 19: Analysis and Modeling of Hydrologic Time Series," *Handbook of Hydrology* (D. R. Maidment, ed.), New York: McGraw-Hill, 1993.

SANDBERG, J., and P. MANZA, *Evaluation of Central Valley Project Water Supply and Delivery Systems,* U.S. Bureau of Reclamation, Mid-Pacific Regional Office, Sacramento, CA, September 1991.

SCHNITTER, N. J., *A History of Dams, the Useful Pyramids,* Rotterdam: A. A. Balkema, 1994.

SCHUSTER, R. J., "Colorado River Simulation System, Executive Summary," U.S. Bureau of Reclamation, Engineering and Research Center, Denver, CO, April 1987.

SCHWEIG, Z., and J. A. COLE, "Optimal Control of Linked Reservoirs," *Water Resources Research,* AGU, Vol. 4, No. 3, 1986.

SHANE, R. M., and K. C. GILBERT, "TVA Hydro Scheduling Model: Practical Aspects," *Journal of Water Resources Planning and Management,* ASCE, Vol. 108, No. 1, March 1982.

SHANE, R. M., E. A. ZAGONA, D. MCINTOSH, T. J. FULP, and H. M. GORANFLO, "Project Object," Civil Engineering, ASCE, Vol. 66, No. 1, January 1996.

SIGVALDASON, O. T., "A Simulation Model for Operating a Multipurpose Multireservoir System," *Water Resources Research,* AGU, Vol. 12, No. 2, April 1976.

SIMONOVIC, S. P., and M. A. MARINO, "Reliability Programming in Reservoir Management, 1, Single Multipurpose Reservoir," *Water Resources Research,* AGU, Vol. 16, No. 5, 1980.

SIMONOVIC, S. P., "Reservoir Systems Analysis: Closing Gap between Theory and Practice," *Journal of Water Resources Planning and Management,* ASCE, Vol. 118, No. 3, May 1992.

SIMONS, T. J., "Development of Three-Dimensional Numerical Models of the Great Lakes," Scientific Series No. 12, Canada Centre for Inland Waters, Burlington, CA, 1973.

SINGH, V. P., *Elementary Hydrology,* Englewood Cliffs, NJ: Prentice Hall, 1992.

SINGH, V. P., Editor, *Computer Models of Watershed Hydrology,* Water Resources Publications, Highland Ranch, CO, 1995.

SMITH, N., *A History of Dams,* Secaucus, NJ: Citadel Press, 1971.

STEFAN, H. G., J. J. CARDONI, and A. Y. FU, "RESQUAL II: A Dynamic Water Quality Simulation Program for a Stratified Shallow Lake or Reservoir: Application to Lake Chicot, Arkansas," Report No. 209, St. Anthony Falls Hydraulic Laboratory, University of Minnesota, Minneapolis, MN, 1982.

STEFAN, H. G., R. B. AMBROSE, and M. S. DORTCH, "Formulation of Water Quality Models for Streams, Lakes, and Reservoirs: Modeler's Perspective," Miscellaneous Paper E-89-1, USACE Waterways Experiment Station, Vicksburg, MS, July 1989.

STRZEPEK, K. M., and R. L. LENTON, "Analysis of Multipurpose River Basin Systems:

Guidelines for Simulation Modeling," Massachusetts Institute of Technology, Civil Engineering Department, Report No. 236, 1978.

STRZEPEK, K. M., L. A. GARCIA, and T. M. OVER, "MITSIM 2.1 River Basin Simulation Model, User Manual," Center for Advanced Decision Support for Water and Environmental Systems, University of Colorado, Draft, May 1989.

STRZEPEK, K. M., M. S. ROSENBERG, D. D. GOODMAN, R. L. LENTON, and D. M. MARKS, "User's Manual for MIT River Basin Simulation Model", Massachusetts Institute of Technology, Civil Engineering Department, Report No. 242, July 1979.

TEJADA-GUIBERT, J. A., S. A. JOHNSON, and J. R. STEDINGER, "The Value of Hydrologic Information in Stochastic Dynamic Programming Models of a Multireservoir System," *Water Resources Research*, AGU, Vol. 31, No. 10, October 1995.

TCHOBANOGLOUS, G., and E. D. SCHROEDER, *Water Quality*, Reading, MA: Addison-Wesley, 1985.

Tennessee Valley Authority, "Water Temperature Prediction Model for Deep Reservoirs (WRMMS)," Report A-2, Norris, TN, 1976.

Tennessee Valley Authority, "Hydrodynamic and Water Quality Models of the TVA Engineering Laboratory," Issue No. 3, Norris, TN, September, 1986.

Texas Water Development Board, "Analytical Techniques for Planning Complex Water Resources Systems", Report 183, Austin, TX, April 1974.

Texas Water Department Board, "Economic Optimization and Simulation Techniques for Management of Regional Water Resource Systems: River Basin Simulation Model SIMYLD-II Program Description," Austin, TX, July 1972.

Texas Water Development Board, "Inventory and Use of Sedimentation Data in Texas," Bulletin 5912, Austin, TX, January 1959.

Texas Water Development Board, "Simulation of Water Quality in Streams and Canals: Theory and Description of QUAL-I Mathematical Modeling System," Report 128, Austin, TX, 1971.

THOMANN, R. V., "Mathematical Model for Dissolved Oxygen," *Journal of Sanitary Engineering*, ASCE, Vol. 89, No. 5, 1963.

THOMANN, R. V., and J. A. MUELLER, *Principles of Surface Water Quality Modeling and Control*, New York: Harper and Row, 1987.

THOMANN, R. V., WINFIELD, R. P., and SEGNA, J. J. "Verification Analysis of Lake Ontario and Rochester Embayment Three-Dimensional Eutrophication Models," Report No. EPA-600/3-79-094, U.S. Environmental Protection Agency, Duluth, MN, August 1979.

TILSLEY, J. M., "Major Dams, Reservoirs, and Hydroelectric Plants—Worldwide and Bureau of Reclamation," U.S. Bureau of Reclamation, Denver, CO, 1983.

TOEBES, G. H., and A. A. SHEPPARD, Eds., "Proceedings of the National Workshop on Reservoir Systems Operations," ASCE, 1981.

TREZOS, T., and W. W.-G. YEH, "Use of Stochastic Dynamic Programming for Reservoir Management," *Water Resources Research*, AGU, Vol. 23, No. 6, June 1987.

U.S. Army Corps of Engineers, "Standard Project Flood Determinations," Engineering Manual 1110-2-1411, Washington, DC, 1952.

U.S. Army Corps of Engineers, Baltimore District, "Metropolitan Washington DC Area Water Supply Study, Final Report," Baltimore, MD, September 1983.

U.S. Army Corps of Engineers, Office of the Chief of Engineers, "Engineering and Design, Hydropower," Engineering Manual 1110-2-1701, Washington, DC, December 31, 1985.

U.S. Army Corps of Engineers, *Reservoir Water Quality Analyses*, Engineering Manual 1110-2-1201, Washington, DC, June 1987.

U.S. Army Corps of Engineers, North Pacific Division, "User Manual, SSARR Streamflow Synthesis and Reservoir Regulation," Portland, OR, August 1987.

U.S. Army Corps of Engineers, *Management of Water Control Systems*, Engineering Manual 1110-2-3600, Washington, DC, November 1987.

U.S. Army Corps of Engineers, *Sedimentation Investigations of Rivers and Reservoirs*, Engineering Manual 1110-2-4000, Washington, DC, December 1989.

U.S. Army Corps of Engineers, *Hydrologic Frequency Analysis*, EM 1110-2-1415, Washington, DC, March 1993.

U.S. Army Corps of Engineers, "Risk-Based Analysis for Evaluation of Hydrology/Hydraulics and Economics in Flood Damage Reduction Studies," EC 110-2-205, February 1994.

U.S. Army Corps of Engineers, "Hydrologic Engineering Requirements for Flood Damage Reduction Systems," EM 1110-2-1419, Draft, September 1994.

U.S. Army Corps of Engineers, "Risk-Based Analysis for Flood Damage Reduction Studies," ETL 1110-20-xxxx, Draft, October 1994.

U.S. Geological Survey, "Water-Loss Investigation: Lake Hefner Studies Technical Report," Geological Survey Professional Paper 269, 1954.

United Nations, "Register of International Rivers," Oxford: Pergamon Press, 1978.

UNNY, T. E., and E. A. MCBEAN, Eds., *Decision Making for Hydrosystems: Forecasting and Operation,* Proceedings of International Symposium on Real-Time Operation of Hydrosystems, Water Resources Publications, Littleton, CO, 1982.

VINCENT, M. K., D. A. MOSER, and W. J. HANSEN, "National Economic Development Procedures Manual-Recreation, Volume I, Recreation Use and Benefit Estimation Techniques," Report 86-R-4, USACE Institute for Water Resources, Alexandria, VA, March 1986.

VOTRUBA, L., and V. BROZA, *Water Management in Reservoirs*, Developments in Water Science, Vol. 33, Amsterdam: Elsevier, 1989.

WAGNER, H. M., *Principles of Operations Research*, Englewood Cliffs, NJ: Prentice Hall, 1975.

WALDROP. W. R. UNGATE, C. D., and HARPER, W. L. "Computer Simulation of Hydrodynamics and Temperatures of Tellico Reservoir," Report No. WR28-1-65-100, TVA Water Systems Development Branch, Norris, TN, 1980.

Water Resources Council, "Principles and Guidelines for Water and Related Land Resources Planning," Government Printing Office, March 1983.

Waterways Experiment Station, "CE-QUAL-R1: A Numerical One-Dimensional Model of Reservoir Water Quality; User's Manual," Instruction Report E-82-1, Environmental Laboratory, USACE, Vicksburg, MS, July 1986.

Waterways Experiment Station, "CE-QUAL-W2: A Numerical Two-Dimensional, Laterally Averaged Model of Hydrodynamics and Water Quality; User's Manual," Instruction Report E-86-5, Environmental and Hydraulics Laboratories, USACE, Vicksburg, MS, August 1986.

Waterways Experiment Station, "CE-QUAL-RIV1: A Dynamic One-Dimensional (Longitudinal) Water Quality Model for Streams; User's Manual," Instruction Report E-90-1, Environmental Laboratory, USACE, Vicksburg, MS, November 1990.

WERICK, W. J., "National Study of Water Management During Drought: Results Oriented Water Resources Planning," *Water Management in the '90s, A Time for Innovation, Proceedings of 20th Annual WRPMD Conference*, ASCE, New York, 1993.

WINDSOR, J. S., "Optimization Model for the Operation of Flood Control Systems," *Water Resources Research*, AGU, Vol. 9, No. 5, 1973.

WURDS, R. A., "Reservoir Management in Texas," *Journal of Water Resources Planning and Management*, ASCE, Vol. 113, No. 1, January 1987.

WURBS, R. A., "Supply Storage in Federal Reservoirs," *Journal of the American Water Works Association*, Vol. 86, No. 4, April 1994.

WURBS, R. A., *Water Management Models: A Guide to Software*, Englewood Cliffs, NJ: Prentice Hall, 1995.

WURBS, R. A., A. S. KARAMA, I. SALEH, and C. K. GANZE, "Natural Salt Pollution and Water Supply Reliability in the Brazos River Basin," Technical Report 160, Texas Water Resources Institute, August 1993.

WURBS, R. A., C. E. BERGMAN, P. E. CARRIERE, and W. B. WALLS, "Hydrologic and Institutional Water Availability in the Brazos River Basin," Technical Report 144, Texas Water Resources Institute, August 1988.

WURBS, R. A., D. D. DUNN, and W. B. WALLS, "Water Rights Analysis Program (TAMUWRAP), Model Description and Users Manual," Technical Report 146, Texas Water Resources Institute, College Station, TX, March 1993.

WURBS, R. A., G. SANCHEZ-TORRES, and D. D. DUNN, "Reservoir/River System Reliability Considering Water Rights and Water Quality," Technical Report 165, Texas Water Resources Institute, College Station, TX, March 1994.

WURBS, R. A., and L. M. CABEZAS, "Analysis of Reservoir Storage Reallocations," *Journal of Hydrology*, Amsterdam: Elsevier, Vol. 92, June 1987.

WURBS, R. A., M. N. TIBBETS, L. M. CABEZAS, and L. C. ROY, "State-of-the-Art Review and Annotated Bibliography of Systems Analysis Techniques Applied to Reservoir Operation," Technical Report 136, Texas Water Resources Institute, June 1985.

WURBS, R. A., "Modifying Reservoir Operations to Improve Capabilities for Meeting Water Supply Needs during Droughts," Research Document 31, USACE Hydrologic Engineering Center, December 1990.

WURBS, R. A., and P. E. CARRIERE, "Evaluation of Storage Reallocation and Related Strategies for Optimizing Reservoir System Operations," Technical Report 145, Texas Water Resources Institute, August 1988.

WURBS, R. A., and S. T. PURVIS, "Military Hydrology Report 20: Reservoir Outflow (RESOUT) Model," MP EL-79-6, USACE, Waterways Experiment Station, April 1991.

WURBS, R. A., and W. B. WALLS, "Water Rights Modeling and Analysis," *Journal of Water Resources Planning and Management*, ASCE, Vol. 115, No. 4, July 1989.

YEH, W. W-G., "Real-Time Reservoir Operation: The Central Valley Project Case Study," *Proceedings of the National Workshop on Reservoir Operations* (Toebes and Sheppard, eds.), ASCE, 1981.

YEH, W. W-G., "Reservoir Management and Operations Models: A State-of-the-Art Review," *Water Resources Research*, AGU, Vol. 21, No. 21, December 1985.

YEH, W. W-G., "State of the Art Review: Theories and Applications of Systems Analysis Techniques to the Optimal Management and Operation of a Reservoir System, Report NSF/CEE-82091, National Science Foundation, June 1982.

YEH, W. W-G., L. BECKER, and W-S. CHU, "Real-Time Hourly Reservoir Operation," *Journal of Water Resources Planning and Management*, ASCE, Vol. 105, No. WR2, September 1979.

YERRAMREDDY, A. K., and R. A. WURBS, "Water Resources Allocation Based on Network Flow Programming," Journal of Civil Engineering Systems, Gordan and Breach, Vol. 13, No. 4, 1995.

YOUNG, G. K., R. S. TAYLOR, and J. J. HANKS, "A Methodology for Assessing Economic Risks of Water Shortages," IWR-72-6, USACE Institute for Water Resources, Alexandria, VA, May 1972.

ZSUFFA, I., and A. GALAI, *Reservoir Sizing by Transition Probabilities: Theory, Methodology, Application*, Littleton, CO: Water Resources Publications, 1987.

INDEX